Design
and Planning
of Engineering
Systems

**PRENTICE-HALL INTERNATIONAL SERIES
IN CIVIL ENGINEERING AND ENGINEERING MECHANICS**

WILLIAM J. HALL, *Editor*

Design and Planning of Engineering Systems

SECOND EDITION

Dale D. Meredith
State University of New York at Buffalo

Kam W. Wong
University of Illinois at Urbana-Champaign

Ronald W. Woodhead
University of New South Wales, Australia

Robert H. Wortman
University of Arizona

Prentice-Hall, Inc.
Englewood Cliffs, New Jersey   07632

Library of Congress Cataloging in Publication Data

Main entry under title:

Design and planning of engineering systems.

   (Prentice-Hall international series in civil
engineering and engineering mechanics)
     Bibliography: p.
     Includes index.
     1. Systems engineering. I. Meredith, Dale Dean.
II. Series.
TA168.D48 1985   620.7′2   83-23025
ISBN   0-13-200189-6

Editorial/production: *Gretchen K. Chenenko*
Cover design: *Ben Santora*
Art production: *Mary Hickman*
Manufacturing buyer: *Anthony Caruso*

Printed in the United States of America

10 9 8 7 6 5 4 3 2

ISBN   0-13-200189-6   01

PRENTICE-HALL INTERNATIONAL, INC., *London*
PRENTICE-HALL OF AUSTRALIA PTY. LIMITED, *Sydney*
EDITORA PRENTICE-HALL DO BRASIL, LTDA., *Rio de Janeiro*
PRENTICE-HALL CANADA INC., *Toronto*
PRENTICE-HALL OF INDIA PRIVATE LIMITED, *New Delhi*
PRENTICE-HALL OF JAPAN, INC., *Tokyo*
PRENTICE-HALL OF SOUTHEAST ASIA PTE. LTD., *Singapore*
WHITEHALL BOOKS LIMITED, *Wellington, New Zealand*

Contents

8 Calculus Methods for Optimization   161

9 Linear Programming   174

Preface

The engineer has professional obligations to clients and to society. In fulfilling these obligations, the engineer must exercise judgment, make decisions, and accept the responsibility for such actions. Society has become increasingly aware of and sensitive to the consequential impacts that develop due to the nature and implementation of engineering works. Therefore, to satisfy their obligations to society, engineers must be able to anticipate a variety of viewpoints and requirements and to incorporate these into decision-making activities. This calls for more than the mere involvement with technical and functional aspects of problems. It also requires the coordination of social goals and objectives and an appreciation for the continuous interactive process that occurs in planning, design, and project implementation.

The engineer must perceive problems in their full environmental context and seek the solution that best satisfies the goals and objectives of clients and society. The approach presented in this book leads the engineer to pose problems in greater depth, to focus on the necessary integration of the factors influencing the problem, and to develop objective rationales for decision making in order to fulfill a professional role in society.

In the design and planning of engineering systems, the engineer must be aware of the environment in which the problem arises and establish relevant goals and objectives. Once the problem has been identified, it is necessary to define and model the problem as well as consider alternative solutions. Finally, provisions must be made for the selection and implementation of the best acceptable solution. The degree to which an engineer is able to perform these activities establishes professional capabilities and credibility. It is hoped that this book will help engineering students to develop an awareness of these activities and to develop an approach that will enable them to meet these professional challenges.

While the professional focus of the material has been maintained, a number of significant changes have been made to this second edition. The book has been reorganized to place more emphasis on problem formulation and the systems theory basis of concept and model development. Chapters dealing with problem formulation and introductory systems theories have been included to give the reader a greater understanding of the conceptual basis for the systems analysis and decision chapters that follow. A stronger emphasis has been placed on the engineer's role in the design and planning process by the addition of a new final chapter that summarizes the focus of the book. Original chapters have been extensively rewritten and strengthened in addition to inclusion of more introductory and concept development problems.

Experience with the initial edition reveals that it has been used by instructors in a number of ways. The material has been organized so that the instructor can select appropriate chapters if desired; however, the general character and presentational focus of the book have not been changed. A course that has a project orientation could place more emphasis on problem definition and formulation as well as on the theory and approach to problem solving. Chapters 1 through 6 and 14 may be useful in such a course. On the other hand, if more emphasis is to be directed to systems analysis methodologies, the instructor may wish to stress Chapters 7 through 13. This particular material is essentially directed at analysis methods and techniques. To strengthen this portion of the book, two new chapters dealing with calculus methods of optimization and nonlinear programming have also been included.

The book is designed to be self-contained. Although it has rigor, it presents elementary but powerful concepts that can be applied to complex professional problems. The concepts are emphasized in such a way as to encourage the student to define, model, and solve problems in a creative and professional manner.

The material developed has been taught successfully at many universities as a one-semester course at the sophomore level. It could be presented at the junior-senior level or as an introductory engineering course in the freshman year. Our experience has been that students gain a better understanding of the concepts when they are confronted with developing a professional approach to an engineering problem. This is best done by incorporating the textual material into a semester project.

The problems at the end of each chapter range from simple reiteration of material covered in the text through a series of professional situations that can readily be handled with the concepts presented. To encourage the use of student projects and to suggest synthesis, the problems have been arranged throughout the book to bear on different aspects of a variety of engineering areas. At the end of each chapter these areas are: transportation (Problem 3), industrial production (Problem 4), high-rise building (Problem 5), water resources (Problem 6), and construction planning and management (Problem 7). The problem set associated with any area can be used as a basis of a semester project or as a guide to the presentation of the material in any engineering discipline.

We gratefully acknowledge extensive debts to our many colleagues who have contributed comments regarding the content and focus of the book.

In a book of this nature there is always a question of what to include and what to leave out. What is included and emphasized here is our view and responsibility.

Dale D. Meredith
Kam W. Wong
Ronald W. Woodhead
Robert H. Wortman

end of each chapter these areas are: transportation (Problem 3), industrial production (Problem 4), high-rise building (Problem 5), water resources (Problem 6), and construction planning and management (Problem 7). The problem set associated with any area can be used as a basis of a semester project or as a guide to the presentation of the material in any engineering discipline.

We gratefully acknowledge extensive debts to our many colleagues who have contributed comments regarding the content and focus of the book.

In a book of this nature there is always a question of what to include and what to leave out. What is included and emphasized here is our view and responsibility.

Dale D. Meredith
Kam W. Wong
Ronald W. Woodhead
Robert H. Wortman

The Design and Planning Process

1

1.1 INTRODUCTION

Engineers are called on to seek solutions to problems that have a far-reaching impact on society. The range of possible alternative solutions must be established and evaluated in terms of improvements to the way of life, public health, safety, and the magnitude of resource commitments. The solution of such problems requires the careful and responsible application of scientific principles, together with a thorough understanding of the social, political, and economic environments in which these problems exist. The challenge to the engineer is to resolve the problems associated with the needs of society within the resources that are available.

The growing awareness of societal complexity has compounded the engineering planning and design process because of the breadth of issues that are involved. At the same time, society is becoming more aware and sensitive to the forces that are molding its form and environment. What is emerging is an awareness that there is a lack of understanding about societal dynamics and the consequences of technical solutions. Thus engineers need to reevaluate their approach to problem formulation and to the manner in which their decisions, designs, and plans are presented to society.

Because of the scope and nature of the planning and design problems that are raised, it is no longer possible for the engineer to rely solely on disciplinary expertise and technical details if the resulting efforts are to meet with success. It is necessary for the engineer to address problems professionally, in a manner that will ensure that broader issues and aspects of a problem are properly treated and considered.

This is evident from the fact that in recent years planners and designers have experienced increasing scrutiny from various sectors of society with respect to all types of proposed development. This scrutiny and review by the general public as well as by other design disciplines has focused on:

1. General questions related to the need for, and consequences of, the proposed facility
2. Specific aspects of the location and design of the particular facility

In some situations, strong opposition has emerged to a proposed solution that has resulted in substantial delays to a project, its modification, or entire rejection. This opposition is often the result of a fear of the consequences that would result from the implementation and use of the proposed facility. In many cases the proposal is unacceptable or inequitable to particular societal groups because of adverse impacts. In either case the result is the same in that the feasibility of the project is questioned and the credibility of the engineer is challenged on the basis of broad societal issues. To meet these challenges the engineer must recognize the comprehensive nature of the problems to be addressed and develop a broad professional approach to the problem solution process.

1.2 THE ROLE OF THE ENGINEER

Engineers traditionally provide a service function to society by addressing and meeting the quality of life and service problems that arise. In practice, the engineer often encounters a societal need in the form of a problem. Either the engineer identifies a problem and brings it to the attention of the responsible authority or client, or the client recognizes a need to be satisfied and presents the engineer with the task of ensuring the satisfaction of that need. The engineer should distinguish between the true problem and the symptoms of a problem. For example, traffic congestion is really a symptom of a traffic problem, not the actual cause of a problem. Traffic congestion can be the result of any number of things, such as inadequate road design, insufficient zoning controls, or lack of alternative forms of transportation.

Well-established procedures exist for initiating and developing engineered facilities. The owner (government, company, or individual) of the proposed facility employs or contracts with a professional engineer and explains the preceived needs, usually in terms of user requirements and the need for a proposed facility. Then the engineer, in cooperation with the owner, should review the owner's desires and the formulation of user requirements to determine if the proposed facility can meet these requirements. After this step, the engineer establishes and evaluates the alternative designs for the proposed facility for presentation to the client accompanied by a professional recommendation on implementation procedures.

When addressing and performing these engineering functions, the engineer is involved in a continuous and cyclic interaction with society in which new needs arise and are identified, the range of technical solutions are formulated and feasible alterations identified, and the preferred alternative is selected, engineered, and implemented. The general role and involvement of the engineer in dealing with problems are graphically depicted in Figure 1.1. This figure indicates the sequential process in which the engineer becomes involved in the identification of the problems, and the nature of the activities that the engineer may engage in in efforts at reducing problems.

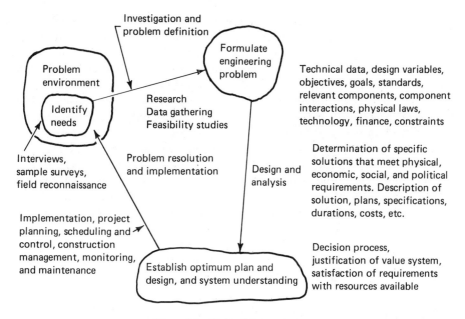

Figure 1.1   Role of the engineer.

The role of the engineer, as outlined above, requires the determination of what is needed; the formulation, design, and detailed specification of feasible alternatives; and decisions relating to the selection of the best solutions and the manner in which they will be built and made available. All these activities involve the engineer in the creative activities of planning and design.

Planning encompasses certain activities that specify how the end product will be achieved. Thus planning includes preliminary investigation, feasibility studies, detailed analyses, the development of specifications for implementation, manufacture and construction, and the formulation of monitoring and maintenance policies and procedures. Planning is concerned with formulating what is to be obtained and how it is to be achieved. Such planning must be consistent with the political, social, environmental, economic, technological, and aesthetic values of the society involved.

Design involves determining the specific form of the end product—its size, shape, properties—and defining the specific emphasis or character of the planning effort that is relevant to the situation: that is, how much investigation, how many feasibility studies. Engineering design is more often concerned with the detailed specification of the facility components and their interrelationships with one another. It requires consideration of the physical laws of nature and the properties of materials and equipment. In many cases, however, the end product of the design process may be a specialized plan, such as a transportation plan, a community development scheme, or the construction plan for a building.

Furthermore, the overall planning activity and professional involvement by the engineer requires consideration of the detailed design aspect of the product. In addition, the design activity in its broadest sense involves determining the extent of the effort and character to be assigned to each step in the planning phase. Thus the planning and design efforts complement each other.

The *design and planning process* that the engineer utilizes in solving societal problems begins with the recognition of an existing *need* or issue that calls for a facility or system of facilities to satisfy that need or issue. The necessary first step in the process is to investigate the environment *surrounding* the need and extract relevant information and data from the environment in order to help define the problem and establish the problem model (see Figure 1.1). Within the context of the problem environment, a problem can involve a collection of needs and interests that may conflict with one another. Hence the engineer may be required to consider the following aspects in relation to a particular facility: its earning capacity and profit potential; its initial, maintenance, and life-use costs; and its marketability, reliability, and anticipated performance. The engineer will also consider the simplicity and elegance of the proposed facility, as well as its political and social acceptability.

Thus, based on the information gathered for the problem, the engineer defines a *model* or representation that depicts the concept of the problem. Using this problem model, the engineer develops an analysis and design procedure that permits the definition of the problem in an objective manner and hence using clearly stated evaluation criteria, selects the best solution to it. This proposed facility or solution will be composed of components, and therefore there will be a concern with establishing the behavior, and gaining an understanding of, the proposed components of the facility. The engineer must then evaluate how well the proposed solution performs in view of the stated user requirements. Finally, the engineer must also be concerned with how the solution is to be implemented. A proposed solution is of no use unless it can be implemented.

At each response stage, the engineer will be using data and models that already exist or that must be acquired or developed. Sometimes a model will require data that must be obtained by observation and measurement. At other times, the available data may limit the types of models that can be developed and used. Finally, there should be concern with the difficulties that are associated with the actual implementation of the conceived solution.

Although Figure 1.1 depicts these efforts in a sequential manner, there may be considerable repetitive cycling within the planning and design process itself. This occurs when:

1. More detailed investigation leads to new insights that require a redefinition of the problem; or

2. Limitations on knowledge and analytical capabilities do not permit the simultaneous consideration of all aspects and implications at any one stage; or

3. Newly discovered technological, financial, social, or political constraints impede the implementation of proposed solutions.

In most cases, the process may require several iterations before a final solution is achieved.

In meeting professional responsibilities, the engineer must utilize approaches that recognize the complexity of the problems that are faced. These approaches must accommodate any interdisciplinary interaction and provide a rational framework for addressing all aspects that are relevant to the problem.

This entire process may be performed by one engineer or it may require a project team to complete the study and evaluation, depending on the size and complexity of the system under consideration. Initially, this would appear to be an overwhelming task, but if comprehensive systematic approaches are utilized, solutions can be achieved that address the broader relevancies of the problems.

1.3 APPLICATION EXAMPLES

A number of major, medium, and minor towns are distributed over a section of the country, as shown in Figure 1.2. Town *A* has been subjected to extensive flooding the past few years and the city council has asked the government for aid in combating the problem. Town *B* has been experiencing a water shortage during the summer months of dry years. The farmers in the area denoted by *E* have combined to form a cooperative to obtain water for irrigation during dry years. The regional conservation club has begun a campaign to inform the population that the quality of water in reach *F* of the stream has deteriorated such that fish can no longer survive there.

These problems have been brought to the attention of the state department of planning and economic development. They may not even be the most critical problems in the region; however, they are the ones that have received attention because of the efforts of those affected to have these problems solved.

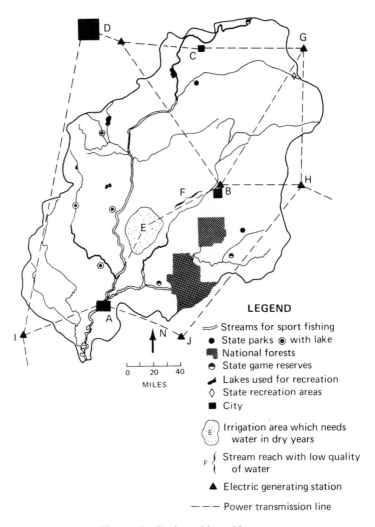

Figure 1.2   Region with problems.

1.3.1 Regional Approach to Problems

The problems stated above are symptoms that indicate that the people of the region have not been able to satisfy their needs. Before any solution is proposed for these problems, what is desired, or to be achieved, should be expressed explicitly. The main requirement in this case may be the maximization of regional welfare, where the term "region" can be employed to denote a geographic area ranging from the size of a small farming field to a large nation. Intermediate stages that will contribute to the satisfaction of the desired end goal may be to generate an increase in income to individuals in the region, develop facilities that provide benefits in excess of their costs, promote and support economic growth, or defend or preserve natural environmental conditions.

If the problems are approached on an individual basis, the proposed solutions may not provide an optimal solution and may even increase the number and severity of the problems rather than solve them. For example, one alternative for reducing the flooding at A is to build a flood control reservoir upstream from the town to store flood waters for release during months when flooding does not occur. One alternative for increasing the water supply for town B might be a water supply reservoir upstream from town B. If these problems are thus approached as individual problems, two reservoirs may be required. However, if they are considered simultaneously, one reservoir might be able to satisfy both the water supply requirements at B and the flood control requirements at A for a much lower cost than two reservoirs.

Therefore, the larger the region considered, the more probable it is that all factors of influence will be within the region being considered. However, the complexity and difficulty of the problem increases with increasing size of the region considered. Thus the region chosen for study, which is shown in Figure 1.2, may be a compromise between including all the influencing factors and keeping the problem small enough to be solved within time, budget, and manpower resources available.

Of the many possible alternatives for achieving the requirements stated above, one is a water resources system along with waste treatment plants to reduce the waste discharge into the stream that results in low water quality. If irrigation water is supplied, it may require extensive pumping and this may increase the power requirements above the present supply capabilities. Thus power supply may become an additional purpose for the development. The points of power demand and supply may be connected in a power network, as shown in Figure 1.2. Hence the influence of power supply and demand from outside the region must be incorporated into any model constructed.

Next, criteria must be developed that can be used to measure how well various proposed alternatives satisfy the stated requirements. One criterion that might be applicable is dollars. The flood damage reduction might be measured in dollars. Power, which can be sold, may also be measured in dollar terms. However, the effects of pollution may be difficult to evaluate in dollar terms. Often when commensurate criteria cannot be developed, the benefits for one of the purposes may be expressed as a standard; that is, as a level that must be provided. Hence this standard becomes a constraint that must be met.

The next step is to identify as many alternatives as possible that might satisfy the stated requirements. At this step, the main concern is identifying possible alternatives, not the optimal or best alternatives. These possible alternatives should include not only those possible under existing technology, but also those possible under technology expected to exist at the time the alternative is needed in the system. The important point is that the only alternatives that will be considered are those that are proposed and explicitly stated.

Therefore, the possible alternatives for the region shown in Figure 1.2 may be a series of multiple-purpose reservoirs, power plants, waste treatment plants, levees, and irrigation canals, as shown in Figure 1.3. A preliminary screening is now performed to determine if each component of the system is technologically and economically feasible. This may involve such things as checking to see if the geological conditions are such that a dam can be constructed at the chosen site, whether there would be enough stream flow to drive a power generator, whether environmental conditions will permit the construc-

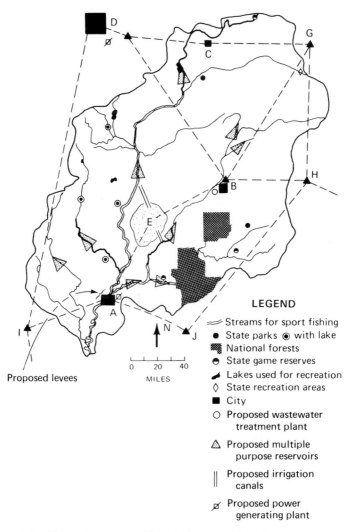

LEGEND

〜 Streams for sport fishing
● State parks ◉ with lake
National forests
● State game reserves
Lakes used for recreation
◇ State recreation areas
■ City
○ Proposed wastewater
   treatment plant
△ Proposed multiple
   purpose reservoirs
‖ Proposed irrigation
   canals
⌀ Proposed power
   generating plant

Proposed levees

0   20   40
MILES

Figure 1.3   Alternatives proposed for solution to problems from a regional view.

tion of a power plant, and so on. This screening may thus reduce the number of possible alternatives that have to be evaluated.

The next step is to develop a model of the system that might be used in an analysis procedure that will aid in determining the components to be included in the final solution and their size. To develop such a model, data must be acquired for the stream flow, water quality, and projections for water use by individuals and industries, crop water requirements, and power use.

One possible model of the system might be stated as:

Maximize   total benefits − total costs

The benefits might include income from power production, flood control, irrigation, and water supply. The costs might include the cost of building and operating the proposed facilities. This criterion function must be maximized subject to the constraints on the system. The following are representative of the constraint relationships that might be developed.

1. Continuity at reservoirs for any time period:

   Outflow ≤ inflow + amount in storage

2. Storage capacity cannot be exceeded:

   0 ≤ amount in storage ≤ maximum capacity of reservoir

3. Flow through turbines:

   Flow through turbines ≤ capacity of turbines

4. Water quality:

   Water quality ≥ minimum acceptable water quality

In addition, there may be budgetary constraints; that is, only so much money is available for developing the system.

After the model has been adequately formulated and solved, a sensitivity analysis may be performed to determine the incremental cost of satisfying the water quality standard. This will then allow the decision maker to make a subjective judgment as to whether it might be appropriate to try a different standard.

It may also be desirable to state a minimum level for flood protection or water supply. Only what is explicitly stated in the model can influence the solution process and the relevance of the answer obtained with the model. However, implicit assumptions may affect the answer; that is, a constraint such as water quality may increase the cost several-fold over that for a slightly lower quality. Thus the engineer should carefully consider, and evaluate, the impact of each model variable and constraint on the model and the resulting decisions. Thus it may become necessary to establish the relationships for the benefits and costs versus size for each of the components in the system in order to conduct a proper analysis.

1.3.2 The Component View

Once the overall regional problem has been defined and a scheme proposed for its resolution, each facility component involved in the scheme can be considered in its own right. The desired performance of each component facility, however, is to contribute toward the requirements of the total scheme. That is, one requirement of the total scheme may be to provide flood control, which may then become the desired performance requirement for the reservoir facility component. Its functional purpose is to reduce the outflow during a flood period, and the measure that may be used to determine how efficiently it achieves this functional purpose may be measured in dollars.

The complete specification of a dam or reservoir requires determining a large number of attributes. However, during the initial formative design stages, the engineer is interested in determining a relationship between the size of the dam and its benefits and costs such that this relationship can be used in the criterion function. This might be done by estimating benefits and costs for different sizes of the reservoir.

To provide the relevant technical data, the engineer needs hydrologic data for the river system for each seasonal period of the year. He or she also needs projections of social, technological (power, ecology, etc.), and use (drinking, recreation, etc.) requirements that may develop for the component. The engineer must develop relationships of total cost versus reservoir size from an estimation of cost based on similar projects using historical costs updated by price index figures and methods.

The initial planning stages for specific facilities necessarily focus on engineering considerations and requirements. However, once these broad details have been established, further planning must be compatible with the problems and resources available for actually implementing the facility.

Although topographic and geological conditions at the dam site may influence the selection of a specific dam type, the availability, quality, and economics of procuring suitable aggregates may decide whether the dam will be entirely concrete or dominantly rock or earth with minimum use of concrete in a spillway and apron.

If the facility component under consideration is the reservoir that has as one of its purposes the storing of water to provide a water supply to town B, one of its physical components will be a distribution system that will allow the water to be transported from the reservoir to the town and then to the users in the town. Although this is a part of the overall regional scheme, it may be studied almost independently of the larger scheme by an engineer for the town.

1.3.3 Implementation Problems

Once a project has been defined and authorized for construction, the project engineer must set up a planning and scheduling model for managing and controlling the project. For each of the project activities, this must establish the necessary resources and initiate the material flows required to complete the activities. A proper attack on each of these activities may require extensive planning studies and design which in their own right may become comprehensive problem formulation efforts.

As an illustrative example, one particular project activity in the construction plan for a reservoir may be "construct concrete spillway," which is a concreting operation of considerable size and may cost millions of dollars and continue over a period of many months. The engineer is therefore faced with the problems of establishing the material flows of aggregate, sand, cement, and additives, and locating and constructing a complete concrete batching plant. This effort must also determine capacity, stockpile sizes, and so on, to ensure uninterrupted production, as well as define the concrete delivery and placement processes.

The extent and use of concrete in the reservoir will have been decided by the locational characteristics and potential and economic sizes of suitable quarries. The costs associated with installing fixed or mobile rock-crushing plants, plus the haulage fleets required to meet the concrete production requirements, must now be investigated. In some cases, cableways or belt conveyors may be more economical, especially in heavily broken country.

If suitable materials are not available in large quantities, the design will tend to minimize the concrete requirements. In some cases, it may become feasible to purchase ready-mixed concrete under contract from a neighboring town and design construction joints and schedule concrete pours to suit the locally available delivery fleet.

However, if the dam requires a large concrete spillway, the production and delivery of concrete becomes a major consideration in the design and economics of the facility. In these cases, considerable attention must be given to planning the entire concrete procurement process. Major considerations relate to the location of the concrete batching plant relative to the concrete site; its rated production capacity affects the duration of the concrete activity and depends on the total concrete quantities involved.

In order to determine overall construction durations so necessary for financial and contract planning purposes, it may become necessary to establish the time required for spillway construction. The engineer may find it necessary to develop a model for the batching plant operation and concrete truck fleet size. Although the haulage distances for aggregate and sand are functions of the availability of suitable sources and economics, that for concrete involves additionally the important features of limited time before its first initial set, the quality deterioration associated with haulage, and its higher unit cost. Consequently, batching plants are usually located as close to the actual concreting site as possible or adjacent to the main materials delivery cable system. These requirements of continuous production at the concrete plant demand a large inventory of storage bins for sand, aggregate, and cement. The engineer may develop models to resolve locational and transportation problems.

Defining the concrete procurement system requires considering many alternatives and system models. The use of cableways for materials delivery is very efficient, unaffected by terrain, weather conditions, and the type of material conveyed, but requires a heavy initial financial investment for steel tower construction. Road haulage systems have the advantage of cheaper initial construction costs and time and permit a staged development in haulage volume, but are either weather dependent or require heavy all-weather preparation. Although haulage roads may have limited final use, cableways have higher salvage potential.

Models can be developed to determine the optimal development of quarry sites and haulage distances and costs. Model studies can help determine stockpile sizes, inventory replacement policies, and the consequences associated with different batch plant sizes and concreting rates.

Extensive initial data must be collected before any decisive actions and designs can be undertaken. Data are required about topographic, soils and geological conditions, road location, and dam-site work areas and access. Extensive comparative estimates must be produced for a large number of alternative designs before final details are defined. Thus the engineer is often called upon to prescribe and authorize extensive studies in all phases of the engineering focus relating to the problem being addressed.

1.4 THE SYSTEMS APPROACH

When accepting a professional assignment it is important that the engineer establish its scope and the extent of the role that must be carried out in its performance. In many cases the engineer has the responsibility of both formulating the problem to be addressed and of implementing its solution. The engineer may have considerable knowledge and experience in accomplishing these tasks and as a result may have developed successful approaches. However, whenever new and complex problems arise, there may be a need to develop new approaches and new techniques to assist in these efforts.

The *systems approach* is such a technique and represents a broad-based systematic approach to problems that may be interdisciplinary. It is particularly useful when problems are complex and affected by many factors, and it entails the creation of a problem model that corresponds as closely as possible in some sense to reality. Its usefulness increases with problem complexity because it permits the engineer to take a broad overall view of the problem under consideration. Thus a clearer understanding of constraints, alternatives, and consequences that are associated with the problem may be obtained.

To ensure that an overall view is attempted, the systems approach discourages the engineer from *initially presenting* a specific problem definition or from *initially adopting* a particular model, mathematics, or solution algorithm.

Instead, the systems approach emphasizes that the *problem environment* be defined in broad terms so that a wide variety of needs can be identified that have some relevance to the problem. These needs should reflect the complex factors, relationships, and conflicts implicit in the problem environment, and which exist in the real-life context, or reality of the problem.

This initial phase of the systems approach stresses the often neglected fact that the nature of the problem should be critically examined and explored. In too many situations, a problem statement is too quickly assumed, or too simply specified, so that the validity of a solution is very questionable and may have little or no relevance to the facts of reality. In such cases, a solution that is based on an erroneous problem statement may not solve the problem under investigation or will have unanticipated consequences. These consequences can be of more concern than the original problem because of potential adverse effects on other systems.

In its elementary form the systems approach stresses the need for the engineer to look for all the relevant factors, influences, or components of the environment that surrounds the problem. In this way the engineer may discover that environmental aspects and impacts can be explicitly included in the problem formulation so that he or she can, and possibly should, address a broader problem statement. In addition, in following the systems approach the engineer will gain a greater understanding of the environment in which the problem is embedded and will become more sensitized to the manner in which the proposed solution can be introduced and in evaluating the resultant impact on the problem environment. Thus the systems approach corresponds to a comprehensive attack on a problem and to an interest in, and commitment to, formulating a problem in the widest and fullest manner that can be professionally handled.

1.5 ENGINEERING PROBLEM SOLVING

In applying the systems approach and developing an attack on any problem, a series of problem-solving steps can be identified. These steps represent specific activities that should be carried out in seeking a professional solution to a problem. These steps are depicted in Figure 1.4. As was indicated in the discussion of the role of the engineer, there may be an iteration in the problem-solving process as the problem analysis and solution progresses. Major *milestones* relative to the status of the effort are also shown in Figure 1.4. These milestones reflect the types of questions that must be addressed and answered at that particular point in the process.

The entire problem-solving process is initiated with a step involving *problem definition*. This step is initiated in response to the recognition of some need that has been demonstrated. Problem definition requires the development of a concise statement of the problem and the constraints that limit the scope of the

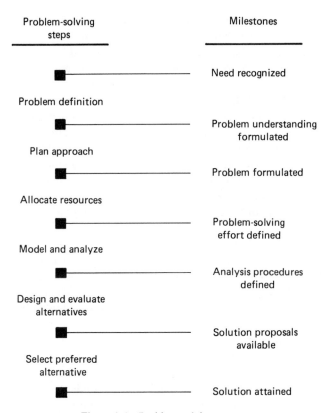

Problem-solving steps	Milestones

Need recognized

Problem definition

Problem understanding formulated

Plan approach

Problem formulated

Allocate resources

Problem-solving effort defined

Model and analyze

Analysis procedures defined

Design and evaluate alternatives

Solution proposals available

Select preferred alternative

Solution attained

Figure 1.4 Problem-solving process.

problem to be considered. Out of this step emerges a formulation and understanding of the problem associated with the need.

Following problem definition, it is necessary to outline and plan an approach to the solutions to the problem. This step includes the establishment of what is to be achieved, the general nature of the alternatives that will be considered, and how the effectiveness of the alternatives will be measured. The product at this point in the process is a statement of the overall problem and its diverse elements, together with an indication of the alternatives to be considered.

An important step in the entire process is the examination and allocation of the engineering-oriented resources that can be applied in attempting to solve the problem. These resources are associated with the amount of time in which the solution must be achieved as well as the manpower resources that can be directed to achieving a solution. This step is critical because resource constraints may affect the approach taken to the attack on the problem area or even force the scope of the problem to be reconsidered. In both cases, the problem formulation may have to be considered.

Up to this point in the process, the general focus has been on the definition of the conceptual approach or attack to be taken in addressing the need. The next step, model and analyze, reflects an effort to begin the detailed technical analysis that is the groundwork for the later evaluation of alternatives. At this point it is necessary to formulate or define the specific technical models or techniques that will be utilized in the design and analysis step. Based on the models or techniques selected, the appropriate data base should be acquired or collected. The result of this step should be that the analysis procedures and models are defined.

The next step in the process is the design and evaluation of alternatives. In this step each alternative should be subjected to a thorough analysis of the costs, benefits, and consequences that would result. This analysis should provide a basis for ranking the alternatives in terms of preference. Also, each alternative should be viewed relative to the *sensitivity* of that problem solution to changes in the *design parameters* or conditions that may be met during the life of the facility. The identification of the preferred alternative should emerge from this step.

1.6 SUMMARY

Engineers have a professional commitment to seeking solutions to a broad spectrum of problems that are identified. This involvement on the part of the engineer may range from addressing problems dealing with large-scale systems of interacting facilities to those that may be considered as a small component of a much larger entity. In all cases it is important that the engineer approach each problem with the intent initially of developing a comprehensive understanding of the situation and the environment in which the problem is contained. The systems approach conceptually represents such an approach to problems. In the process of seeking a solution to a problem, a series of problem-solving steps can be identified. These steps represent a logical sequence of efforts that are necessary for completing a professional attack on a problem. Later chapters contain more detailed material and concepts that are associated with these steps.

PROBLEMS

P1.1. For any specific engineering effort, determine the following.
  (a) Required technical data
  (b) Standards utilized
  (c) Financing requirements and methods

  (d) Necessary approvals

  (e) Constraints

P1.2. Review a set of engineering plans and specifications. For those documents, identify the content and ascertain the engineering meaning of each.

P1.3. Review a transportation-oriented project in your local area, such as a transportation study or a street improvement project. For the project determine the scope of the effort and the issues that are addressed.

P1.4. For a given industrial facility, determine the rationale for its location. For that facility, define how engineers were involved in its development.

P1.5. Identify city policies and ordinances relative to the location of a large building.

P1.6. Figure P1.6 shows the general land use for a river system. Identify the various social, economic, political, and technical issues and conflicts involved.

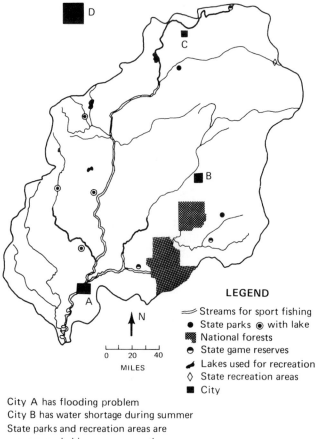

LEGEND

≈ Streams for sport fishing
● State parks ◉ with lake
▓ National forests
◖ State game reserves
◢ Lakes used for recreation
◊ State recreation areas
■ City

City A has flooding problem
City B has water shortage during summer
State parks and recreation areas are
  overcrowded in summer months
City D is large city near basin
Rest of basin is mainly in agricultural use

Figure P1.6

P1.7. Review a publication such as *Civil Engineering* or *Engineering News-Record.* For several projects discussed in these publications, discuss the problems, difficulties, and issues associated with each.

P1.8. Review the code of ethics for a professional group such as the American Society of Civil Engineers. Discuss the role, obligations, and responsibilities of an engineer in terms of dealing with societal problems.

Problem
Formulation

2

2.1 THE SYSTEMS NATURE OF ENGINEERING PROBLEMS

In approaching and solving a problem, one of the initial challenges the engineer faces is to understand the nature of the problem, the environment in which it exists, and the *generating* and *response phenomena* that are associated with the development of the problem and the impact of a proposed solution on the environment. The nature of the problem can provide an indication of the intrinsic factors that are involved in creating the problem and in establishing its scope; thus the solution of the problem must also consider these factors. The *environment* consists of the setting that contains, generates, or surrounds the problem. The *response phenomena* are indicative of the way that the problem area and its environment will respond to some disturbing stimulus. This stimulus could be a modification that is brought about as the result of the introduction of the facility designed as a solution to the specific problem that has arisen.

If the engineer is to undertake and achieve a rational approach to the solution of the problem, it is imperative that this definition and understanding adequately represent the problem in its total context. There must be a recognition of the complexity of the total problem environment and accommodation of the natural as well as the sociopolitical forces that are involved. Certainly, there are limitations on the scope of the problem that can be investigated, and these limitations must also be recognized in formulating a solution. This aspect of the problem is an important consideration and is discussed further in later sections of this chapter.

There are numerous examples that demonstrate the highly complex nature of engineering problems as well as the interaction and response of the problem area and its environment.

1. A temporary interruption to freeway traffic flow in a city can have not only an immediate impact on the use of that freeway, but also a snowballing and often prolonged impact on the city's entire transportation system. By the same token, a freeway improvement could affect the capacity and performance of the entire transportation system.
2. Failure to expedite and receive material that is required on a construction site can delay construction of a facility and result in a drop in field productivity and the temporary unemployment of construction workers.
3. Failure of a machine part can cause a malfunction or limit the operation of that machine such that production is greatly reduced or stopped.
4. Damage to or modification of some part of a building's foundation can have widespread implications on the loads the building can sustain, and therefore limit use of the building.

In all cases, the effect of the interaction, response, or malfunction may ultimately be traced to the point where social, political, or economic consequences can be defined. Furthermore, all the problems are characterized by the fact that a disturbing stimulus produces an effect on the environment.

This effect is important and interesting to the engineer in two ways. First, the disturbing stimulus defines the cause of a problem; thus the identification of this stimulus permits the engineer to direct the solution at the cause. Second, when a feasible *solution* to a problem is posed, the engineer must consider the consequences that will result throughout the environment if the proposed solution is implemented. Part of the professional challenge to the engineer is the fact that the analysis and design must not only consider the immediate problem but also the various consequences. Thus the engineer must treat the problem and the environment as a whole because of their interactive aspects rather than deal with a fragmented aspect in an isolated context.

A facility with which an engineer is involved is in reality part of a larger entity. In addition, that facility contains a number of parts or components that must be integrated to fulfill a purpose or function. In essence, the engineer is dealing with a system or a system of systems, and the approach that he or she utilizes must recognize and deal with the problem on this basis.

A *system* can be defined as a collection of components, connected by various types of interactions or interrelationships which collectively respond to a stimulus or demand and fulfill some specific purpose or function. In a system, each component responds to stimulation according to its intrinsic nature, but the actual stimulation it receives and its subsequent actual behavior is conditioned by the presence and interaction of the other system components. Accordingly, the loads or demands placed on a system call into play the individual behavior of the system components. The *system response* that results from the applied loads and demands develops from the synthesized composit behavior of that of the individual components.

In dealing with a system, it is possible to identify the following characteristics:

1. There are specific purposes or functions that must be, or are being, fulfilled or performed.
2. There are a number of components (at least two) that can be identified as necessary ingredients or fundamental parts of the system. Furthermore, each component has a variety of attributes that implicitly, physically, and behaviorally are necessary for its effective descriptions.
3. The components are interrelated in a manner satisfying interface consistency between the components.
4. There are constraints that restrict the system's behavior and the individual component response.

These characteristics apply to a wide range of general system types that are likely to be encountered in engineering design and planning. The system types reflect the breadth of engineering involvement and might be classified in terms of physical, organizational, and process-oriented systems. In all the system types, the engineer is involved with the conceptualizing, planning, design, implementation, and operation of the facility that results.

If a system were being considered in which the components were actually physical parts, the system would be classified as a *physical system*. Engineering examples of physical systems include an automobile, a building, a dam, a piece of construction equipment, an elevator, a water supply system, and a highway or street system.

More specifically, and by way of an example, a highway interchange could be considered as a physical system. Although it is a subcomponent of the highway or street system, it might be defined with the following characteristics:

System purpose	To allow the orderly flow of traffic at the intersection of two highways
System components	Two highways, ramps, traffic controls, signing, or motorist communication
System structure	Physical layout of the interchange
System constraints	Traffic volume, human reactions, traffic regulations, vehicle characteristics, design standards, and so on.

It may be noted that each of the components listed above may be subdivided further. The highways, for example, might be considered with roadway, structures, and drainage components.

In some cases it may be more convenient to characterize or define a system in terms of the interaction of *process components*. For example, a sewage disposal system can be considered as being made up of collection, treatment, and disposal processes. The selection and design of a particular technology for a process will lead to a unique set of physical components, whereas the selection and design of a different technology for the *same* process will lead to a completely different set of physical components. The household septic tank treatment works process, for example, is completely different from the sewage treatment works process of a sewage system. As with physical systems, the process approach to systems definition also has hierarchical levels. The treatment process, for example, in a city sewerage system may have screening separation, settling, digestion, drying, and incineration processes, and each of these have component processes.

The components of an *organizational system* would consist of activities or actions that would be required to accomplish some project or task. For example, the activities of selecting, purchasing, procurement, site handling, and installation of the physical components of a process establish a construction support organizational system. Also, the organizational design and handling of the construction support organizational system is itself an organizational system. Finally, the organizational design, setting up, and staffing of an organization that will design, implement, and manage the organizational systems can also be considered an organizational system. In each case it is possible to define the functional purpose of the organizational system as well as components, structure, and constraints.

Because of the systems nature of engineering problems, the engineer must develop a relevant system formulation for the problem. The problem formulation process includes the initial determination of a statement of the needs that generate the problem and identification of the relevant systems and their components and hierarchical nature as necessary steps toward identifying the environment of the problem and concludes with the definition of the problem scope and constraints.

2.2 THE NEEDS STATEMENT

The *needs statement* is an expression of an unfulfilled requirement. It provides a means for abstracting from a complex environment a specifically focused requirement that can be addressed as a way of providing a solution.

The essential feature of a needs statement is that it portray a disparity between what is wanted and what is available. The resolution of a need implies a reallocation of resources so that what is wanted is made possible. *What is available* is an indication of the resources that are actually and potentially available for the problem's solution. *What is wanted* is an unconstrained statement of an ultimate state of the problem's solution. The disparity is an indication that a limited resource problem exists that can be solved with the aid of a *decision process* and the commitment of additional resources to meeting a possibly compromised agreement as to what is wanted and what is to be made available.

A schematic illustration of the need statement that highlights the manner in which needs may be satisfied is presented in Figure 2.1. An ultimate goal is to satisfy or eliminate the need, which in essence corresponds to the elimination of the disparity gap between what is wanted and what is available. As indicated, the need statement can be evaluated in terms of studies and model formulations that in some way focus on a quantification of the magnitude of the resource disparity. The resolution of the need requires initially an evaluation of the extent to which resource demands and availabilities can be equated. If they cannot be fully equated, the manner in which either or both demands and availability positions can be changed should be established and trade-off solutions obtained. The need statement thus leads to the consideration of *goals* and the identification of those sets of *objectives* that once obtained assist in the attainment of desired goals. The extent to which an objective is obtained can be measured once a criteria is established. Thus the need statement formulation of Figure 2.1

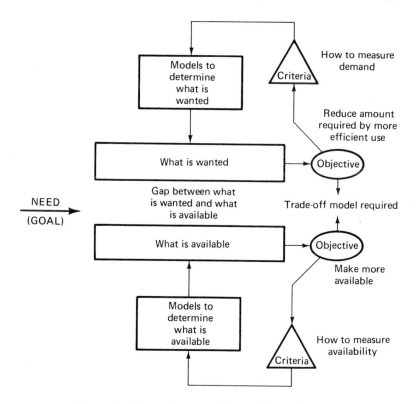

Figure 2.1 Schematic representation of the need statement.

introduces into the engineering design and planning process a professional decision-making role and increases the professional responsibility of the engineer.

Goals should reflect the desired end state that is to be achieved. Normally, the end state is related to the purpose function that is to be fulfilled by the solution to a problem. The objectives indicate the ways in which a goal can be achieved. For any given problem statement, there may be a number of objectives that would serve to meet the goal fully or partially. Although Figure 2.1 depicts a single objective relative to what is available and a goal for what is needed, probably a set of goals and objectives for a problem will be considered. Finally, *criteria* are measures of the manner by which the objectives can be quantified and measured. The definition of criteria is particularly important because of the implications regarding the evaluation models and techniques that are ultimately used.

Considerable skill is required of the engineer in the detection and proper formulation of the needs statement(s) relevant to the problem being addressed. Consequently, it is important that the problem formulation process be developed in terms of the needs statement approach. However, before this can be done it is necessary to develop further system concepts.

2.3 THE HIERARCHICAL NATURE OF SYSTEMS

The initial task in the design and planning process is the formulation and identification of the problem to be addressed. In this step it is necessary to delineate a general problem environment that permits the general system that is associated with the problem to be identified. Furthermore, the components that are contained in the system should be identified. This step is depicted graphically in Figure 2.2. In this figure, four components are schematically depicted for the purpose of representing the concept. In reality, the fundamental parts of the system under investigation must be identified; and the definition of

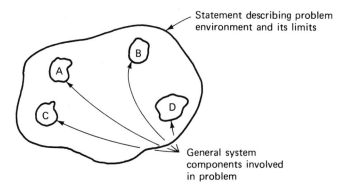

Figure 2.2   Problem environment and component identification.

these elements will specify the number of components. The identification of the component parts should also include a description and quantification of the functional characteristics and properties of each component.

In many cases the definition of system components is not an easy task since it requires some degree of search and investigation in order to understand the system and its makeup. Given a systems-oriented problem that is unfamiliar, the initial effort would be to review available material on the knowledge related to that system. Where the state of knowledge is limited, it may be necessary to undertake a search for component definition in terms of the attributes or characteristics of each of the parts so far identified. In both cases, it is mandatory that a thorough understanding of the extent and nature of the system be achieved.

The identification and nature of the interactions between the components must now be determined for each component identified for the system. Again, the identification of the interactions requires a thorough understanding of how the components function and interact. Of course, the components must in some way either simultaneously or sequentially interact in order for the collection of components to operate as a system. The attributes of the components themselves may provide clues as to the occurrence and nature of an interaction. With the unfamilar system, it may be necessary to examine each component for potentially compatible component attributes to determine the presence or possibility of an interaction. Taken compositely, the interactions serve to form a *system structure* that depicts the relation of the system components and the way they function with respect to other components. In essence, the system structure defines the system and influences its behavior. For the hypothetical system shown in Figure 2.2, the system structure has been added and is shown schematically in Figure 2.3.

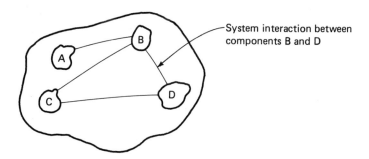

Figure 2.3   Problem environment and system components with structure.

The engineer must recognize that the problem may involve or be related to other systems. Thus, in defining the problem, these other systems must be identified. For example, a system with which an engineer may be directly concerned is in reality a component of some larger system. Thus, for a comprehensive view of the problem, the engineer also must define the components and structure of the larger system. It can be noted that the components of the larger system would have broader functions or purposes. Also, the attributes of the larger component may differ from the attributes or characteristics of each of the subcomponents. This relationship of the system under investigation to a large system is shown in the top portion of Figure 2.4. Viewing the system under consideration in the context of the larger system has the following benefits. First, a clearer definition of system function can be ascertained; and second, the consequences of modifying the system can be established in the broader context.

A microscopic investigation of the system under study can be made by subdividing each of the components into component parts. This permits a detailed investigation of that parent component to determine possible design modifications. In contrast to examining the broader system levels, an examination of the subcomponents reveals more restrictive functions and impacts. This microanalysis of the system components is depicted graphically in the bottom portion of Figure 2.4. An investigation of each component reveals that it is a system in itself.

Thus a *hierarchical system structure* can be determined that permits analysis of the systems at the various levels. The hierarchical structure is shown in Figure 2.4 and compositely depicts the system components. It must be recognized that the hierarchy of systems shown in Figure 2.4 depicts an *hierarchical strand.* Obviously, the full problem definition would include similar strands for each of the components being considered. An example of a hierarchical strand for a physical system is shown in Figure 2.5.

The hierarchical nature and structure of systems permits the analysis of a system in terms not only of the system under consideration, but also in terms of both higher- and lower-level systems. Furthermore, this structure provides the framework for analyzing the comprehensive aspects of the problem as well as the technical details that must be addressed and considered. Certainly, the framework reflects the understanding of the overall system that has been developed in the course of the problem formulation. In this respect, the system structure permits an examination of the response to any modification or change of a part of the system. This is vital to the planning and design effort because it defines those components that are sensitive to any particular problem solution.

To illustrate the multilevel systems concept, a transportation example has been chosen that has an emphasis on motor vehicle parking. Initially it might seem that parking is not a complex problem area; however, a thorough investigation would reveal that parking can have significant effects in terms of a breadth of design and planning questions. As the scope of a parking problem changes, different design and planning questions result that require the development and use of different modeling and analysis rationales.

Motor vehicle parking can be considered as a basic component of the high-

Macrosystem

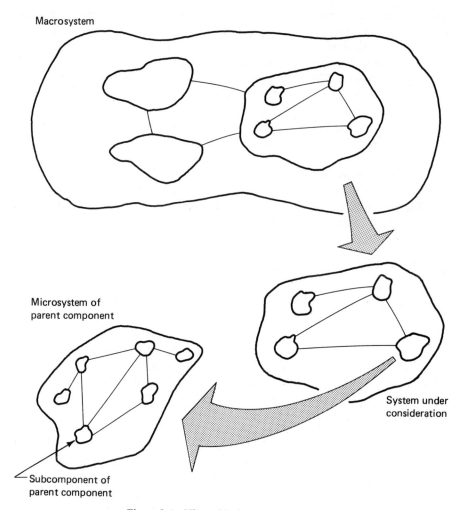

Figure 2.4   Hierarchical system structure.

way system, together with the vehicle and roadway components. Each of these three components have attributes or characteristics that may also be identified. For example, the parking component has physical characteristics that reflect the layout of the parking spaces. Other characteristics may be associated with the overall supply of parking and possibly attributes related to *policies* governing the use or management of parking in general. Similar characteristics of the vehicle and roadway components may also be identified.

For this particular system, each component may have an interaction with both of the other component elements, and these interactions may be expressed in terms of the component attributes. The modification of vehicle dimensions, for example, potentially affects the physical dimensions of the parking spaces.

TRANSPORTATION SYSTEM

| Vehicle | Residential street | Collector road | Rural road | Interchange | Freeway |

Driveway   Intersection   Frontage road   Access road

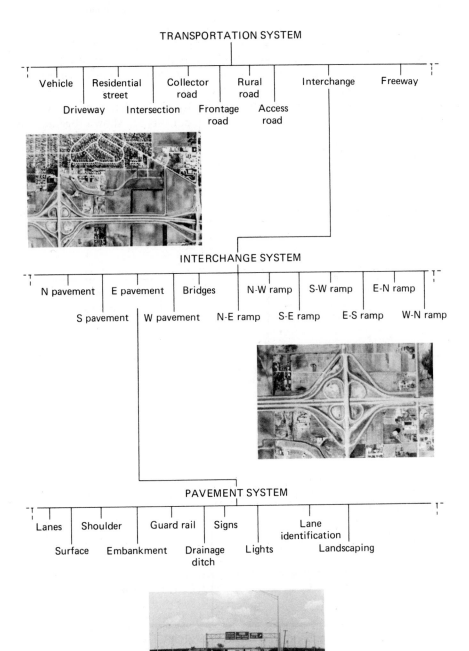

INTERCHANGE SYSTEM

| N pavement | E pavement | Bridges | N-W ramp | S-W ramp | E-N ramp |

S pavement   W pavement   N-E ramp   S-E ramp   E-S ramp   W-N ramp

PAVEMENT SYSTEM

| Lanes | Shoulder | Guard rail | Signs | Lane identification |

Surface   Embankment   Drainage ditch   Lights   Landscaping

Figure 2.5 Levels of systems. (Aerial photographs courtesy of Chicago Aerial Survey.)

A change in vehicle weight will affect the design of safety barriers on a roadway or possibly the pavement structure. Also, the modification of a roadway network may affect the demand for parking. Finally, policies on vehicle use certainly interact with parking and roadway requirements.

Figure 2.6 depicts the representation of this formulation of the highway system. Typical attributes of each of the components are shown together with

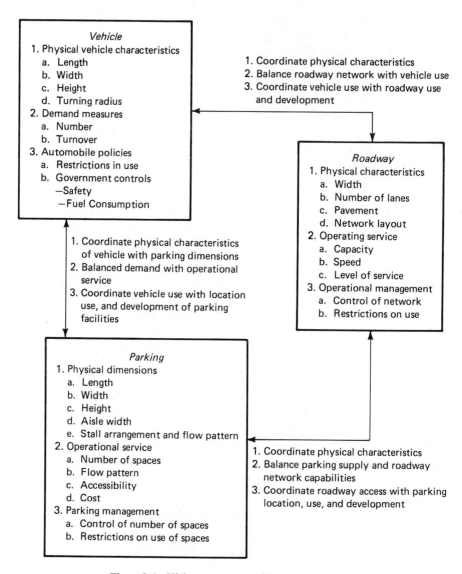

Figure 2.6 Highway system problem formulation.

several component interactions. It may be noted that different design and planning approaches, techniques, and models would be required to address problem areas posed by the differing types of interactions. For example, in the case of the interaction of the vehicle and parking components, the coordination of physical characteristics of the automobile and parking would focus on techniques to analyze the actual layout of spaces. At the next level, the balance of supply with demand would require planning forecast models that estimate the total number of required spaces given some level of vehicle use. Finally, at the management level of interaction, the models and analysis would be directed to trade-offs and consequences of providing or not providing new facilities or imposing restrictions on vehicle use.

Decisions with respect to the provision of parking facilities would also have implications on the need for roadway access; thus the roadway component would be affected as well. Users of the highway system could be attracted to new parking facilities, and a different use pattern of the roadway network would evolve. On the other hand, failure to provide adequate parking could result in people choosing other destinations of travel that could increase the use and wear of the vehicle and the roadway. In essence a change in one of the components of even this simple system requires careful thought because of the consequences.

The highway system is a part of an overall transport system, which is comprised of multiple forms of modes of movement; thus a second-level system emerges that has an increased problem scope. In the case of a particular design problem, it would be necessary to identify the specific modes that would be involved. This next-higher-level system is depicted schematically in Figure 2.7. For the sake of simplicity, the other modes or forms of transportation have been generalized. In addition, transportation is a service function that provides mobility for societal activities or objectives; thus transportation can be characterized as a component of an urban or societal system. This expansion of the system to the next-higher level is also shown in Figure 2.7. This figure reveals a composite view of the hierarchical system that would be involved and portrays the full system environment in which the engineer can embed the problem and examine the extent of the system that should be considered. Certainly, the higher-level systems represent an increase in problem scope, whereas the lower-level systems reflect an increase in technical detail. It must be recognized that it would be possible to extend the hierarchy by further subdivision of the highway system components, even though it was not included in this particular example.

As has been mentioned, the hierarchical system provides the framework of defining the consequences of a given action relative to a component in the hierarchy of systems. For example, assume that a city council is considering the adoption of a policy that limits the amount of parking in a downtown area. In this case, a solution has been specified and the engineer must evaluate the consequences of the imposition of such a solution. Using the framework depicted in Figure 2.7, the components at the various systems levels can be defined for

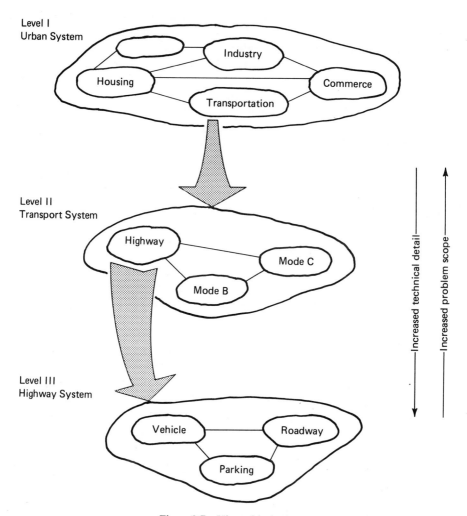

Figure 2.7 Hierarchical system.

which there would be the need to consider potential consequences. If parking were modified, the immediate impact would be related to the vehicle and roadway components. Typical questions to be addressed at this point would examine roadway and vehicle use or the need for roadway network changes. The impact can then be traced to the consequences to the various forms of transportation and ultimately to the consequences to the urban system. For example, at the transportation system level, the impact could be related to the demand or use of other forms of transport. At the urban system level, consequences could be associated with impacts on location of housing, business, and industry.

2.4 THE HIERARCHICAL NATURE
OF THE PROBLEM ENVIRONMENT

A basic relationship exists between the need statement as presented in Section 2.2 and the formulation of a problem in terms of a hierarchical system structure. Recalling that a goal is a desired end state that is to be achieved, and the goal can be expressed as the function that is to be fulfilled by the solution to a problem. An examination of a particular system level in a hierarchy of systems reveals that the purpose or function can be ascertained by defining the role of a system in relation to the next-higher-level system. The objectives may be expressed as a function of the components of a particular system level. In essence, the components indicate the potential specific modifications that may be taken to improve the functioning of a particular system. Based on this concept a hierarchical relationship emerges that is conceptually depicted in Figure 2.8. In order to maintain some degree of simplicity, the figure is shown with a single strand of goals and objectives. Of course, the number of objectives at any level

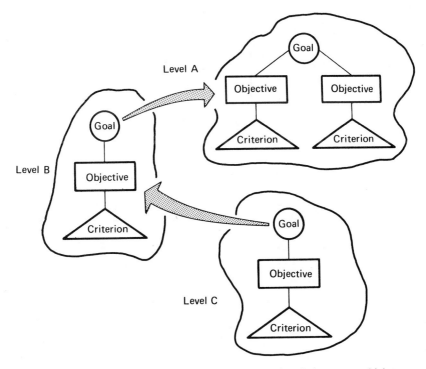

Figure 2.8 System goal as defined by role of system in relation to next-higher system.

would be a function of the number of components at that level. Constraints that exclude a component from being considered would also eliminate that component from being included in the objectives.

The need model or statement concept can also be related to the hierarchical system model. The need statement requires that the engineer resolve the conflict between what is needed and what is available. When applied to the hierarchical system, the need statement can be expressed in terms of the function of a system.

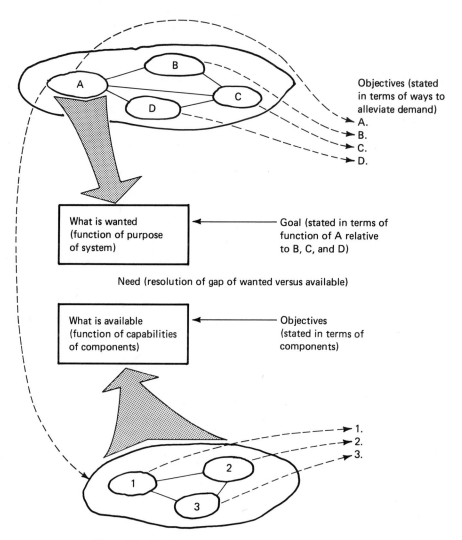

Figure 2.9   Need model related to the hierarchical system.

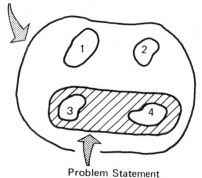 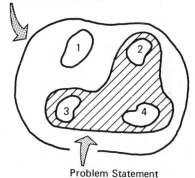

Problem Environment   Problem Environment

Problem Statement   Problem Statement

(a) Problem statement based   (b) Problem statement based
     on components 3 and 4   on components 2, 3, and 4

Figure 2.11 Different problem statements based on components that are included.

...oblem environment. Several alternatives may be considered, which include ...e following:

1. When a component provides input or a lower-level system is involved, it may be treated as either a constant or variable input.
2. Where quantifiable relationships cannot be utilized, the influences outside the defined limits may be considered in a subjective manner.
3. In some cases, components or other levels of systems that lie beyond the defined limits may have to be ignored. This alternative is not desirable, but may be necessary because of the limits that are imposed.

The problem definition, then, will delineate the problem environment and ...ll make clear what goals are to be achieved, what difficulties are to be over-...me, what resources are available, what constraints will exist for an acceptable ...lution, and what criteria will be used to judge the validity of a possible ...lution.

Having developed a statement of the problem, the engineer must now ...cide how the problem will be resolved and the manner in which the components ...e to be considered in this solution. In order to initiate the analysis and design, ...e engineer must develop a model that represents the problem as defined by ...e problem statement.

In modeling the problem that is defined by the problem statement, the ...gineer must examine each of the components associated with the problem. ...r each component that is not included in the problem statement, a decision ...ust be made with respect to how or if the component will be included in ...alysis. In addition, the relationship and ordering of each component must

The question of what is available is related to the components at that system level. The need or what is required is related to the next-higher-level system. Although the components at the higher level create the need or demand for fulfillment, those same components represent options or alternatives for reducing the demand or possibly alleviating what is wanted.

The need statement as it is related to the hierarchical system, goals, and objectives is shown in Figure 2.9. The importance of the need statement in the design and planning process lies in the fact that the challenge is not only to examine how to provide for a demand, but also to consider the alternatives for reducing or even eliminating the demand. In this way, a comprehensive examination of the problem area can be ensured.

For example, in the highway system the goal might be considered in terms of a specific mobility that is required. The highway system alternatives for improving mobility are embedded in the vehicle, roadway, and parking components. The objectives, therefore, would be associated with improving the service offered by these three components.

On the other hand, the alternatives for alleviating the mobility requirements on the highway system is to improve or modify the services provided by the other forms of transport. The need would be to resolve the differences between the requirements for mobility and the capabilities of the highway system.

In some cases, several need statements may be developed that indicate the definition of multiple goals. For example, using a transportation hierarchy, assume that a city wishes to develop an improved transport system that will serve to:

1. Stimulate downtown business
2. Reduce air pollution
3. Alleviate traffic congestion

Figure 2.10 illustrates how this could be depicted. Implicit in the statement of the problem is the task of a transportation solution. The same solution may not satisfy all three stated needs. In such a case it would be necessary to establish a priority with respect to the satisfaction of the goals and the resolution of the need.

Although the problem statement focuses on determining a transportation solution, the problem should also be examined in terms of other alternatives at the urban system level. For example, changes in the location of housing, business, and industry may represent a better solution than one oriented to transportation. A comprehensive problem formulation should include such alternatives.

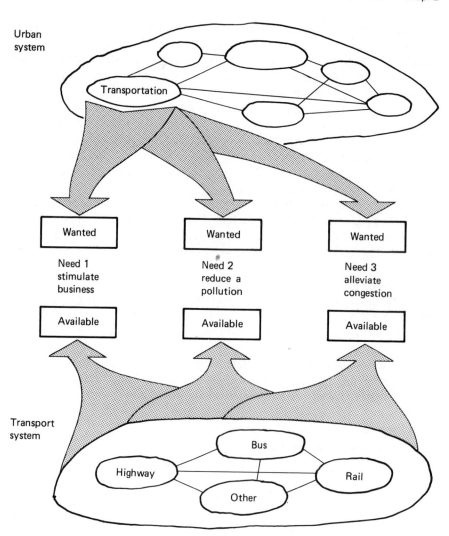

Figure 2.10 Description of multiple need statements.

2.5 DEFINITION OF PROBLEM SCOPE AND CONSTRAINTS

In utilizing a systems approach to design and planning problems, a basic difficulty lies in determining how much of the problem environment is to be considered in the problem definition. Obviously, the investigation increases considerably with each higher system level that is considered. Also, the solution that is achieved depends on that portion of the environment with which the engineer can actually deal.

As the formulation of the hierarchical system is developed, a n limitations will probably become apparent with respect to the design ning effort. These limits may be associated with jurisdictional control of knowledge, or resource constraints.

In the first case, a limit may be reached when some aspect of th lies beyond the influence or control of the professional or professiona dealing with the problem. This type of limit would normally be enco the breadth or scope of the problem increases.

The lack of knowledge concerning a system and its functioning the definition of the problem environment. This lack of knowledge ca to both the breadth and depth of the investigation. In both cases, created by encountering a system in which the components, struct havior are not understood. This type of limit can be overcome to a cel by including pertinent professionals from other disciplines in the s problem. In addition, the inclusion of higher-level systems in the structure will require that persons with a higher level of decision-mal sibility be involved in the problem and its solution. These persons not represent the same professional discipline; thus a problem may is related to the transfer of technical knowledge. Furthermore, the lower levels of systems from the hierarchical structure requires grea technical expertise concerning specific aspects of the problem.

Finally, the time requirements for the study may also pose a c both the breadth and depth of the investigation. As some part of th either considered more broadly or in more detail, the time requi increase considerably. Normally, the engineer faces some time co dictates when an answer to the problem must be attained. This tir certainly must be recognized and considered in defining the scope gation and the resources that are available to accomplish the inv

Thus, in addition to determining the hierarchical system st ciated with the problem, the limitations of influence, knowled and time also must be defined. From this, a statement of the pro that defines the aspects of the problem environment that will be i investigation.

It must be recognized that various problem statements or be generated, depending on how many and which system levels an are incorporated into the engineer's view of the problem. For ex 2.11 depicts conceptually two different problem definitions that a including different components. Figure 2.11a indicates that the ment is based on two components; Figure 2.11b indicates the incl er entire component. The influence of a component may, in fact, some limited or partial manner.

With limits imposed on the study, the engineer must decic components or other system levels that are outside the constra

be made with respect to other components, and the behavior of this interaction must be explicitly defined. If this cannot be accomplished, it indicates that the engineer is unable to solve the problem as it is defined, and the problem statement must be reexamined.

For the parking example, Figure 2.12 represents three problem statements that would result from considering varying degrees of problem scope. It should be noted that for each problem statement the design problem changes due to the fact that different components must or can be considered. Figure 2.12a

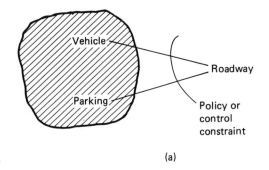

(a)

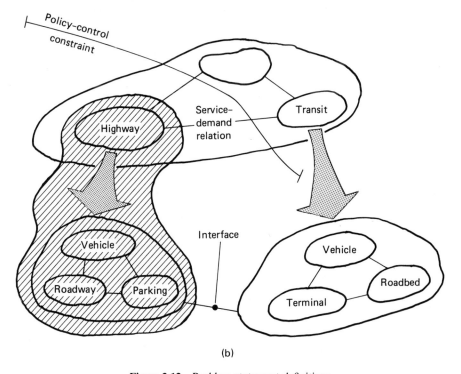

(b)

Figure 2.12 Problem statement definitions.

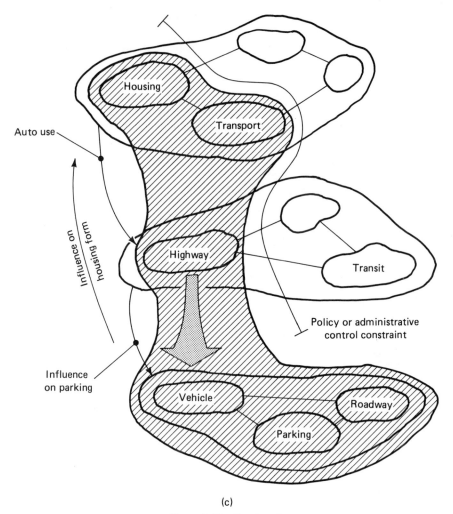

(c)

Figure 2.12 Continued.

depicts the rather simple problem of coordinating the design of the vehicle and the parking facility. This type of problem focuses on the geometric aspects of space design and layout and excludes consideration of the roadway or street network aspects. In essence, this has been the somewhat traditional view of parking and the policies for facility development.

The coordination of roadway considerations and parking raises broader design questions which involve roadway network capacity and parking supply information. Figure 2.12b reveals a problem statement with this increased scope; however, it does not include consideration of impacts on other forms of transport. Parking demand ratios implicitly treat the problem in this context by relating parking needs to the number of trips being made or to area population estimates. A parking analysis approach has been proposed for use in transportation planning studies which incorporates the evaluation of parking adequacy with trip distribution, traffic assignment, and modal split. This approach permits an evaluation of parking not only in terms of the highway network but provides a basis for examining the impact on the entire transportation system.

Finally, Figure 2.12c depicts a problem statement in which parking is indirectly linked with questions related to urban form and structure. This view of the problem addresses the broader management questions and involves strategic parking policy.

In all cases it may be noted that the scope of a specific design statement is a function of a constraint that exists. Frequently, the constraint is of a jurisdictional nature. The immediate acceptance of a jurisdictional constraint represents to some extent the influence of current social and political values and policies. Often, however, decision makers for higher-level systems are not aware of the full consequences of such constraints and of the need for a broader view of a problem. The engineer has a professional responsibility to ensure that decision makers are well aware of the ramifications of current jurisdictional constraints.

2.6 SUMMARY

The complexity of design and planning problems requires that a definitive framework be developed for examining the full spectrum of component involvement and the consequences that result. The multilevel system structure serves this purpose and provides the engineer with a rational basis for addressing a problem. In fact, the use of a multilevel system framework results in a number of implications that affect the engineer in the conduct of a design and planning effort. In a complex problem, these implications generally work to the advantage of the engineer in achieving an acceptable design.

The first implication that might be cited is related to the formulation of policy. Certainly, the understanding of the multilevel hierarchical system permits the engineer to question the nature of the policies that govern and constrain his design. Furthermore, it permits him to advise the policymakers of the implications of a particular policy.

Second, the engineer may use the hierarchical system to aid in the resolution of design and planning decisions. At each system level, the function of a particular system may be evaluated in terms of the larger system and its function. Thus the view of the engineer is broadened because of the recognition of the problems of the larger system and the information requirements of those who may choose between various alternatives. Also, the hierarchical model of the problem statement enables the more realistic and natural definition of goals and objectives as a means of obtaining more effective solutions to a given problem.

A third implication is associated with required research. The systems framework provides a guide to the areas where the design is constrained because of limitations of knowledge. The framework therefore serves to indicate the research that is needed to address that particular aspect of design.

Finally, the framework permits an evaluation of different designs and design strategies. The consequences of a particular design can be traced to determine the impacts that may be involved. Thus it is a guide for the determination of the consequences that need to be considered and the way they are to be considered.

An examination of a hierarchical system further reveals that planning and design involve more than the achievement of compatibility of the physical components. A system is involved and it is necessary for design to address the management aspects of the system. The management implications raise some rather far-reaching questions as the manner in which the planning and design effort will be carried out, and in how the solution facility will be implemented and managed in practice. In many cases these management aspects have not been considered in the planning context. Nevertheless, management is a vital issue or facility in design and evaluation.

A view of the problem in this context raises broader issues that must be addressed and requires the establishment of compatible design and development policies at the various decision levels. In this environment, the engineer can properly address the design and evaluation of facilities that provide the required service, and the multilevel system model concept becomes vital.

PROBLEMS

P2.1. For each of the following situations, identify the systems of physical objects, agencies, and social groups involved and the possible interaction conflicts that should be considered in the engineering design and planning process (see Section 2.1).

(a) A city council has passed an environmental act requiring that all existing and future utilities be placed underground.

(b) An area within a high-density city has an extremely high crime rate, and the city engineer has been asked to determine if there are engineering solutions that can help to reduce the incidence of crime.

(c) An underpass within a business district is frequently flooded during storms.

(d) A company is changing its product line and desires to remodel its factory to accommodate this change while old products are being phased out.

(e) An urban renewal plan requires the demolition of a high-density residential area and its conversion into a recreational use facility.

(f) A developmental group desires to develop a recreational resort in a publicly owned primitive area and must meet aesthetic, ecological, and environmental constraints.

P2.2. Develop need models for the various levels and decision makers involved in the situations described in Problem P2.1 (see Section 2.2).

P2.3. Observe a major street intersection in your city. For this intersection, identify the hierarchical system that is involved and the components that are contained in each level of the system. Determine the purpose or function at each level for each system and its components. If you detect problems at this intersection, indicate how an examination of the hierarchical system could be utilized to identify the cause of the problems and possible solutions that might be considered. How is the solution approach for this situation different from one that would be useful for an airport, rail station, large parking lot, or shopping center?

P2.4. Visit a production facility and identify one or more systems that exist. List the components of each system and their purpose or function. To what extent does the system change when the engineer faces the following problems?
(a) Spatial layout of material flow
(b) Sequential operations in the product process line
(c) Choice of inventory level for raw materials and finished product in order to meet order and delivery requirements
(d) Effect on product line flows of introducing a new product requiring the use of both general-purpose and special-purpose machines

P2.5. Develop structured general problem statement models for a high-rise building in terms of need model components and the decision hierarchy, and hence develop the relevant goals, objectives, and criteria for the following problems.
(a) Selecting floor layout
(b) Selecting floor heights as influenced by floor use, desired artificial environments supplied by utilities, costs, and attractiveness to tenants and customers
(c) Selecting material for the building's surface
(d) Selecting the transportation system in the building
(e) Determining size and location of the high-rise building
(f) Establishing zoning requirements for city areas

P2.6. Suggest a possible system that can be used to alleviate the problem issues associated with water supply, flooding, and recreation for the region shown in Problem P1.6.

P2.7. Identify systems, system components, structure, and possible constraints in the following situations.
(a) A concrete highway bridge spanning a small country stream
(b) A steel highway bridge spanning a coastal estuary with a swing span to permit shipping traffic
Can you identify various levels of systems associated with the physical objects, the functions being performed, and the processes that permit, produce, use, and maintain the facilities involved.

P2.8. The hierarchical system structure depicted in Figure 2.4 permits the analysis of systems at various levels. This structure provides the framework for analyzing the comprehensive aspects of the problem as well as the technical details that must be addressed and considered. The problems raised at different levels require different types of answers, models, and decision processes.

(a) What types of answers and models can you identify at different levels?

(b) What degree of detail would have to be incorporated into these models?

(c) What types and amounts of data might be required at different levels?

An Introduction to Systems Theories

3

3.1 INTRODUCTION

Given the problem statement, there is a need to then examine the methods, or manner, in which the problem can be expressed for analysis and solution. This chapter begins to examine alternatives for treating and addressing the problem in terms of its description, methods for predicting model response, and formulation of the design analysis as decision processes. For example, given the formulation of a problem in a systems context, a fundamental question that must be addressed is how the system or even parts of the system will be portrayed, treated, and analyzed.

3.2 SYSTEMS ANALYSIS VIEWPOINTS

A variety of theories and viewpoints have been developed and used for system identification, description, behavior prediction, and management purposes. Each system theory is established on the basis of a certain level of assumed knowledge as to what constitutes the system under consideration, of what is known about its internal structure and workings, of the nature and complexity of the system performance as it reacts in its environment, and on the engineering or management issues under consideration. The manner in which the system problem is perceived and set up is often very subjective and constrained.

Even though the formulation of a problem on the basis of components and interactions was stressed in Chapter 2, there are times when it is not possible to define the operating or organizational system in such detail. These cases often reflect situations where the systems do not have accessible or visible parts. Other *assemblages of parts* may not be classed, or operate, as a system unless the engineer can connect them in a conceptual manner which establishes a systems purpose and structure for the otherwise independent parts. For these cases, the identification of a system relevant to the problem being addressed depends mainly on the ability of the engineer to formulate a systems purpose and to see, or conceptualize, a number of components linked together by a structure of some conceived type, which operates in an environment in which constraints may develop.

To accommodate the spectrum and diversity of problems that are encountered in design and planning, several generic systems viewpoints, or theories, have been identified that permit the necessary portrayal and analyses. These different systems theories can be described, and related to each other, in terms of what is initially known about the systems structure of a system, and on the system nature of the basic problem formulation to be addressed by the systems theory.

Several classifications of theories or viewpoints that are commonly found and applied in design and planning efforts are:

1. *The black box approach.* The black box approach is utilized when little or nothing is known about the internal linkages or composition of a given system. The system is depicted and analyzed in terms of the black box system response to some given input.

2. *The state theory approach.* In state theory, an endeavor is made to describe the internal workings and responses of systems in terms of a minimal consistent set of system indicators. The main interest in this case is in the description of the *state* of the system and on the detection, or prediction, of changes in system state with new inputs.

3. *The component integration approach.* This approach attempts to establish systems response in terms of the behavior of known components and specifically identified or manufactured linkages of these components.

4. *The decision process approach.* An approach to systems theory formulated as sets of integrated decisions and decision processes.

Each system theory or viewpoint necessarily reflects a view of reality in the manner in which the system is portrayed and in the selection of system problems the theory purports to address. A necessary corollary follows in the selection of the criteria to be used in establishing acceptable solutions to systems problems. Thus each systems theory exhibits a unique modeling approach in the manner in which the system itself is portrayed as well as in the way in which a system problem is formulated and solved. To some extent these systems theories can be ranked (1) in terms of what is known about the system nature of a problem, and (2) in terms of the extent to which human beings are involved in the content of the system or the manner in which their decisions are handled, or viewed, by the systems theory. Using these viewpoints as reference points, the following four sections of this chapter discuss in greater detail each of the theories listed above.

3.3 THE BLACK BOX APPROACH

A basic behavioral concept of a system is that it is a device which accepts one or more inputs and generates from them one or more outputs. In this behavioral approach to a system the internal composition and workings of the device may be unknown, known only partially, or ignored. In any case, actual interest is focused on the input–output behavior of the system. The system device itself is, in fact, or is considered as such, a single, all-embracing, impenetrable black box. This simple behavioral approach to systems is generally known as the black box approach and is represented schematically in Figure 3.1.

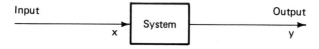

Figure 3.1   Black box.

In fact, the black box problem is a basic problem of science. The general complexity of reality and its phenomena often presents the would-be investigator with a situation that may only be viewed as a black box system. It may be impossible to identify the system components, their attributes, or the constraints. Thus an attempt to link observed system output with observed, or prescribed, system inputs may be the only possible approach to gaining an understanding of the natural system. Indeed, initially there may even be difficulty in ascertaining what is an acceptable input to the system, and little may be known of the magnitude, character, or time delay of the system response or even if one exists. Thus an initial problem to be addressed by the black box approach is that of *system identification*. In such cases, one must attempt to identify the test that can be applied to the system to establish its nature or at least some particular relevant feature.

In many instances, although much may be known of a system, it may be convenient to consider it as a black box system because all that is relevant to the needs of the moment may be the system performance in terms of the relationships between system inputs and system outputs. In a sense this statement of system behavior is often the goal of systems analysis, wherein a systems problem as formulated is solved once an explicit relationship is obtained between what is known (input variables) and what is required (solution output variables).

A typical example of such a situation is the computer program, especially in relation to the features of many hand-held calculators. These calculators frequently will make some type of calculation simply by pushing a button. All that is required is the prior keyed input of data and the selection of the appropriate program button. These built-in features enhance the system complexity of the calculator and the user may not know, or care about, the algorithm involved or the form of the printed logic circuits that make up the calculator.

Mathematically, the black box system can be formulated in terms of the relationships between its inputs (x) and outputs (y). Using this concept the black box system is typically represented by a function, F, and the input–output transformation that it produces by

$$y = F(x)$$

The input x to the black box system may be an element, or a collection of elements taken from the set of permissible input variables as given by a vector $\mathbf{X}$. Similarly, the output y may be an element, or a series of elements in a vector $\mathbf{Y}$ of possible output variables. In this way the behavior of the black box system may be conveniently described in terms of the changing values over time of both the input vector $\mathbf{X}$ and the output vector $\mathbf{Y}$.

In the case of the hand-held calculator, inputs (x) must be keyed in by the selection and touch sequencing of the appropriate buttons from among the total set of buttons (X). The system output may be a graphic display, or printout, so that a similar reasoning can establish the set of possible output variables (Y). Many situations exist, however, where such a simple explanation is not possible.

The identification problem of the black box may be thought of as the problem of finding expressions for the function F in terms of inputs which are considered more manageable or more clearly indicative of certain properties of the system. Inherent in any such approach is the need to ensure that a causal relationship does exist between the system inputs and outputs that is valid over considerable periods of time. Time must always be considered as a potential, or real, input parameter for a black box system. Thus it may be necessary to consider the time history of inputs when attempting to relate system inputs to system outputs. Similarly, an input to a system at a single instant of time may have long-lasting repercussions in the output of the system. In this sense the black box system may have a memory for a long-past input.

An example of a very useful black box approach to complex natural phenomena is the use of the instantaneous unit hydrograph to investigate the relationship between rain storms and the resulting runoff on a catchment area. The instantaneous unit hydrograph for runoff from a watershed basin is a fundamental concept in hydrology in which a single instantaneous system input can produce a prolonged system output. The instantaneous unit hydrograph is postulated as the catchment runoff that would be observed if at a given point in time a storm of 1 in. of rainfall fell simultaneously and uniformly over the entire catchment basin. A schematic illustration of the system is shown in Figure 3.2.

In the unit hydrograph approach no attempt is made to discover the intricacies of the catchment topography and creek–river interaction because

As the formulation of the hierarchical system is developed, a number of limitations will probably become apparent with respect to the design and planning effort. These limits may be associated with jurisdictional control, the state of knowledge, or resource constraints.

In the first case, a limit may be reached when some aspect of the problem lies beyond the influence or control of the professional or professionals who are dealing with the problem. This type of limit would normally be encountered as the breadth or scope of the problem increases.

The lack of knowledge concerning a system and its functioning may limit the definition of the problem environment. This lack of knowledge can be related to both the breadth and depth of the investigation. In both cases, the limit is created by encountering a system in which the components, structure, or behavior are not understood. This type of limit can be overcome to a certain degree by including pertinent professionals from other disciplines in the study of the problem. In addition, the inclusion of higher-level systems in the hierarchical structure will require that persons with a higher level of decision-making responsibility be involved in the problem and its solution. These persons may or may not represent the same professional discipline; thus a problem may result that is related to the transfer of technical knowledge. Furthermore, the inclusion of lower levels of systems from the hierarchical structure requires greater depth in technical expertise concerning specific aspects of the problem.

Finally, the time requirements for the study may also pose a constraint on both the breadth and depth of the investigation. As some part of the problem is either considered more broadly or in more detail, the time requirements may increase considerably. Normally, the engineer faces some time constraint that dictates when an answer to the problem must be attained. This time constraint certainly must be recognized and considered in defining the scope of the investigation and the resources that are available to accomplish the investigation.

Thus, in addition to determining the hierarchical system structure associated with the problem, the limitations of influence, knowledge or ability, and time also must be defined. From this, a statement of the problem emerges that defines the aspects of the problem environment that will be included in the investigation.

It must be recognized that various problem statements or definitions can be generated, depending on how many and which system levels and components are incorporated into the engineer's view of the problem. For example, Figure 2.11 depicts conceptually two different problem definitions that are the result of including different components. Figure 2.11a indicates that the problem statement is based on two components; Figure 2.11b indicates the inclusion of another entire component. The influence of a component may, in fact, be included in some limited or partial manner.

With limits imposed on the study, the engineer must decide how to treat components or other system levels that are outside the constrained or limited

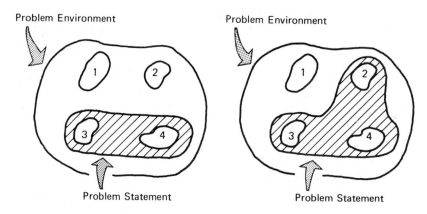

Problem Environment   Problem Environment

Problem Statement   Problem Statement

(a) Problem statement based   (b) Problem statement based
     on components 3 and 4   on components 2, 3, and 4

Figure 2.11 Different problem statements based on components that are included.

problem environment. Several alternatives may be considered, which include the following:

1. When a component provides input or a lower-level system is involved, it may be treated as either a constant or variable input.
2. Where quantifiable relationships cannot be utilized, the influences outside the defined limits may be considered in a subjective manner.
3. In some cases, components or other levels of systems that lie beyond the defined limits may have to be ignored. This alternative is not desirable, but may be necessary because of the limits that are imposed.

The problem definition, then, will delineate the problem environment and will make clear what goals are to be achieved, what difficulties are to be overcome, what resources are available, what constraints will exist for an acceptable solution, and what criteria will be used to judge the validity of a possible solution.

Having developed a statement of the problem, the engineer must now decide how the problem will be resolved and the manner in which the components are to be considered in this solution. In order to initiate the analysis and design, the engineer must develop a model that represents the problem as defined by the problem statement.

In modeling the problem that is defined by the problem statement, the engineer must examine each of the components associated with the problem. For each component that is not included in the problem statement, a decision must be made with respect to how or if the component will be included in analysis. In addition, the relationship and ordering of each component must

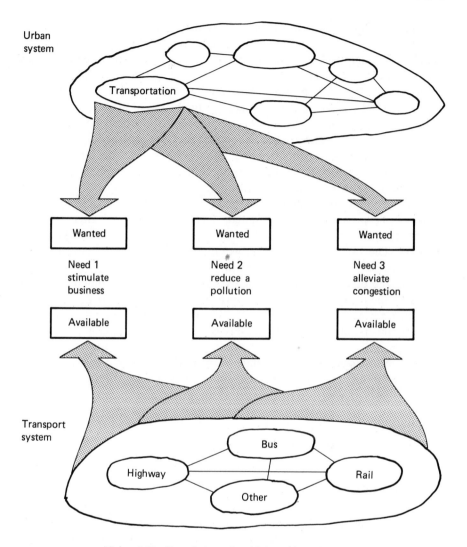

Figure 2.10   Description of multiple need statements.

2.5 DEFINITION OF PROBLEM SCOPE AND CONSTRAINTS

In utilizing a systems approach to design and planning problems, a basic difficulty lies in determining how much of the problem environment is to be
considered in the problem definition. Obviously, the investigation increases
considerably with each higher system level that is considered. Also, the solution
that is achieved depends on that portion of the environment with which the
engineer can actually deal.

The question of what is available is related to the components at that system level. The need or what is required is related to the next-higher-level system. Although the components at the higher level create the need or demand for fulfillment, those same components represent options or alternatives for reducing the demand or possibly alleviating what is wanted.

The need statement as it is related to the hierarchical system, goals, and objectives is shown in Figure 2.9. The importance of the need statement in the design and planning process lies in the fact that the challenge is not only to examine how to provide for a demand, but also to consider the alternatives for reducing or even eliminating the demand. In this way, a comprehensive examination of the problem area can be ensured.

For example, in the highway system the goal might be considered in terms of a specific mobility that is required. The highway system alternatives for improving mobility are embedded in the vehicle, roadway, and parking components. The objectives, therefore, would be associated with improving the service offered by these three components.

On the other hand, the alternatives for alleviating the mobility requirements on the highway system is to improve or modify the services provided by the other forms of transport. The need would be to resolve the differences between the requirements for mobility and the capabilities of the highway system.

In some cases, several need statements may be developed that indicate the definition of multiple goals. For example, using a transportation hierarchy, assume that a city wishes to develop an improved transport system that will serve to:

1. Stimulate downtown business

2. Reduce air pollution

3. Alleviate traffic congestion

Figure 2.10 illustrates how this could be depicted. Implicit in the statement of the problem is the task of a transportation solution. The same solution may not satisfy all three stated needs. In such a case it would be necessary to establish a priority with respect to the satisfaction of the goals and the resolution of the need.

Although the problem statement focuses on determining a transportation solution, the problem should also be examined in terms of other alternatives at the urban system level. For example, changes in the location of housing, business, and industry may represent a better solution than one oriented to transportation. A comprehensive problem formulation should include such alternatives.

be made with respect to other components, and the behavior of this interaction must be explicitly defined. If this cannot be accomplished, it indicates that the engineer is unable to solve the problem as it is defined, and the problem statement must be reexamined.

For the parking example, Figure 2.12 represents three problem statements that would result from considering varying degrees of problem scope. It should be noted that for each problem statement the design problem changes due to the fact that different components must or can be considered. Figure 2.12a

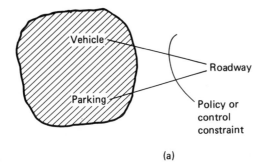

(a)

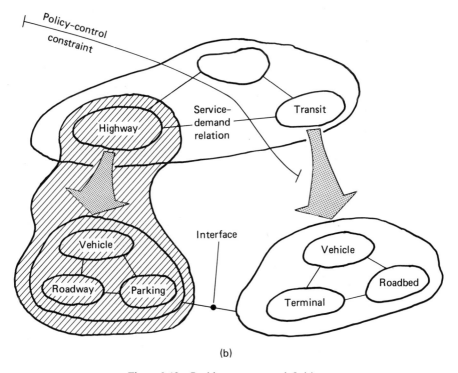

(b)

Figure 2.12   Problem statement definitions.

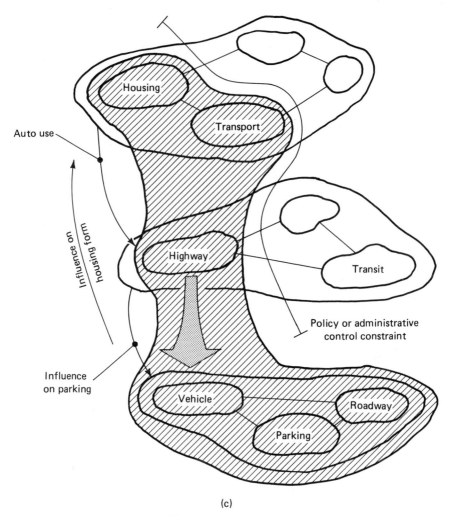

(c)

Figure 2.12   Continued.

depicts the rather simple problem of coordinating the design of the vehicle and the parking facility. This type of problem focuses on the geometric aspects of space design and layout and excludes consideration of the roadway or street network aspects. In essence, this has been the somewhat traditional view of parking and the policies for facility development.

The coordination of roadway considerations and parking raises broader design questions which involve roadway network capacity and parking supply information. Figure 2.12b reveals a problem statement with this increased scope; however, it does not include consideration of impacts on other forms of transport. Parking demand ratios implicitly treat the problem in this context by relating parking needs to the number of trips being made or to area population estimates. A parking analysis approach has been proposed for use in transportation planning studies which incorporates the evaluation of parking adequacy with trip distribution, traffic assignment, and modal split. This approach permits an evaluation of parking not only in terms of the highway network but provides a basis for examining the impact on the entire transportation system.

Finally, Figure 2.12c depicts a problem statement in which parking is indirectly linked with questions related to urban form and structure. This view of the problem addresses the broader management questions and involves strategic parking policy.

In all cases it may be noted that the scope of a specific design statement is a function of a constraint that exists. Frequently, the constraint is of a jurisdictional nature. The immediate acceptance of a jurisdictional constraint represents to some extent the influence of current social and political values and policies. Often, however, decision makers for higher-level systems are not aware of the full consequences of such constraints and of the need for a broader view of a problem. The engineer has a professional responsibility to ensure that decision makers are well aware of the ramifications of current jurisdictional constraints.

2.6 SUMMARY

The complexity of design and planning problems requires that a definitive framework be developed for examining the full spectrum of component involvement and the consequences that result. The multilevel system structure serves this purpose and provides the engineer with a rational basis for addressing a problem. In fact, the use of a multilevel system framework results in a number of implications that affect the engineer in the conduct of a design and planning effort. In a complex problem, these implications generally work to the advantage of the engineer in achieving an acceptable design.

The first implication that might be cited is related to the formulation of policy. Certainly, the understanding of the multilevel hierarchical system permits the engineer to question the nature of the policies that govern and constrain his design. Furthermore, it permits him to advise the policymakers of the implications of a particular policy.

Second, the engineer may use the hierarchical system to aid in the resolution of design and planning decisions. At each system level, the function of a particular system may be evaluated in terms of the larger system and its function. Thus the view of the engineer is broadened because of the recognition of the problems of the larger system and the information requirements of those who may choose between various alternatives. Also, the hierarchical model of the problem statement enables the more realistic and natural definition of goals and objectives as a means of obtaining more effective solutions to a given problem.

A third implication is associated with required research. The systems framework provides a guide to the areas where the design is constrained because of limitations of knowledge. The framework therefore serves to indicate the research that is needed to address that particular aspect of design.

Finally, the framework permits an evaluation of different designs and design strategies. The consequences of a particular design can be traced to determine the impacts that may be involved. Thus it is a guide for the determination of the consequences that need to be considered and the way they are to be considered.

An examination of a hierarchical system further reveals that planning and design involve more than the achievement of compatibility of the physical components. A system is involved and it is necessary for design to address the management aspects of the system. The management implications raise some rather far-reaching questions as the manner in which the planning and design effort will be carried out, and in how the solution facility will be implemented and managed in practice. In many cases these management aspects have not been considered in the planning context. Nevertheless, management is a vital issue or facility in design and evaluation.

A view of the problem in this context raises broader issues that must be addressed and requires the establishment of compatible design and development policies at the various decision levels. In this environment, the engineer can properly address the design and evaluation of facilities that provide the required service, and the multilevel system model concept becomes vital.

PROBLEMS

P2.1. For each of the following situations, identify the systems of physical objects, agencies, and social groups involved and the possible interaction conflicts that should be considered in the engineering design and planning process (see Section 2.1).

  (a) A city council has passed an environmental act requiring that all existing and future utilities be placed underground.

  (b) An area within a high-density city has an extremely high crime rate, and the city engineer has been asked to determine if there are engineering solutions that can help to reduce the incidence of crime.

  (c) An underpass within a business district is frequently flooded during storms.

  (d) A company is changing its product line and desires to remodel its factory to accommodate this change while old products are being phased out.

  (e) An urban renewal plan requires the demolition of a high-density residential area and its conversion into a recreational use facility.

(f) A developmental group desires to develop a recreational resort in a publicly owned primitive area and must meet aesthetic, ecological, and environmental constraints.

P2.2. Develop need models for the various levels and decision makers involved in the situations described in Problem P2.1 (see Section 2.2).

P2.3. Observe a major street intersection in your city. For this intersection, identify the hierarchical system that is involved and the components that are contained in each level of the system. Determine the purpose or function at each level for each system and its components. If you detect problems at this intersection, indicate how an examination of the hierarchical system could be utilized to identify the cause of the problems and possible solutions that might be considered. How is the solution approach for this situation different from one that would be useful for an airport, rail station, large parking lot, or shopping center?

P2.4. Visit a production facility and identify one or more systems that exist. List the components of each system and their purpose or function. To what extent does the system change when the engineer faces the following problems?
(a) Spatial layout of material flow
(b) Sequential operations in the product process line
(c) Choice of inventory level for raw materials and finished product in order to meet order and delivery requirements
(d) Effect on product line flows of introducing a new product requiring the use of both general-purpose and special-purpose machines

P2.5. Develop structured general problem statement models for a high-rise building in terms of need model components and the decision hierarchy, and hence develop the relevant goals, objectives, and criteria for the following problems.
(a) Selecting floor layout
(b) Selecting floor heights as influenced by floor use, desired artificial environments supplied by utilities, costs, and attractiveness to tenants and customers
(c) Selecting material for the building's surface
(d) Selecting the transportation system in the building
(e) Determining size and location of the high-rise building
(f) Establishing zoning requirements for city areas

P2.6. Suggest a possible system that can be used to alleviate the problem issues associated with water supply, flooding, and recreation for the region shown in Problem P1.6.

P2.7. Identify systems, system components, structure, and possible constraints in the following situations.
(a) A concrete highway bridge spanning a small country stream
(b) A steel highway bridge spanning a coastal estuary with a swing span to permit shipping traffic
Can you identify various levels of systems associated with the physical objects, the functions being performed, and the processes that permit, produce, use, and maintain the facilities involved.

P2.8. The hierarchical system structure depicted in Figure 2.4 permits the analysis of systems at various levels. This structure provides the framework for analyzing the comprehensive aspects of the problem as well as the technical details that must be addressed and considered. The problems raised at different levels require different types of answers, models, and decision processes.

(a) What types of answers and models can you identify at different levels?

(b) What degree of detail would have to be incorporated into these models?

(c) What types and amounts of data might be required at different levels?

An Introduction
to Systems Theories

3

3.1 INTRODUCTION

Given the problem statement, there is a need to then examine the methods, or manner, in which the problem can be expressed for analysis and solution. This chapter begins to examine alternatives for treating and addressing the problem in terms of its description, methods for predicting model response, and formulation of the design analysis as decision processes. For example, given the formulation of a problem in a systems context, a fundamental question that must be addressed is how the system or even parts of the system will be portrayed, treated, and analyzed.

3.2 SYSTEMS ANALYSIS VIEWPOINTS

A variety of theories and viewpoints have been developed and used for system identification, description, behavior prediction, and management purposes. Each system theory is established on the basis of a certain level of assumed knowledge as to what constitutes the system under consideration, of what is known about its internal structure and workings, of the nature and complexity of the system performance as it reacts in its environment, and on the engineering or management issues under consideration. The manner in which the system problem is perceived and set up is often very subjective and constrained.

Even though the formulation of a problem on the basis of components and interactions was stressed in Chapter 2, there are times when it is not possible to define the operating or organizational system in such detail. These cases often reflect situations where the systems do not have accessible or visible parts. Other *assemblages of parts* may not be classed, or operate, as a system unless the engineer can connect them in a conceptual manner which establishes a systems purpose and structure for the otherwise independent parts. For these cases, the identification of a system relevant to the problem being addressed depends mainly on the ability of the engineer to formulate a systems purpose and to see, or conceptualize, a number of components linked together by a structure of some conceived type, which operates in an environment in which constraints may develop.

To accommodate the spectrum and diversity of problems that are encountered in design and planning, several generic systems viewpoints, or theories, have been identified that permit the necessary portrayal and analyses. These different systems theories can be described, and related to each other, in terms of what is initially known about the systems structure of a system, and on the system nature of the basic problem formulation to be addressed by the systems theory.

Several classifications of theories or viewpoints that are commonly found and applied in design and planning efforts are:

1. *The black box approach.* The black box approach is utilized when little or nothing is known about the internal linkages or composition of a given system. The system is depicted and analyzed in terms of the black box system response to some given input.

2. *The state theory approach.* In state theory, an endeavor is made to describe the internal workings and responses of systems in terms of a minimal consistent set of system indicators. The main interest in this case is in the description of the *state* of the system and on the detection, or prediction, of changes in system state with new inputs.

3. *The component integration approach.* This approach attempts to establish systems response in terms of the behavior of known components and specifically identified or manufactured linkages of these components.

4. *The decision process approach.* An approach to systems theory formulated as sets of integrated decisions and decision processes.

Each system theory or viewpoint necessarily reflects a view of reality in the manner in which the system is portrayed and in the selection of system problems the theory purports to address. A necessary corollary follows in the selection of the criteria to be used in establishing acceptable solutions to systems problems. Thus each systems theory exhibits a unique modeling approach in the manner in which the system itself is portrayed as well as in the way in which a system problem is formulated and solved. To some extent these systems theories can be ranked (1) in terms of what is known about the system nature of a problem, and (2) in terms of the extent to which human beings are involved in the content of the system or the manner in which their decisions are handled, or viewed, by the systems theory. Using these viewpoints as reference points, the following four sections of this chapter discuss in greater detail each of the theories listed above.

3.3 THE BLACK BOX APPROACH

A basic behavioral concept of a system is that it is a device which accepts one or more inputs and generates from them one or more outputs. In this behavioral approach to a system the internal composition and workings of the device may be unknown, known only partially, or ignored. In any case, actual interest is focused on the input–output behavior of the system. The system device itself is, in fact, or is considered as such, a single, all-embracing, impenetrable black box. This simple behavioral approach to systems is generally known as the black box approach and is represented schematically in Figure 3.1.

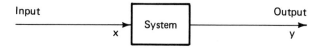

Figure 3.1   Black box.

In fact, the black box problem is a basic problem of science. The general complexity of reality and its phenomena often presents the would-be investigator with a situation that may only be viewed as a black box system. It may be impossible to identify the system components, their attributes, or the constraints. Thus an attempt to link observed system output with observed, or prescribed, system inputs may be the only possible approach to gaining an understanding of the natural system. Indeed, initially there may even be difficulty in ascertaining what is an acceptable input to the system, and little may be known of the magnitude, character, or time delay of the system response or even if one exists. Thus an initial problem to be addressed by the black box approach is that of *system identification*. In such cases, one must attempt to identify the test that can be applied to the system to establish its nature or at least some particular relevant feature.

In many instances, although much may be known of a system, it may be convenient to consider it as a black box system because all that is relevant to the needs of the moment may be the system performance in terms of the relationships between system inputs and system outputs. In a sense this statement of system behavior is often the goal of systems analysis, wherein a systems problem as formulated is solved once an explicit relationship is obtained between what is known (input variables) and what is required (solution output variables).

A typical example of such a situation is the computer program, especially in relation to the features of many hand-held calculators. These calculators frequently will make some type of calculation simply by pushing a button. All

that is required is the prior keyed input of data and the selection of the appropriate program button. These built-in features enhance the system complexity
of the calculator and the user may not know, or care about, the algorithm
involved or the form of the printed logic circuits that make up the calculator.

Mathematically, the black box system can be formulated in terms of the
relationships between its inputs (x) and outputs (y). Using this concept the black
box system is typically represented by a function, F, and the input–output
transformation that it produces by

$$y = F(x)$$

The input x to the black box system may be an element, or a collection of
elements taken from the set of permissible input variables as given by a vector
$\mathbf{X}$. Similarly, the output y may be an element, or a series of elements in a vector
$\mathbf{Y}$ of possible output variables. In this way the behavior of the black box system
may be conveniently described in terms of the changing values over time of both
the input vector $\mathbf{X}$ and the output vector $\mathbf{Y}$.

In the case of the hand-held calculator, inputs (x) must be keyed in by the
selection and touch sequencing of the appropriate buttons from among the total
set of buttons (X). The system output may be a graphic display, or printout, so
that a similar reasoning can establish the set of possible output variables (Y).
Many situations exist, however, where such a simple explanation is not possible.

The identification problem of the black box may be thought of as the
problem of finding expressions for the function F in terms of inputs which are
considered more manageable or more clearly indicative of certain properties of
the system. Inherent in any such approach is the need to ensure that a causal
relationship does exist between the system inputs and outputs that is valid over
considerable periods of time. Time must always be considered as a potential,
or real, input parameter for a black box system. Thus it may be necessary to
consider the time history of inputs when attempting to relate system inputs to
system outputs. Similarly, an input to a system at a single instant of time may
have long-lasting repercussions in the output of the system. In this sense the
black box system may have a memory for a long-past input.

An example of a very useful black box approach to complex natural
phenomena is the use of the instantaneous unit hydrograph to investigate the
relationship between rain storms and the resulting runoff on a catchment area.
The instantaneous unit hydrograph for runoff from a watershed basin is a fundamental concept in hydrology in which a single instantaneous system input
can produce a prolonged system output. The instantaneous unit hydrograph is
postulated as the catchment runoff that would be observed if at a given point
in time a storm of 1 in. of rainfall fell simultaneously and uniformly over the
entire catchment basin. A schematic illustration of the system is shown in
Figure 3.2.

In the unit hydrograph approach no attempt is made to discover the
intricacies of the catchment topography and creek–river interaction because

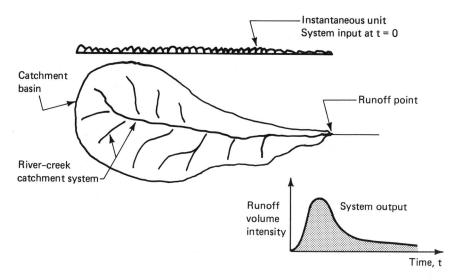

Figure 3.2 Instantaneous unit hydrograph for a catchment basin: input–output black box model.

these aspects of the natural system are automatically captured in the shape, magnitude, and delay characteristics of the unique unit hydrograph that results. In effect, the history of each raindrop as it hits a specific spot on the catchment basin, its subsequent runoff movements, and eventual discharge at the catchment basin output position contributes to the natural system integration that produces the resultant unit hydrograph runoff output response to the initial rainfall input.

In many design situations, interests are focused on the behavior of the facility as a whole rather than on the complex interrelationships among the system components and their individual behavior. If simple functional performance criteria can be established, the need for an internal knowledge of the mechanics and structure of the components can be bypassed, and system design can be performed through direct application of the input–output behavior characteristics.

A simple example is in engineering design involving plates, sheathing, and shells. Consider, for instance, a thin, square sheet of steel acting as a simply supported horizontal plate subjected to vertical loading. The displacement responses of the plate to various loading patterns can be readily measured independently of a knowledge of the internal stress patterns and the manner in which its internal elements behave. The simple behavioral approach to structural elements and structural systems can be viewed as special cases of the black box system methodology.

Finally, many situations exist in engineering where system response is not directly, or linearly, related to systems input. In many cases the engineer deals with a facility that has a threshold that must be reached before the facility is in a permissive mode to react to system inputs. Many black-box-type systems exist in engineering where the same system input may produce different or no outputs. An example of such a situation is the relationship between inputs and outputs in a reservoir. Initially, on completion of the reservoir, inflow establishes the dead storage of the reservoir and only after this dead storage has been formed will system outputs be possible.

In summary, the fundamental problem of the black box system is to establish the functional relationship between system inputs and outputs. As indicated in the examples above, it is not easy to establish these relationships. Fortunately, in most practical engineering situations all that is required in system description, design, or behavior prediction is to relate known internal system behavior through to the point where system inputs and outputs can be functionally and explicitly related. Once this has been achieved, full knowledge of the complete system can be conveniently ignored and the black box system formulation becomes adequate for engineering purposes.

3.4 THE STATE THEORY APPROACH

A relatively simple, and intuitively appealing conceptual approach to the description and understanding of the behavior of a system is to devise a set of reference system variables which provide a unique and meaningful descriptive measure of the system state at an instant in time. Such a set of reference variables is called the *state vector* of the system and each reference variable in the state vector is called a *state variable* of the system. The system configurations and behavior is then captured and portrayed in terms of the set of the changing values of each system state variable during the life of the system. Whether the state variables are continuously read or evaluated, or measured only at specific discrete times depends mainly on the nature of the system and the purpose behind the system study.

A common example of the state variable approach to monitoring or describing the behavior of a system is the instrumentation panel in motor vehicles. The fuel tank volume status, vehicle speed, engine revolutions per minute, turning light indicator status, battery charging, door lock displays, and so on, provide a ready means of portraying and evaluating vehicle status during driving operations. A more comprehensive check would require readings on tire pressures and oil, radiator, battery, and brake fluid levels plus indicators of tire wear. Finally, most drivers constantly monitor engine and body noises as early indicators of potential malfunctions. An illustration of a possible vehicle state vector and its component state variables is given in the following list:

    Car identification variables
      Year
      Make

Engine number

Instrument panel variables
  Gas tank
  Speed
  Rpm
  Various indicator lights

Engine performance check variables
  Oil level
  Battery level
  Radiator level
  Brake fluid level

Vehicle performance characteristics
  Tire
  Braking
  Stearing
  Ride quality

Appearance indicators
  Body condition
  Interior condition
  Paint quality

A number of basic problems must be addressed when developing a state theory approach to systems. A fundamental problem arises in the determination of whether a particular variable can qualify as a system state variable. If it does, it is necessary to consider whether it is redundant in terms of explaining the system theory purpose or its value can be deduced from other state variables. Another related fundamental problem is the determination of what constitutes a minimally acceptable set of state variables in the state vector capable of serving the needs of portraying system configuration and behavior and/or the purpose of the system modeler. The nature of a system state variable and the number of entries in the required state vector are intimately tied to the systems purpose being modeled and the complexity of the systems theory being formulated.

Another fundamental issue affecting the state formulation of a system is whether the intent is to portray the configuration and status of the system or whether a predictive systems model is required capable of predicting future behavior patterns. In the simpler case of portraying system configuration and status, the choice of system state variables and the size of the state vector depend on the accuracy with which the system is to be described.

The description of a system is therefore intimately linked with the construction and characteristics of a system model. Selecting a specific set of variables out of the large number that may exist enables a specific system description to be developed. An entirely different selection of variables may permit another and entirely different system description. Clearly, it is important to consider carefully the purpose of the system description and to include all the dominant variables that affect the problem under consideration. The system behavior is then portrayed by the succession of states that the system assumes at particular points of time.

In the more difficult case of formulating a system predictive model, the state vector must contain state variables that enable the future states of the system to be predicted in terms of the current state vector and the current and future system inputs. The state vector may have to contain information on the set of possible new states that can be reached from the current state. In this situation the state that does next occur is, so to speak, selected by the occurrence and magnitude of the next system input. In this way, changes to what will normally follow as a consequence of the current state and past histories of states and inputs can be detected as the consequence of new and future system inputs. Thus a whole variety of state theory models can be developed.

As an example of the development of a state model of an engineering problem consider the water supply system shown in Figure 3.3, which basically portrays the flow and use of water in a typical city–farmland situation. To describe the water distribution system in detail would require describing the system components and their interaction: river, dam, pipeline, chlorination plant, town retriculation systems, drainage system, waste treatment plant, irrigation canal, irrigation areas, seepage system flows, and so on. The individual component state vectors for each of these components would require many descriptions: physical, economic, social, and the like.

Suppose that the basic problem is concerned with water management and more specifically with the management of the water impounded by the reservoir; that is, the owner wants to determine how much water to release each

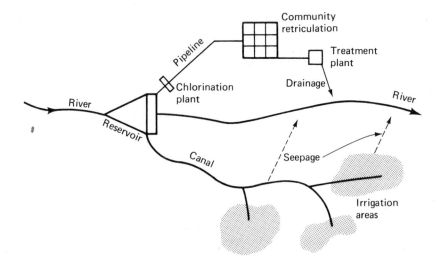

Figure 3.3 Typical city–farmland water supply system.

month to maximize her income from the sale of water. In this case, attempts might be made to model the reservoir in terms of the reservoir size; depth of water; surface area exposed to evaporation; stream-flow characteristics; the community's annual, seasonal, and daily demands; agricultural demands; minimum stream-flow requirements to maintain ecological, salinity, and boating conditions; and so on.

If, for example, too much water is released for irrigation, future flows may not materialize, thus possibly jeopardizing the community's water supply. If, on the other hand, too little water is released, sudden future flows may cause the reservoir to fill and cause flooding. Clearly, a management problem is involved in handling the storage volume.

The development of a major management control model for the reservoir requires, as a first step, the development of a suitable model for the reservoir storage volume. A simple model can be developed considering the reservoir purely from a storage volume point of view. For this model the volume of impounded water is the state variable, because the state of the reservoir can be described in terms of how full it is at any time, as shown in Figure 3.4. The state can change depending on the nature of the inflow and outflow. Although inflow can easily be considered an environmental input to the system, the outflow, which is primarily the result of human action, can be considered a negative input from the storage volume point of view.

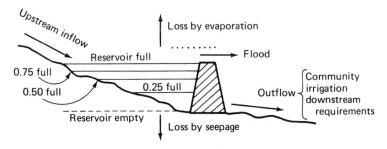

Figure 3.4   Water reservoir system.

Because of its simplicity, a graphical concept will be used, as shown in Figure 3.5. Incremental volumes of quarter reservoir capacity are used to differentiate the various states of the reservoir; Therefore, $S_i = 1, \frac{3}{4}, \frac{1}{2},$ and $\frac{1}{4}$ means that the reservoir is full, three-fourths full, one-half full, and one-fourth full, respectively. $S_i = 1$ and $w_i = \frac{1}{4}$ means that the reservoir is overfull as a result of a storm and that a one-fourth unit volume of water will spill over, resulting in a minor flood during the next period. In Figure 3.5b a typical system response is shown. The actual seasonal inflow from the stream can be obtained from historical data and modeled as a chance node with probabilities associated with the various different inflows for each seasonal time increment.

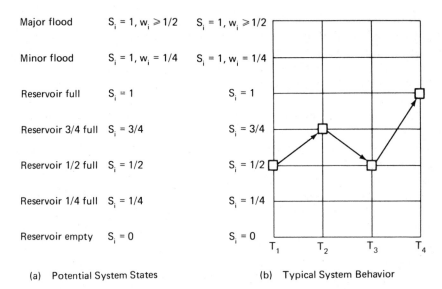

Major flood	$S_i = 1, w_i \geqslant 1/2$
Minor flood	$S_i = 1, w_i = 1/4$
Reservoir full	$S_i = 1$
Reservoir 3/4 full	$S_i = 3/4$
Reservoir 1/2 full	$S_i = 1/2$
Reservoir 1/4 full	$S_i = 1/4$
Reservoir empty	$S_i = 0$

(a) Potential System States   (b) Typical System Behavior

Figure 3.5 Potential system states and typical system behavior for reservoir system model.

In spring, for example, there is a probability of P_1 that the inflow will be a three-fourths volume, P_2 that there will be a one-half reservoir volume, and P_3 that there will be a one-fourth volume, as shown in Figure 3.6. Similar possible influences for summer, fall, and winter are also shown in Figure 3.6.

If the reservoir is full (i.e., $S_i = 1$) and there is an inflow of one-half reservoir volume with no scheduled release, the succeeding state, S_{i+1}, is 1, which means that a major flood has resulted from a flood overflow and that the reservoir is still full. A minor flood results from a flood overflow when the inflow plus the amount in storage minus the scheduled release exceeds the reservoir capacity by one-fourth of the reservoir volume.

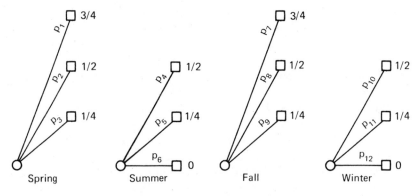

Figure 3.6 Seasonal stream flow.

The transformations involved in this state system model are simply scalar additions or subtractions such that

$$S_{i+1} = S_i - x_i + I_i - w_i$$

where x_i is the quantity of water scheduled for release, I_i is the inflow, and w_i is the flood overflow during system time T_i to T_{i+1}. In general, however, the transformations correspond to the evaluation, or interpretation, of equations of state for the model and represent a major task. Further development of this model to include possible management decisions related to current reservoir system state, local needs, and anticipated seasonal river behavior will be considered later in this chapter.

As another example of a schematic graphical representation of a state system situation, consider the modeling of the parking lot of a ready-mix plant in terms of the availability of concrete transit mix trucks. Assuming that the plant owns five trucks, the parking lot status is related directly to the number of trucks it contains. Because the company owns five trucks, the parking lot can be only one of six possible states: empty (S_0) or with one (S_1), two (S_2), three (S_3), four (S_4), or five (S_5) trucks parked. Figure 3.7 illustrates these possible states and a schematic state model.

Number of trucks in parking lot	Parking lot state	Linear graph state model
0	S_0	⓪
1	S_1	①
2	S_2	②
3	S_3	③
4	S_4	④
5	S_5	⑤

Figure 3.7 Parking lot states for ready-mix plant.

The state transformations for the parking lot are simply related to the arrival or departure of trucks. Figure 3.8 shows the parking lot state transformations for single trucks and Figure 3.9 the transformations associated for a batching plant call for two trucks. Other state transformation models can be developed for batching plant requests for three-, four-, and five-truck-full orders. In each case, however, the parking lot state is equal to the number of trucks parked, and the transformations effect an addition or subtraction to the number of parked trucks.

A more elaborate and descriptive state vector representation of the operation of the ready-mix plant can be developed. It could, for example, consider

Transformation effect of batching plant
calling for one truck

Transformation effect of single
truck return to parking lot

Figure 3.8 Parking lot state transformations for order for one truck for ready-mix plant.

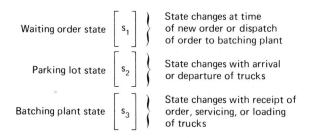

Figure 3.9 Parking lot state transformations for order for two trucks for ready-mix plant.

the number of orders currently unfilled and held in the order receiving office. Again, the state transformation is related simply to the additions of an incoming order or the subtraction of an order released to the batching plant. Similarly, the batching plant can be empty awaiting a mix order, mixing, filling a truck, or full awaiting a truck.

A total state vector can be developed for the entire operational side of the ready-mix-plant operation and corresponds to the aggregation of the system's component state vectors, as shown in Figure 3.10. The monitoring of the ready-mix-plant operation would then be accomplished by evaluating each state variable during the period of operation.

Generally speaking, the state vector representation of a system will contain different types of state variables or variables that monitor or are required for different purposes in a system theory approach. An illustration of the range of classes of state variables that may be relevant to a system theory is given by the following list:

Waiting order state	$\begin{bmatrix} s_1 \end{bmatrix}$	State changes at time of new order or dispatch of order to batching plant
Parking lot state	$\begin{bmatrix} s_2 \end{bmatrix}$	State changes with arrival or departure of trucks
Batching plant state	$\begin{bmatrix} s_3 \end{bmatrix}$	State changes with receipt of order, servicing, or loading of trucks

Figure 3.10 State vector for ready-mix plant.

1. System characteristics relating to number of components, system structure
2. Variables describing the current system configuration, components currently in operation, spatial and relative location
3. Variables describing past system configuration(s)
4. Current input parameters
5. Variables describing past system input(s)
6. Variables indicating the current set of potential next system states
7. Variables indicating past history of potential next system states, plus history of states, plus history of states actually achieved
8. Special-purpose or integrated summary variables

A full description of a complex engineering system is beyond human capabilities because of the infinite complexity of physical materials and forces, and the nature of the environment in which the system is embedded. Nevertheless, engineers are called on to plan, design, build, monitor, and maintain systems that perform specific engineering and societal functions over long periods of time. They must, therefore, in particular cases, gain an understanding of a system; develop adequate descriptions of the system, its environment, and its use; and thus be able to identify the changing status and behavior of the system over time.

3.5 THE COMPONENT INTEGRATION APPROACH

In most engineering problems the desired solution facility is obtained by the manufacture or construction, or both, of a variety of individual components that are physically linked or connected in a specific manner and configuration. The manufactured system facility is a specially integrated assemblage of physical system components. Usually, the system components have known characteristics or specifically designed and built-in functions and the pattern or configuration that links and integrates the components establishes the intrinsic structure and behavior potential of the system as a whole. A systems approach that addresses the various design and planning process problems associated with component integration must be capable of handling explicit component identification and behavior specifications, and exhibit specific linkage configuration statements.

A number of basic design and planning problems must be addressed by the component integration approach. A fundamental problem is to predict system behavior and response in terms of a total input knowledge of component properties and configuration patterns. Ancillary problems of specific design interest focus on the prediction of incremental changes in system behavior and response for given changes in component behavior or linkages.

The design process can be simply visualized as the progressive selection and design of components to achieve desired overall system objectives. A more difficult formulation of the design process encompasses, in addition, the initial problems of establishing desired system component configurations and the resulting number of components prior to individual component design.

A simple example of the physical linkage of components is shown in Figure 3.11a. As indicated, the structural system is made up of a number of structural components simply pinned to each other and to a pinned support at the bottom and a tie connection at the top. When subjected to a vertical load as shown, the various system components, pins, and supports respond to system input loads

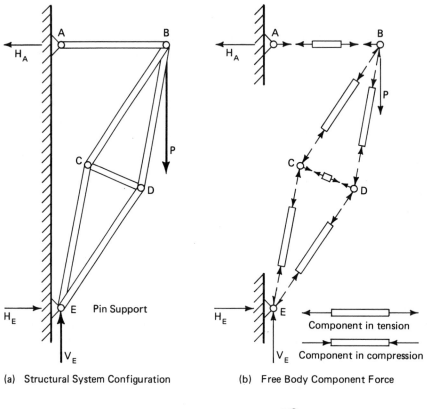

(a)   Structural System Configuration

(b)   Free Body Component Force

(c)   Use for Structural System

Figure 3.11   Structural system model.

by taking up loads and deformations so that the overall system response of pin displacements and component member strains reaches some equilibrium state. Figure 3.11b, for example, indicates in exploded form the nature of the component member forces on each individual component element that results from the applied vertical load. The system analysis procedures for the determination of reaction forces, component member forces, displacements, and resulting system deflections must explicitly take into account the spatial location, orientation, and linkage configurations of the individual system components.

The component integration approach, however, is not limited to purely physically linked components but can be applied to situations involving planned sequential actions, work processes, and other engineering-focused problem areas.

For example, operational system models can be developed for the description of construction and process operations if component work tasks and potential idle resource states are modeled as system components. In this approach the technological structure of the operation provides the system structure via directed linkages which indicate the direction of resource flows and the sequencing of the component work tasks. The construction operation or process is then seen as a specifically ordered and integrated set of component work tasks.

As a specific example, consider an earth-moving operation involving scrapers and bulldozers. The scrapers are push-loaded and boost-accelerated by a pusher dozer, and then haul and dump loads in an embankment area before returning for another load. A system model that portrays the technological structure of this earth-moving operation is shown in Figure 3.12. As indicated, work tasks are schematically modeled as square node components, and the locations of potential idle resource states in the process by queue nodes. The technological sequence of these system components is established by the resource flows involved in the process and indicated by directed arrows. As a modeling requirement, work tasks that require resource combinatorial logic are shown shaded. Hence since a dozer and a scraper must be available before the loading work tasks can begin, the "couple" work task is shown shaded. If the system model is viewed as a system map on which resource "trains" flow, the dozer resource is constrained to the dozer cycle, the various scrapers to the scraper cycle, and the common track must be traveled (if at all) by joint resources (i.e., a dozer and a scraper resource) simultaneously.

3.6 DECISION PROCESSES APPROACH

A *decision* can be viewed as the selection of a preferred alternative, or course of action, from among those alternatives then known and freely available to the decision maker for selection. Decision making implies that the decision maker is aware of a range of alternatives and of the potential consequences that could follow the selection of an alternative. In addition, the decision maker is

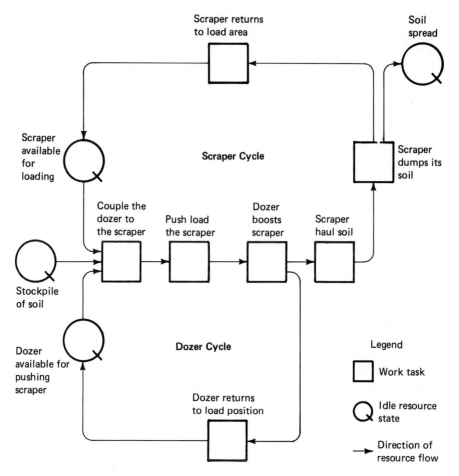

Figure 3.12   Earth-moving operation system model.

capable of making a decision when the desirability of each potential outcome is evaluated in some way according to one or more criteria. If alternatives, outcomes, and criteria are missing at a particular decision point in a decision process, no valid decision situation exists.

Decision processes addressing major problem areas involve a number of linked decisions. These decisions are usually sequentially related to each other but often interact in that several passes through the decision process are required before a satisfactory and consistent set of decisions have been made. Thus decision processes can be viewed as component-integrated systems in much the same way as physical systems are made up from physically integrated components.

A great variety of engineering decision processes exist in the design and planning process. They can be grouped into different classes depending on the extent to which the decision process is formalized and structured. As an illustration, consider the following examples.

Many computer programs incorporate into their logical structure rigorous *preference* ranking wherever certain conditions arise. If the computer program is used in a decision process, the user, knowingly or more often in ignorance, implicitly accepts the preference ranking built into the computer program by the programmer. Many engineering design and planning processes have specific built-in preference structures. These formalized preferences arise from *policy decisions*, from past experience which indicates that some preferences best meet design objectives or public approval, or from the explicit preferences of the original structure of the decision process or computer program. In more critical decision processes the computer program must present the decision maker with the range of alternatives and require decision input before proceeding to the next phase of the calculations. This is often accomplished by the programmer establishing an interactive human–machine interface procedure whenever preference ranking criteria are required. In these cases it is not possible to build into the decision process formulation rigorous preference sequences and structures. Instead, the decision process must include only logical relationships and constraints, whenever these restrictions apply, and alert the decision maker to the need to enter personalized preferences, criteria, and so on.

As an example of the modeling of an engineering decision process, consider the state model previously developed for the reservoir problem. It can be readily extended to incorporate possible reservoir management decisions. Suppose that at a certain time the decision has been made to release water equivalent to a one-fourth reservoir volume and that subsequently, local rainfall produces a stream inflow equal to one-half the reservoir volume. Then assuming an initial reservoir content state S_i, Figure 3.13 models the interaction of the management decision to meet a commitment to supply water during the next time interval and the actual inflow that may occur due to natural stream flow and storms.

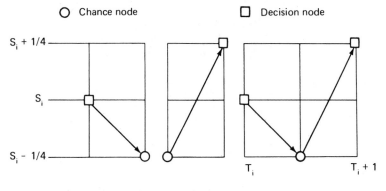

(a) Decision Phase   (b) Chance Phase   (c) System Time Increment

Figure 3.13   State transformation for reservoir system model.

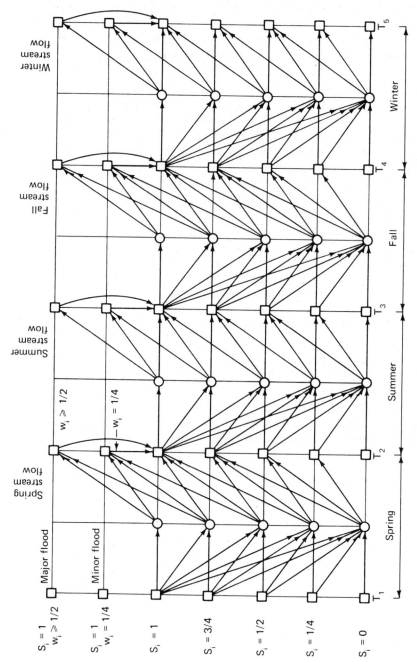

Figure 3.14 Reservoir system management model.

Using the state model for seasonal inflow of Figure 3.6, the basic model for reservoir management is shown in Figure 3.14. Notice that at each decision node all possible decisions have been shown. Thus, if the reservoir is at the three-fourths level, four possible decisions can be made: to release 0, 1, 2, or 3 one-fourth volumes of water. Also, for each chance node, all possible inflows have been shown.

The graphical model presented is similar in characteristics to those to be introduced in Chapter 11 and similar solution techniques can be employed. Before any solution technique can be used, the model must be further developed to include the monetary values to be received from the sale of water to the city and farms, the monetary penalties that will be assessed if minor and major floods occur, the penalty cost to be assessed if the city water requirements are not met, and so on. These values may depend on the use of the water, how much is used, and when it is used.

Figure 3.14 portrays the range of management decision alternatives available to the reservoir manager for all possible reservoir states. Given a specific initial reservoir state and a set of regional needs for water and resulting benefits together with the differing ranges of possible stream inflows in different seasons and the penalty flooding costs that may arise naturally or through bad reservoir management, a variety of reservoir management strategies can be developed (see Chapter 11). The decision maker must establish desired goals and determine the reservoir management strategies that best enable the goals to be achieved.

3.7 SUMMARY

Four generic system viewpoints or approaches have been identified and discussed. These approaches permit the portrayal and analysis of systems for different purposes and at various levels. The examples indicate that to utilize any one of these approaches, it is often necessary to develop models of the system under study. The next chapter provides insight into how to approach model building.

PROBLEMS

P3.1. Systems can be described by state vectors in terms of the values adopted by state variables. Different descriptive state vectors can be developed depending on the complexity of the system and the purpose of the description. Identify a number of compatible purposes and state vectors for the areas listed below. How would you establish the size of the descriptive vector and the order of importance of the state variables?
(a) A moving pendulum
(b) A drawbridge

(c) A storm water discharge pipe with flap valve

(d) A house telephone

(e) A highway toll collection area

(f) A shipping harbor

(g) A railroad switching yard

P3.2. Review a number of engineering analysis methods and techniques. For each, develop a classification scheme in terms of the theory concepts presented in this chapter.

P3.3. Figure P3.3 illustrates the conditions at Washington and Anderson streets. Left turns are not permitted at this intersection. You have been assigned to develop a state theory model for the auto traffic going east on Washington at this intersection.

(a) Why would the time that the light is green not be a good state variable?

(b) If the number of cars at the light waiting to cross is chosen as the state variable, write the state equation; be sure to define terms in equations.

(c) How would you obtain data for the terms in your state equation?

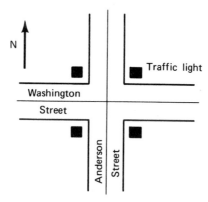

Figure P3.3

P3.4. Given a manufacturing or production assembly process, indicate the appropriate systems viewpoints that can be applied to planning, design, monitoring, or controlling such an operation. Explain why such viewpoints would be applicable.

P3.5. For a high-rise building, why is the component integration viewpoint applicable to the planning or design of the building? Indicate the components that would be involved.

P3.6. Given a water resources problem such as the river system described in Problem P1.6, identify how each of the systems theories presented in this chapter can be applied.

P3.7. Examine some type of construction activity. For that activity, develop a model in a form similar to that shown in Figure 3.12 which highlights the systems nature and resource interaction requirements of the construction activity.

P3.8. Review several planning or design activities that have been undertaken in your locality. For each, determine the systems analysis viewpoints that were or could have been utilized.

An Introduction to Systems Modeling

4

4.1 THE NEED FOR MODELING

One of the major steps in the design and planning process is the modeling of the problem that has been formulated. The engineer must decide how the problem will be resolved and the manner in which the conflicting components, objectives, and constraints are to be considered in developing a solution. To initiate the design and analysis process, the engineer must develop one or more models that can, either individually or collectively, depict all the relevant aspects of the problem.

Models are used extensively by engineers: as aids in the description, analysis, and design phases of problem solving; to facilitate the communication of ideas to others; and as a means of storing information for future reference and use. A *model* is a conceived image of reality; it normally will be a simplification of reality. This does not mean that all models are simple. The complexity of the model depends on the object or process being represented and the purpose of the inquiry. Models may have little resemblance to the actual appearances of the original, but in symbolic terms, reproduce the essential elements of reality.

The modeling technique to be used for a given problem depends on the nature of the problem, the results that are desired, and the resources and capabilities that are available. In the modeling process, the engineer must examine each of the components associated with the problem. For each component that is not included in the problem model, a decision must be made with respect to how or if the component will be included in the analysis. If this cannot be accomplished, it means that the engineer is unable to solve the problem as it is defined; and either the problem statement must be reexamined, or a new modeling approach must be attempted.

This chapter introduces a broad classification of modeling types and purposes and then considers the graphical and mathematical modeling of engineering problems. Analysis and solution techniques are left to later chapters. The modeling process is illustrated by examples.

4.2 MODELING TYPES AND PURPOSES

The engineer can use a variety of different types of modeling languages and formats. These might include:

1. Physical models, which are actual physical representations
2. Graphical models, which are line or schematic drawings
3. Mathematical models, which represent the problem in mathematical terms

Each of these types of models has applications and uses in the design and planning process. Physical models are frequently used to provide a scaled representation of a proposed or completed design. For example, a physical model of a dam, a building, or an airport may be used to assess various design features.

Some of the most commonly used graphical models in engineering include engineering drawings, copies of which are commonly referred to as blueprints; histograms or charts of data and statistical results; curves and graphs showing the interrelationships between two or more parameters; flowcharts portraying the logic of a computer algorithm; perspective drawings of complex mechanisms or buildings; and topographical maps of land surfaces.

Mathematical models, using special symbols and notation, are often used to depict laws of nature and relationships between parameters and problem constraints. Typical examples are models for the paths of planets through space or of a satellite in the earth's atmosphere, the behavior of a gaseous substance under changing temperature, and the performance and fuel consumption of a piece of construction equipment over a rough terrain. Economic models and social behavior models can also be expressed in mathematical terms. In its simplest form, a mathematical model may have only one mathematical expression. On the other hand, a mathematical model of a complex system, such as the structured network of a high-rise building, may require thousands of mathematical statements. With the increasing availability of electronic computers, mathematical models play an increasingly important role in engineering design and planning.

In addition to classifying models by type, models may be grouped on the basis of their use or purpose. Following is a general classification based on use:

1. *Descriptive models.* The descriptive model is used to present the detailed specifications of what is involved and what is to be accomplished. It is, in fact, a concise description of some aspect of the problem, and if properly formulated, provides the framework for design and planning efforts in problem solution.

2. *Behavioral models.* Behavioral models are used to represent the response characteristics of a system. In analysis and design, they are used either to design components to achieve specified response characteristics, or to determine the system response for a given set of components and system structure.

3. *Decision models.* Decision models are used to select the most favorable solutions from among the alternatives that are available according to criteria that the engineer establishes. They are used to investigate and resolve conflicts and to select the best alternatives and strategies.

The following sections of this chapter focus on the graphical and mathematical modeling processes. However, as mentioned above, each illustrative example may incorporate descriptive, behavioral, and decision modeling components.

4.3 LINEAR GRAPH MODELING CONCEPTS

A common feature of many engineering systems is that they are composed of a number of components physically interwoven in the form of a network. Typical examples are the interstate highway system, a city sewage collection and disposal system, a local telephone system, the retriculation lines and storage reservoirs of a water supply system, and the steel framework of a skyscraper.

In addition, there are many engineering decision problems and organizational systems that, although they do not have the physical appearance of networks, can be usefully interpreted as such; for example, the flow of decisions and authority within an industrial firm can be described by a network system. Similarly, the schedule of jobs in a construction project may be viewed as a network of activities. The flow of cash among the agencies within a state government or among the departments within a construction company may be considered as a network of cash flow.

The linear graph is a simple, but powerful graphical technique that is particularly useful for the modeling and analysis of network systems or for the representation of system relationships. Linear graphs are composed of *nodes* (usually represented as a point) and *branches* (usually represented by lines between nodes). Nodes, for example, can be used to represent system components, and the branches can be used to represent the interrelationships between these components. Alternatively, for example, in the case of networks, the branches can be used to represent network components, with the nodes indicating the manner in which network components are connected. Additionally, it may be convenient for the purpose of realistically modeling a particular problem to assign attributes of various qualities to either nodes or branches, or both.

Linear graphs are not drawn to scale, and the relative positions of the nodes and branches do not necessarily represent the actual relative positions of the system parameters or components. Therefore, linear graphs should not be confused with geometric graphs and the physical properties of lines and points. Linear graph theory is an abstract concept and is called *linear* because of its concern with the connectivity of lines and nodes. The primary purpose of a linear graph is to present a concise, graphical representation of the interrelationships of the system that are relevant to the system problem at hand, and to provide a logical structure as a framework that can be suitably annotated and labeled to capture relevant problem parameters. It also provides a conceptual way of portraying the environment or system in which a problem is embedded.

Figure 4.1 shows two linear graph models of a system consisting of four components. Each component is represented by a node, and the interrelationships between components are represented by branches linking the related component nodes. Because linear graphs focus on the connectivity of nodes and branches the two models (a) and (b) in Figure 4.1 are equivalent. The concentration of linear graph theory on *connectivity* provides simple tools and concepts for model construction.

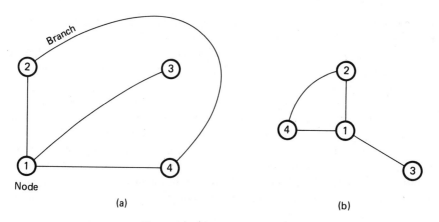

Figure 4.1   Linear graph models.

The modeling capability of linear graphs is shown in the following examples.

Example 4.1

Consider the problem of a construction contractor who is choosing a route to move some heavy construction equipment from Chicago to Urbana, Illinois. The contractor has no preference between two-lane or four-lane highways, but does want to avoid traveling on secondary roads because of load limitations. The contractor wishes to find the route that requires the shortest travel time. After studying the highway map and consulting the highway authorities, the choices are narrowed down to three, as indicated in Figure 4.2a. Figure 4.2b is a linear graph model of the three feasible routes.

In the figure, the nodes denote the cities and towns, and the branches, the highway connections. The numbers along the branches indicate the driving time in minutes. From this graph, the best route is Chicago–Kankakee–Paxton–Rantoul–Urbana, which takes 160 minutes. The second best choice is the Chicago–Dwight–Urbana route, which requires 170 minutes. The route going through Bloomington requires 200 minutes.

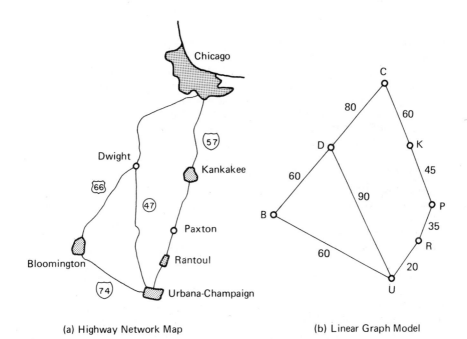

(a) Highway Network Map   (b) Linear Graph Model

Figure 4.2   Linear graph representation of a highway network.

Suppose that an engineer is preparing a plan for improving the highway network in the same area as that illustrated in Figure 4.2. The graph model may now include the following additional features: feasible routes for new highways, existing highways for which additional lanes can be added, towns that require better highway connections because of industrial developments, recreational sites, and so on. The engineer may also choose to use several different symbols among the nodes to differentiate towns, cities, and recreational sites, as well as different symbols among the branches to differentiate the different classes of highways.

Example 4.2

Figure 4.3 is a linear graph representing the scheduling problem in assembling a prefabricated house. Each arrow models a work activity and is labeled with its estimated duration in hours. The directions of the arrows indicate the order in which the

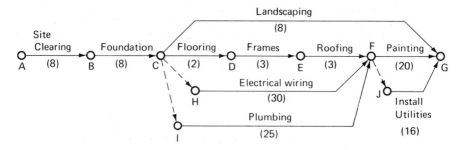

Figure 4.3 Linear graph model of a construction plan.

activities must be performed. For example, frames cannot be erected until flooring is completed, but can be started before completing electrical wiring and plumbing. The model clearly shows the order, or precedence, for all the tasks and the length of time needed to complete each task. Using such a diagram, the total project duration and the amount of extra or slack time available for each work activity can be computed. In this example, the shortest possible time the project can be completed is 66 hours. Such a diagram is called a CPM (critical path method) diagram and is being used extensively for planning and controlling construction progress.

Example 4.3

Consider the problem encountered by a contractor who is performing some work with special equipment in a river flats area that has been subjected in the past to high-water conditions and occasional destructive flooding. There is a 4-month period during which there is no use for the equipment either on this job or on others. The contractor can keep the equipment on the job in the river flats, or else move it out, store it, and then move it back, at a total cost of $1800.

If the contractor keeps the equipment in the river flats area, there is the option of building a platform for the equipment at a cost of $500, which will protect it against high water, but not against a destructive flood. The damage that would be caused by high water amounts to $10,000 if there is no platform. A destructive flood would entail a loss of $60,000, regardless of whether or not a platform is built. The probability of high water in the 4-month period is 0.25; the probability of a destructive flood is 0.02. The contractor has to choose from three possible options:

1. Move equipment
2. Leave equipment in the area and build a protective platform
3. Leave equipment in the area unprotected

Figure 4.4 is a linear graph model of this decision problem. This type of linear graph model is called a *decision tree*. A branch represents either an alternative available to the decision maker or a possible outcome that can result from an uncontrollable chance event. A node denotes either the occurrence of an event at which a decision must be made, the occurrence of an event where outcome is determined by chance, or a final condition.

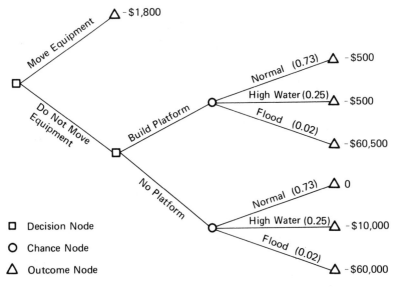

Figure 4.4   Linear graph model of a contractor's decision problem.

The decision nodes, denoted by squares, are those at which the decision maker must choose one course of action from among the alternatives available. The chance nodes, indicated by circles, are those at which the decision maker has absolutely no control. The actual specific outcome from a chance node, of course, depends on nature, but in the prior situation for the contractor, is modeled in Figure 4.4 as a probability outcome using past observations of nature. The numbers on the chance outcome branches indicate the probability that these outcomes would occur. The final outcomes, shown as triangular nodes at the tips of the tree, show the cost to the decision maker for each possible combination of events.

The preceding examples of the modeling of physical, organizational, and sequential decision systems problems illustrate the use of linear graphs as an aid in modeling complex problems. The actual modeling of problems using linear graphs requires creativity and an intimate knowledge of the problem on the part of the investigator. Methods of analysis using linear graphs are presented in Chapter 7 for path and flow problem networks, in Chapter 11 for decision analysis, and in Chapter 13 for network planning and project scheduling.

4.4 MATHEMATICAL MODELING CONCEPTS

Mathematical models contain information expressed in certain specific forms about the problem they address and, require interpretation according to certain predefined rules. Consequently, mathematical formulations often need the explicit support of both descriptive and schematic modeling components as aids in interpreting the symbols and variables used in the model formulation. Usually, the modeling medium is the symbolism of a specific branch of mathematics and requires the logic and understanding of that mathematical area for its interpretation.

A mathematical model can therefore be considered to be made up of a number of modeling components. Typical modeling components are:

1. *Descriptive components*. These modeling components establish or define the problem in words and pose or formulate the purpose to be achieved, or sought for, in its solution. Descriptive components are concerned with problem formulation and solution goals.

2. *Schematic components*. These modeling components complement the descriptive component models in establishing visual understanding of problem definition and are often used to illustrate the interpretation to be made for symbols and interrelationships among system components.

3. *Symbolic definition components*. These modeling components list and describe all the symbols, variables, parameters, and constraints used in the model formulation.

4. *Mathematical components*. These modeling components establish the mathematical reasoning and processes involved in the problem definition and solution processes, as well as formulating the decision processes in a manner that best relates problem definition input to problem solution output.

As indicated in Figure 4.5, the first three modeling components define the manner in which the final mathematical component is to be interpreted.

These four modeling components can be readily recognized in most

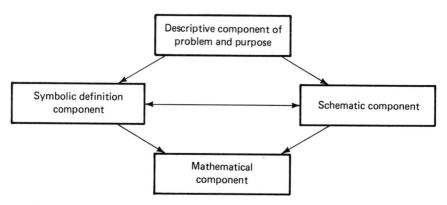

Figure 4.5 Basic components of a mathematical model.

mathematical formulations of problems. In some cases the schematic modeling component is omitted through personal choice, and in others the symbolic definition modeling component is not explicitly formulated especially if the symbols used in the mathematical modeling component have a universal acceptance and recognition.

The mathematical modeling process is concerned with the definition and formulation of a consistent and compatible set of descriptive, schematic, symbolic, and mathematical modeling components that reflect the nature of the engineering problem addressed, required solution goals, and the mathematical knowledge and skill of the modeler involved.

As an illustration of the mathematical modeling process and the various kinds of modeling components used in problem solving, consider the following example.

Modeling Component 1: Description of Problem and Purpose.   A straight, uniform bar of specified length is to be made from a light elastic material. The bar is to hang vertically from a support and carry a specific axial load at its free bottom end within a specific deflection range.

This modeling component is a statement of the modeler's view and evaluation of the problem. It formulates the model purpose and system constraints; its definition is a critical step since it provides the requirements and objectives for the final model.

Modeling Component 2: The Symbolic Definition

Let A = connection point of the bar to the support
B = bottom end point of the bar
L = length of the bar, in inches
A_x = cross-sectional area of the bar, in square inches
E = Young's modulus of the material, in pounds per square inch
P = axial force along the bar member, in pounds
σ = bar member stress, in pounds per square inch
ϵ = bar member strain
ΔL = bar member elongation, in inches
P'_B, P'_A = applied axial load at end B, and the reaction load at A, respectively, in pounds
u'_B, u'_A = displacements of the points B and A, respectively, in inches
Δ = maximum permissible movement of the applied load at B, in inches

Modeling component 2 is the interface between the modeler's knowledge of the real problem and the requirements of a specific technological model. These symbols, in fact, represent the system components and their attributes that are relevant to the problem.

Modeling Component 3: The Schematic Representation

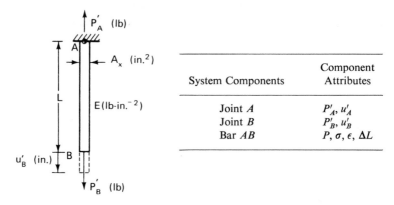

System Components	Component Attributes
Joint A	P'_A, u'_A
Joint B	P'_B, u'_B
Bar AB	$P, \sigma, \epsilon, \Delta L$

Modeling Component 4: Mathematical Formulation of the Bar Behavior

Bar assumed weightless

$P = $ constant

$$\sigma = \frac{P}{A_x}$$

$$\epsilon = \frac{\Delta L}{L} = \frac{\sigma}{E} = \frac{P}{A_x E}$$

mathematical logic and derivation (i.e., program segment for deriving mathematical relationships)

Therefore,

$$\Delta L = \frac{PL}{A_x E}$$

mathematical relationship

Modeling component 4 is a mathematical formulation for the technological description of the bar behavior. Its selection depends on the level of technology available and known to the modeler and the relevant depth considered necessary to meet the problem requirements. This modeling component identifies the five parameters that describe the system response and structure of the bar. If any four parameters are known, the fifth can be found directly from the desired mathematical relationships. It can be used to study the bar response to the design problem described later.

Modeling Component 5: Mathematical Formulation of the System Component Interaction

$$\Delta L = u'_B - u'_A$$

$$u'_A = 0$$

$$P = P'_B$$

$$P'_A = P'_B$$

Modeling component 5 defines the system structure relating the component bar parameters and the node parameters. It normally requires a unique creative effort on the modeler's part, since a specific technological formulation for the problem may not exist.

Modeling Component 6: Mathematical Derivation of the System Response

From component 4: $\Delta L = \dfrac{PL}{A_x E}$ where P'_B is the system input

From component 5: $\Delta L = u'_B$

$\dfrac{L}{A_x E}$ is the system mode of response

$P = P'_B$

u'_B is the system response and system output

Therefore,

$$u'_B = \frac{P'_B L}{A_x E}$$

Modeling component 6 is a statement indicating how the modeler intends to *read* the system. It focuses on the input–output description of the system and represents the solution process for the systems model.

Modeling Component 7: Mathematical Formulation of the Design Problem

$$u'_B = \frac{P'_B L}{A_x E} \leq \Delta$$

Therefore,

$$A_x E \geq \frac{P'_B L}{\Delta} = \text{a constant}$$

Modeling component 7 is a mathematical formulation exposing the design variables A_x and E. The modeler can now establish a design criterion to test whether a particular design is suitable or not. Since the design criterion is bounded on one side only, the designer must establish a value system for departures of $A_x E$ from the constant $P'_B L / \Delta$.

The total mathematical model for the system design problem is summarized in Figure 4.6. The seven model components portray in various ways such things as the embodiment of model purpose, scope of investigation, reality constraints on conditions and choices, and the modeler's bias and knowledge of technology. They do not, of course, model the design process and selection of specific attributes for the model components.

The model itself can be checked for internal consistency; for level of accuracy relevant to the real-world problem, with problem identification, needs, and resources; and for the impact of the modeler's bias.

Modeling component 1, for example, prescribes that the structure be a bar (not a truss or combination of structural elements) and that it be uniform, straight, and be made from a light elastic material. The inclusion of all (or any one) of these descriptions severely limits the design choices potentially available and, in effect, plays a dominant role in the specific forms of component models 2 to 7.

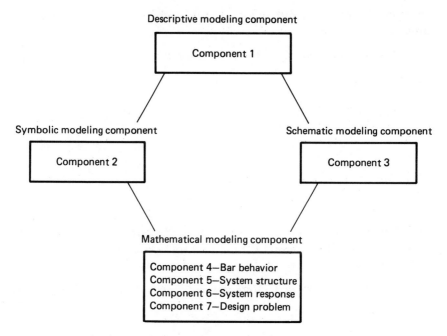

Figure 4.6   Mathematical model of system design problem.

The requirement for a light material (the validity of which should be considered) tentatively supports the assumption of a weightless bar in modeling component 4 and the vague selection criterion imposed in modeling component 7. Existing technology permits modeling component 4 to be updated to include the effect of the self-weight of the bar. Whether this is done or not reflects a decision made by the modeler. If the material selection aspect of the problem is to be elaborated, at least models 1 and 7 are affected and require replacement by new or updated versions.

Similarly, an investigation can be directed to the necessity for the imposition of constraints such as straightness and uniformity on the bar. If the load is applied dynamically or suddenly, model 4 may not be accurate or suitable, in which case, modeling components 1, 4, 6, and 7 must be changed.

Suppose that the cost of the structure and the time to manufacture and erect the structure enter into consideration and affect the model's purpose. The requirement for internal consistency in the modeling components demands that if costs enter into modeling component 1, a consideration of costs must appear in at least one of the following modeling components. The specific form of its inclusion will expose the modeling and design rationale of the modeler, which can then be examined and critiqued.

The development of useful and valid models requires a high level of creative ability, technical and professional knowledge, and an understanding of the forces, constraints, and values at work in the environment in which the problem exists.

4.5 MATHEMATICAL MODELING: A COMPOUND BAR SYSTEM

If a single equation can be considered as a model of a condition or component, then a set of equations intuitively represents a model of a system. In fact, many system models take the form of a set of simultaneous equations, either algebraic, functional, differential, or integral.

In some cases, each individual equation introduces a new condition initiated by the consideration of another component in the system model. In other cases, each equation introduces a new constraint on the relationships among the set of system variables.

Great insight into mathematical models can be gained if the various equations are grouped into sets in which each set of equations focuses on a specific type of system constraint (i.e., component behavior, system structure, system equilibrium).

In the mathematical modeling of engineering systems, each component of the system may require modeling components as defined above and as illustrated in Figure 4.6.

As an example, consider the compound bar system shown in Figure 4.7. Although this particular compound bar system may be considered to be over-simplified and have little value in practical applications, the example will serve to illustrate an approach that is applicable to the modeling of many engineering systems. In fact, some of the most sophisticated computer programs that have been developed for the analysis of structural networks are based on mathematical models which are simple extensions of the one to be presented here.

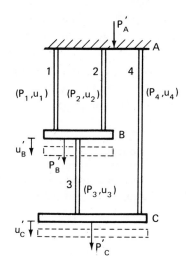

Figure 4.7 Compound bar system.

4.5.1 Statement of the Problem

The compound bar system shown in Figure 4.7 consists of four bar components, 1, 2, 3, and 4, connected to three short horizontal members, A, B, and C, in which member A is a fixed support. The construction is such that these horizontal members can be considered as rigid cross pieces so that the vertical bar members are constrained to axial deformations only; that is, the horizontal members remain horizontal and can be treated as individual joints in the structure.

The system can be loaded with joint forces P'_B and P'_C at B and C, developing the reaction force P'_A. The joint forces P'_B and P'_C can be considered as system inputs.

As a result of the applied system loading, the bars will be loaded, producing axial forces $P_1, P_2, \ldots, P_4$, and the bars will deform by the amounts $u_1, u_2, \ldots, u_4$. This internal loading and deformation of the system will cause joints B and C to displace amounts u'_B and u'_C. Since joint A is a fixed support, the displacement u'_A is zero. The joint displacements u'_B and u'_C can be considered as measurements of the system response (i.e., output) to the system loading input.

The problem is to develop a mathematical behavior model that can be used to analyze the system response under various combinations of loading at B and C. This is realized if a set of equations can be developed to express the system response parameters u'_B and u'_C as functions of the system input parameters P'_B and P'_C.

4.5.2 System Constraints

The following constraints in the structure are recognized:

1. *Component behavior.* The response of a bar member when subjected to an axial load is governed by a physical law, commonly referred to as an *equation of state.*
2. *System compatibility.* Geometric relationships exist between the elongation of the bar members and the physical displacements of the joints.
3. *System equilibrium.* The components are in stable equilibrium under static loading.

4.5.3 Component Behavior

It has already been shown that the following relationship exists between the axial force P_i acting on a bar and the corresponding elongation u_i:

$$P_i = \frac{E_i A_i}{L_i} u_i = k_i u_i \qquad (4.1)$$

where E_i is Young's modulus of the material used in bar i in pounds per square inch, A_i is the cross-sectional area of bar i in square inches, and L_i is the length of bar i in inches. In structural engineering, k_i is called the stiffness coefficient of bar i and is expressed in units of bar force per unit displacement (lb/in.). The stiffness coefficient k_i thus reflects the "stiffness" of bar i and is a function of the material, the cross-sectional area, and its length.

To complete the model on bar behavior, a sign notation must be developed for Equation 4.1. Since the bar is elongated when the axial force is in tension, and shortened when the axial force is in compression, it is intuitively obvious from Equation 4.1 that P_i is positive when it is in tension and negative when it is in compression. Similarly, u_i is positive when the bar is elongated and negative when it is shortened.

One equation can be used to describe the behavior response of each of the four members in the given structure. Thus

$$P_1 = k_1 u_1$$
$$P_2 = k_2 u_2$$
$$P_3 = k_3 u_3 \qquad (4.2)$$
$$P_4 = k_4 u_4$$

In matrix notation, Equations 4.2 become

$$\begin{bmatrix} P_1 \\ P_2 \\ P_3 \\ P_4 \end{bmatrix} = \begin{bmatrix} k_1 & 0 & 0 & 0 \\ 0 & k_2 & 0 & 0 \\ 0 & 0 & k_3 & 0 \\ 0 & 0 & 0 & k_4 \end{bmatrix} \begin{bmatrix} u_1 \\ u_2 \\ u_3 \\ u_4 \end{bmatrix} \qquad (4.3)$$

that is,
$$\mathbf{P}_{(4 \times 1)} = \mathbf{K}_{(4 \times 4)} \mathbf{u}_{(4 \times 1)} \qquad (4.4)$$

where $\mathbf{P}_{(4 \times 1)}$ = column matrix of component axial forces
  $\mathbf{u}_{(4 \times 1)}$ = column matrix of component axial deformations
  $\mathbf{K}_{(4 \times 4)}$ = square diagonal matrix of component stiffnesses; its diagonal form indicates (and establishes through matrix multiplication $\mathbf{u}_{(4 \times 1)}$) the independence of the four equations

4.5.4 System Compatibility

Connecting the bars together to form the system structure requires the development of mathematical expressions to enforce the geometric compatibility between member elongations and the joint displacements. System compatibility is ensured if the individual members continue to be connected to the joints whatever the joint displacements. This implies that the members must elongate just enough to suit the joint displacements. Let a downward joint displacement be represented by a positive value, and an upward displacement by a negative value. Then the condition of geometric compatibility between joint displacements and member elongations can be completely described by the following set of equations:

$$u_1 = -u'_A + u'_B$$
$$u_2 = -u'_A + u'_B$$
$$u_3 = -u'_B + u'_C$$
$$u_4 = -u'_A + u'_C$$

(4.5)

Since joint A is assumed fixed, $u'_A = 0$, and the equations above can be simplified to the following:

$$u_1 = u'_B$$
$$u_2 = u'_B$$
$$u_3 = -u'_B + u'_C$$
$$u_4 = u'_C$$

(4.6)

In matrix notation, Equation 4.5 becomes

$$
\begin{bmatrix} u_1 \\ u_2 \\ u_3 \\ u_4 \end{bmatrix}
=
\begin{bmatrix} -1 & 1 & 0 \\ -1 & 1 & 0 \\ 0 & -1 & 1 \\ -1 & 0 & 1 \end{bmatrix}
\begin{bmatrix} u'_A \\ u'_B \\ u'_C \end{bmatrix}
$$

(4.7)

that is,

$$\mathbf{u}_{(4 \times 1)} = \bar{\mathbf{A}}_{(4 \times 3)} \mathbf{u}'_{(3 \times 1)}$$

(4.8)

where $\mathbf{u}'_{(3 \times 1)}$ = column matrix of joint displacements

$\bar{\mathbf{A}}_{(4 \times 3)}$ = rectangular matrix of coefficients that yield relative displacements on matrix multiplication with $\mathbf{u}'_{(3 \times 1)}$

4.5.5 System Equilibrium

Since the applied loads P'_B and P'_C are static, the system is in equilibrium if the algebraic sum of the forces acting on any joint in the structure is equal to zero. Figure 4.8 illustrates the forces acting on the present structure. Bar axial forces are shown as tensile (i.e., positive) forces, and joint forces as downward. Let a downward joint force be considered positive, and an upward joint force be negative. The following equilibrium equations can be derived.

At joint A: $P'_A + P_1 + P_2 + P_4 = 0$

At joint B: $P'_B - P_1 - P_2 + P_3 = 0$ (4.9)

At joint C: $P'_C - P_3 - P_4 = 0$

Rearranging terms to separate the applied joint loads (cause) from the bar component member loads (effect) results in

$$P'_A = -P_1 - P_2 \qquad\quad - P_4$$
$$P'_B = +P_1 + P_2 - P_3$$
$$P'_C = \qquad\qquad\quad\; + P_3 + P_4$$

(4.10)

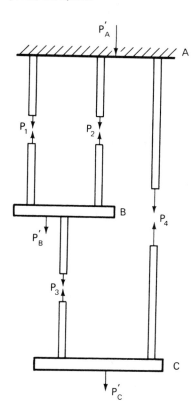

Figure 4.8 Member forces and joint loads.

In matrix motation

$$
\begin{bmatrix} P'_A \\ P'_B \\ P'_C \end{bmatrix} = \begin{bmatrix} -1 & -1 & 0 & -1 \\ 1 & 1 & -1 & 0 \\ 0 & 0 & 1 & 1 \end{bmatrix} \begin{bmatrix} P_1 \\ P_2 \\ P_3 \\ P_4 \end{bmatrix} \tag{4.11}
$$

By comparing the coefficient matrix of Equation 4.11 with the coefficient matrix $\bar{\mathbf{A}}$ in Equation 4.7, we have

$$
\mathbf{P}'_{(3 \times 1)} = \bar{\mathbf{A}}^T_{(3 \times 4)} \mathbf{P}_{(4 \times 1)} \tag{4.12}
$$

in which $\mathbf{P}'$ is the column matrix of joint loads and $\bar{\mathbf{A}}^T$ is the transpose of $\bar{\mathbf{A}}$; that is, the rows and columns have been interchanged.

4.5.6 System Model

A complete mathematical model of the compound bar system is obtained by grouping Equations 4.4, 4.8, and 4.12 together as follows:

Four equations of state:

$$
\mathbf{P}_{(4 \times 1)} = \mathbf{K}_{(4 \times 4)} \mathbf{u}_{(4 \times 1)} \tag{4.4}
$$

Four compatibility equations:

$$\mathbf{u}_{(4 \times 1)} = \bar{\mathbf{A}}_{(4 \times 3)}\mathbf{u}'_{(3 \times 1)} \tag{4.8}$$

Three equilibrium equations:

$$\mathbf{P}'_{(3 \times 1)} = \bar{\mathbf{A}}^T_{(3 \times 4)}\mathbf{P}_{(4 \times 1)} \tag{4.12}$$

The 11 equations contain the 14 system variables $P_1, P_2, P_3, P_4, u_1, u_2, u_3,$ $u_4, P'_A, P'_B, P'_C, u'_A, u'_B,$ and u'_C. For any solution, three variables must be specified. In this particular example, $u'_A = 0$ and P'_B and P'_C are known from the given loading condition.

Furthermore, by performing some simple substitutions, a mathematical model can be developed to express the relationship between the system response vector $\mathbf{u}'$ and the system input vector $\mathbf{P}'$. Thus, by substituting Equation 4.4 into Equation 4.12, the following results:

$$\mathbf{P}' = \bar{\mathbf{A}}^T\mathbf{K}\mathbf{u}$$

Then, substituting Equation 4.8 into this equation yields

$$\mathbf{P}' = (\bar{\mathbf{A}}^T\mathbf{K}\bar{\mathbf{A}})\mathbf{u}' \tag{4.13}$$

which is the desired mathematical behavior model.

Equation 4.13 expresses the relationships between $\mathbf{P}'$ and $\mathbf{u}'$ in the given compound bar system. If $\mathbf{u}'$ is specified, $\mathbf{P}'$ can be computed directly from Equation 4.13. If $\mathbf{P}'$ is specified, $\mathbf{u}'$ can be computed from the following matrix expression:

$$\mathbf{u}' = (\bar{\mathbf{A}}^T\mathbf{K}\mathbf{A})^{-1}\mathbf{P}' \tag{4.14}$$

In longhand form, Equation 4.13 becomes

$$\begin{bmatrix} P'_A \\ P'_B \\ P'_C \end{bmatrix} = \begin{bmatrix} -1 & -1 & 0 & -1 \\ +1 & +1 & -1 & 0 \\ 0 & 0 & +1 & +1 \end{bmatrix} \begin{bmatrix} k_1 & 0 & 0 & 0 \\ 0 & k_2 & 0 & 0 \\ 0 & 0 & k_3 & 0 \\ 0 & 0 & 0 & k_4 \end{bmatrix} \begin{bmatrix} -1 & +1 & 0 \\ -1 & +1 & 0 \\ 0 & -1 & +1 \\ -1 & 0 & +1 \end{bmatrix} \begin{bmatrix} u'_A \\ u'_B \\ u'_C \end{bmatrix}$$

By completing the matrix multiplications,

$$\begin{bmatrix} P'_A \\ P'_B \\ P'_C \end{bmatrix} = \begin{bmatrix} -k_1 & -k_2 & 0 & -k_4 \\ k_1 & k_2 & -k_3 & 0 \\ 0 & 0 & k_3 & k_4 \end{bmatrix} \begin{bmatrix} -1 & +1 & 0 \\ -1 & +1 & 0 \\ 0 & -1 & +1 \\ -1 & 0 & +1 \end{bmatrix} \begin{bmatrix} u'_A \\ u'_B \\ u'_C \end{bmatrix}$$

there results

$$\begin{bmatrix} P'_A \\ P'_B \\ P'_C \end{bmatrix} = \begin{bmatrix} k_1 + k_2 + k_4 & -k_1 - k_2 & -k_4 \\ -k_1 - k_2 & k_1 + k_2 + k_3 & -k_3 \\ -k_4 & -k_3 & k_3 + k_4 \end{bmatrix} \begin{bmatrix} u'_A \\ u'_B \\ u'_C \end{bmatrix} \tag{4.15}$$

Equation 4.15 in effect contains the following three algebraic equations:

$$P'_A = (k_1 + k_2 + k_4)u'_A - (k_1 + k_2)u'_B - k_4u'_C \qquad (4.16)$$

$$P'_B = -(k_1 + k_2)u'_A + (k_1 + k_2 + k_3)u'_B - k_3u'_C \qquad (4.17)$$

$$P'_C = -k_4u'_A - k_3u'_B + (k_3 + k_4)u'_C \qquad (4.18)$$

Since $u'_A = 0$, these equations may be simplified as follows:

$$P'_A = -(k_1 + k_2)u'_B - k_4u'_C \qquad (4.19)$$

$$P'_B = (k_1 + k_2 + k_3)u'_B - k_3u'_C \qquad (4.20)$$

$$P'_C = -k_3u'_B + (k_3 + k_4)u'_C \qquad (4.21)$$

Thus, given P'_B and P'_C, u'_B and u'_C can be determined by solving Equations 4.20 and 4.21. The reaction joint load P'_A can, in turn, be computed from Equation 4.19. In matrix notation, Equations 4.20 and 4.21 may be written as

$$\begin{bmatrix} P'_B \\ P'_C \end{bmatrix} = \begin{bmatrix} k_1 + k_2 + k_3 & -k_3 \\ -k_3 & k_3 + k_4 \end{bmatrix} \begin{bmatrix} u'_B \\ u'_C \end{bmatrix} \qquad (4.22)$$

and the solution is

$$\begin{bmatrix} u'_B \\ u'_C \end{bmatrix} = \begin{bmatrix} k_1 + k_2 + k_3 & -k_3 \\ -k_3 & k_3 + k_4 \end{bmatrix}^{-1} \begin{bmatrix} P'_B \\ P'_C \end{bmatrix} \qquad (4.23)$$

which represents the final input–output response model. This inverse matrix embodies all three types of system constraints in the problem: component behavior, system compatibility, and system equibrium, and reflects the materials used for the bar members, the geometry and equilibrium of the structure. Thus, for a given design of the structure, this inverse matrix remains constant and the mathematical model in Equation 4.23 can be used to solve for the joint displacements u'_B and u'_C for various sets of joint loads P'_B and P'_C.

This behavior model can be used in an iterative procedure to design a structure. The properties of the structural components and configuration can be chosen and the response tested via the behavior model. If the response does not meet the chosen criteria, the properties of the structural components and configuration must be changed and the analysis repeated.

4.5.7 Numerical Example

As an illustration, consider the specific case shown in Figure 4.9. According to Equation 4.23,

$$\begin{bmatrix} u'_B \\ u'_C \end{bmatrix} = \begin{bmatrix} 12 & -4 \\ -4 & 5.5 \end{bmatrix}^{-1} \begin{bmatrix} 5 \\ 8 \end{bmatrix}$$

Therefore,

$$\begin{bmatrix} u'_B \\ u'_C \end{bmatrix} = \begin{bmatrix} 0.11 & 0.08 \\ 0.08 & 0.24 \end{bmatrix} \begin{bmatrix} 5 \\ 8 \end{bmatrix} = \begin{bmatrix} 1.19 \text{ in.} \\ 2.32 \text{ in.} \end{bmatrix}$$

and $P'_A = -8u'_B - 1.5u'_C = -13$ lb.

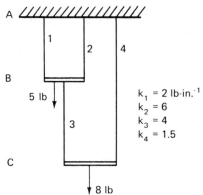

$k_1 = 2$ lb-in.$^{-1}$
$k_2 = 6$
$k_3 = 4$
$k_4 = 1.5$

Figure 4.9   Compound bar example.

If the loads are changed so that $P'_B = 6$ lb and $P'_C = 10$ lb, the new solution can be computed as follows:

$$\begin{bmatrix} u'_B \\ u'_C \end{bmatrix} = \begin{bmatrix} 0.11 & 0.08 \\ 0.08 & 0.24 \end{bmatrix} \begin{bmatrix} 6 \\ 10 \end{bmatrix} = \begin{bmatrix} 1.46 \text{ in.} \\ 2.88 \text{ in.} \end{bmatrix}$$

and $P'_A = -8u'_B - 1.5u'_C = 16.00$ lb.

A linear graph and topological matrix formulation of the compound bar problem is possible and is very useful as the logical basis for the automatic formulation of problems for solution by digital computer.

4.6 A MATHEMATICAL MODELING APPROACH
TO THE CONCRETE BATCHING PLANT

In many engineering management and decision situations no clear cut and/or rigorous problem formulation may exist. In these cases the engineer must build up a problem formulation and system model in an ad hoc manner if informed decisions are to be made. As an example of this approach, consider a mathematical modeling approach to the concrete batching plant problem area mentioned previously as a portion of the regional planning problem introduced in Chapter 1.

The necessity for carefully considering the design and operation of a concrete batching plant is directly related to the magnitude of the concrete requirements of the project. If large quantities of concrete are required over a considerable period of time, opportunities may exist for both increasing production capacities and for realizing economies in concrete procurement. Engineers may then be called on to develop mathematical, economical, and simulation models of entire processes, and to develop management policies for their operation.

In general, mathematical models will be required for the following system features:

1. *System component and technology models.* These models will define the number, type, size, and characteristics of the plant entities involved in the concrete procurement process associated with the batching plant. In addition, the component technology models will define the transformations produced on the materials processed.

2. *The procurement process structure definition models.* These models define the routes taken by the various materials (sand, aggregates, cement, additives, concrete, etc.) through the various system components of the batching plant. Depending on the issues involved, the routes will be defined from sources (quarries, stockpiles, factories, bins, etc.) to the final destination (loading hopper, ready-mix truck, concrete spillway and apron, etc.).

3. *The management policy models.* These models will define the various actions to be taken when certain conditions exist, or provide criteria to select an alternative when several options become available. The models should focus on inventory features of material stockpiles, initiating and terminating concrete mixing, the rates at which quarries will be excavated and stockpiles replenished, and so on.

4. *System response models.* These models will determine the characteristics of the total procurement system in the batching plant that address the problems under consideration. Some will focus on the time characteristics of the system, such as the time response for changing concrete mixes, to deplete and replenish stockpiles, and idle time. Some, however, will focus on management problems associated with production costs and the feasibility of additions and alterations to eliminate system bottlenecks or to increase productivity.

The development of formal mathematical models will then permit analysis of the batching plant by providing the basis for functional steps in a computer program. The following material is intended to illustrate an approach to the mathematical modeling of the batching plant but does not attempt to either justify the level of modeling or present a complete model.

The concrete batching plant consists of a set of hardware component items connected by a system of belt conveyors, feed chutes, and hoppers. A typical layout for a plant is shown in Figure 4.10.

The plant components can be described in symbolic terms with associated vectors of attributes to describe the physical and functional characteristics of the component. For example, the cement storage bin can be denoted by the symbol CEMENTSTORAGEBIN. Various attributes of the bin might be of interest: its physical volume, the full-capacity cement load, and the current status of the bin in terms of its cement content. These attributes can then either be assigned positional entries in a $N \times 1$ vector called CEMENTSTORAGEBIN(N) or given unique symbolic descriptors. The concrete plant mixer will have a capacity MIXVOLUME, a loading time MIXLOADTIME, a mixing time MIXTIME, and an emptying

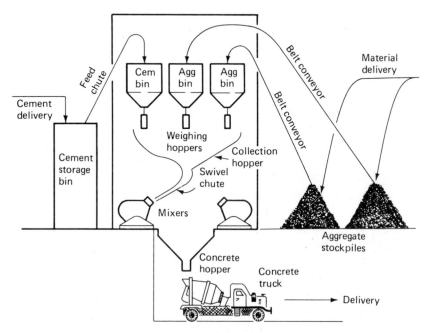

Figure 4.10   Concrete plant layout and operations.

time, MIXDUMPTIME. The mixer times may be assumed to be constants or they may be probabilistic time distributions depending on the nature of the mixer operations and the level of modeling detail and problems requiring solutions. In a similar manner, each component shown in Figure 4.10 can be described so that a set of formal symbolic labels and vector attributes result.

The material flow processes are associated physically with the properties of the belt conveyors, feed chutes, and other mechanical parts. The flows can be expressed as time dependent rate equations with capacity limits. For example, the Ith aggregate belt conveyor BELTCONVEYOR(I) will have characteristics such as physical length, width, and inclination, as well as delivery rates in volume or weight units per unit of time as functions of belt speed, and the available transmission horsepower. Due to changing efficiency of equipment, aggregate bulking conditions on the belt, feed, and so on, the delivery rate may well be a variable with a specific probabilistic distribution that can be expressed in terms of the mean and standard deviation. Field observations on similar equipment may provide the raw data for defining specific model characteristics. For the cement and concrete delivery processors, discontinuous flow properties must be assumed as a function of time.

Management policy statements can be related to the conditional occurrence of events or system states. For example, if the level of cement in the cement bin, CEMENTBINLEVEL(T), falls below a certain value, CEMENTBINLEVEL(A), the feed chute, CEMENTFEEDCHUTE, from the cement storage bin must be activated by the operator for a certain period of time, DELTATREFILL, to insure that the cement bin is almost filled. The policy statement for this situation may take the following form:

IF (CEMENTBINLEVEL(T) $\leq$ CEMENTBINLEVEL(A)) THEN ACTIVATE
CEMENTFEEDCHUTE FOR TIME DELTATREFILL

Similar policy statements can be developed for each decision point in the batching plant operation.

The concrete plant is intended to produce a total volume of concrete, TOTALCONCRETE, over a construction period of time OPERATIONLIFE. It receives orders of size ORDERSIZE for a mix type, such as MIXTYPE(I). In reality, ORDERSIZE is a function of time and the various concrete mix types require different proportions of aggregate, sand, and cement.

Once sufficient variables and items have been identified, a variety of relationships can be developed linking the symbolic labels into a system of functional equations.

For example, from continuity considerations,

$$\int_{T=0}^{T=\text{OPERATIONLIFE}} \text{ORDERSIZE}(T)dT = \text{TOTAL CONCRETE}$$

$$= \text{ESTIMATED TOTAL CONCRETE}$$

$$+ \text{ADDITIONAL CONCRETE}$$

The amount of detail required and the system model scope that is developed will be a function of the problem definition and objectives. Various alternatives exist for the model both in the level of detail and the manner in which the parameters are characterized (e.g., by deterministic or probabilistic times). The result of the modeling efforts becomes a formal statement of the system and the system problem.

An initial modeling segment for the mixer production time, MIXERPRODUCTIONTIME, for the Nth order of ORDERSIZE(N) may be given by

$$\text{MIXERPRODUCTIONTIME} = \frac{\text{ORDERSIZE}(N)}{\text{MIXVOLUME}} \times (\text{MIXLOADTIME}$$

$$+ \text{MIXTIME} + \text{MIXDUMPTIME})$$

$$+ \text{IDLETIME}$$

where IDLETIME is the amount of time during the process time for the order that the batching plant is not being serviced by delivery trucks at the loading dock. Consequently, IDLETIME is a function of the truck fleet size, TRUCKFLEETSIZE, and characteristics of the delivery cycle and concrete pouring operation in the field.

The actual model that the engineer develops must be compatible with an available system analysis procedure and supported by the specific data relevant to the particular problem under consideration.

4.7 SUMMARY

All system models must portray in some manner the number and behavior of its components, the system structure of the component interactions, and system conditions introduced by the exposure of the system to physical laws and requirements imposed by organizational considerations. In addition, the relationship and use of the model by the modeler in the design process may require that the model exhibit decision aspects.

The essential difference between the mathematical and graphical modeling of systems is that the mathematical model must provide explicit structural statements in its formulation, whereas the graphical model analysis can build on an existing graph structure. Consequently, the mathematical model and the individual equations it uses often exhibit the mathematical characteristics of the system graph structure. The use of mathematical models is usually more convenient than a graphical analysis when analyzing large systems.

PROBLEMS

P4.1. For each of the following situations, identify system components, linear graph structure, and characteristics. To what extent can linear graphs aid in representing and understanding the situation?
(a) A school bus route
(b) A river system
(c) A variety of road and highway intersections
(d) A plant layout
(e) A city water distribution system
(f) A bridge span

P4.2. Develop descriptive, symbolic, schematic, and mathematical models for each of the following situations. To what extent would your model formulation change if your purpose is:
(a) To describe the physical situation?
(b) To design and manufacture the facility?
(c) To use the facility?
  1. A man standing on a ladder that makes an angle of 20 degrees with a vertical wall
  2. The flow of electricity through an insulated cable
  3. A 300-lb weight resting on the free end of a horizontal cantilever
  4. A steel cable supported at both ends
  5. A vehicle moving along a curve of 900-ft radius at a speed of 60 mph
  6. The foyer entrance to a high-rise building

P4.3. In Example 4.1, the modeling of a road network is presented. Indicate how the following can be represented on such a graph model.
(a) Turn restrictions at the roadway intersections

(b) One-way roadways

(c) Other descriptors that would indicate the travel conditions on any particular roadway segment

P4.4. A plant production line consists of five processing units, as shown in Figure P4.4a. Processing unit P4 is an assembly process requiring the assembly of two items from P3 on line *B* and one item from P2 in line *A*. All other processing units perform operations on individual items.

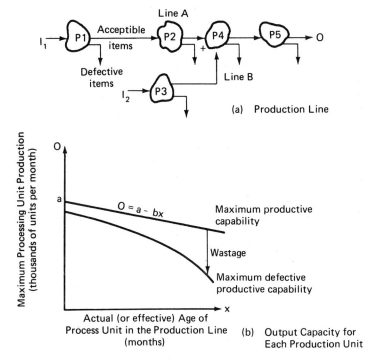

Figure P4.4

The production from each processing unit measured in terms of items processed depends on both the input supply to the unit and on the age of the processing unit in terms of actual (or effective) months of production spent in the production line. In addition, while the productive output of a unit decreases with time, the percentage of defective items produced increases with time, so that the effective production of acceptable items is a function of both unit age and quality.

Assume that the maximum output capacities of each unit decrease linearly with time and that the percentage of defective items increases parabolically with time, as illustrated in Figure P4.4b and quantified in Table P4.4. Thus in any specific unit the number of defective items produced is given by $(c + dx^2)$, where x is the actual or effective age (in months) of the process unit in the production line.

The supply rates I_1 and I_2 for lines *A* and *B* have been set at 120 and 140

TABLE P4.4

	a	b	c	d	Initial Age of Machine (Months)
P1	120	7	0.85	0.5	1
P2	100	6	0.75	0.5	0
P3	150	7	0.95	0.5	2
P4	80	3	0.75	0.5	1
P5	80	4	0.95	0.5	1

(thousands per month), although the factory could deliver at the maximum rates of 150 (thousands per month) to both lines.

(a) What is the monthly output of acceptable finished products for the first 6 months of operation?

(b) What are the optimal values of I_1 and I_2 for this production rate, and what bottlenecks exist?

(c) What single processing unit should be replaced now at the start of the production run? Find the new monthly output for the first 6 months and the optimal values of I_1 and I_2.

P4.5. Develop a mathematical formulation for the broad characteristics of a high-rise building with a rectangular cross section.

(a) What would you choose for an objective, and how would you incorporate into your model the following considerations?

   1. Local zoning ordinances restricting the building height and percentage of ground cover

   2. Floor space required for tenants and services

   3. Structural, mechanical, and elevator costs that increase with building height

   4. Differential building surface and heating costs for the east-west and north-south faces

(b) How could you incorporate into the model considerations relating to each of the following?

   1. Rental income as a function of type of tenant

   2. Maintenance cost

   3. Total cost

P4.6. The linear graph in Figure P4.6 represents a water pipe network. The lengths of the pipe sections are shown in the graph, and all pipes are 12 in. in diameter. The flow (q_i) along the ith pipe may be related to the pipe diameter (d_i), length

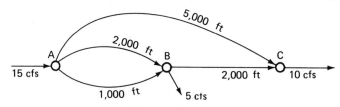

Figure P4.6

of pipe (L_i), and head loss along the pipe (h_i) by the equation

$$q_i = 44.8 \frac{d_i^{2.5}}{L_i^{0.5}} h_i^{0.5}$$

(a) What are the three different sets of physical constraints that must be satisfied by the network in a steady-state flow condition?

(b) Develop a mathematical model of the system by deriving equations to describe all the physical constraints.

(c) By means of the mathematical model developed in part (b), compute the flows (q_i) and the hydrostatic head at joints B and C.

P4.7. Develop a mathematical formulation for the compound bar problem discussed in Section 4.5 which includes the following considerations.

1. Bar lengths, sizes, and material types are to be explicitly included.
2. Costs must be considered as a function of material type, quantity used, and joint costs as a function of bar perimeter attached to the joint.

How is the solution process affected, and what might be the objectives and constraints, that you must now consider?

P4.8. The models that have been developed above are the types that are useful for components in the need model shown in Figure 2.1. What is needed is a model that incorporates several need models into a general model for the entire system, as shown in Figure 2.9. How would you approach the problem of combining the types of models developed above into the general model for the entire system area of a problem?

Optimization Concepts

5

5.1 THE OPTIMIZATION PROCESS

In most engineering problems, many feasible solutions exist. It is the professional responsibility of the engineer to seek out the best possible solution that can be achieved according to the goals and objectives for the total system, commensurate with the availability of resources and time. The processes by which the best solution is determined is called *optimization*.

Optimization is an essential and necessary phase of the design and planning process. After the problem environment has been carefully defined and some initial solutions have been found, system design models are developed to describe the design characteristics and the interactions of the system requirements. These models may be graphical, mathematical, or physical. They provide a vehicle to analyze the behavior and response of the design when it is exposed to a wide range of environmental conditions. As a result of the understanding gained from these analyses, design modifications may be made or completely new designs may be needed. This cyclic process of design and analysis is reiterated until a set of most promising solutions to the system problems emerges. These solutions must then be evaluated and ranked according to how well they fulfill the goals and objectives that have been established for the proposed system. During the ranking and evaluation phase, new understanding may be gained and new system constraints may be discovered. It is then necessary to return again to the design and analysis cycle.

The optimization process is a cyclic process that consists of a continuous interplay of design, analysis, and ranking of alternative solutions. Within the framework of the systems approach, several methods and procedures have been developed to aid the decision maker in the search for optimal solutions. Systems theories and linear graph and mathematical modeling methods help to describe the interactions of system components and to provide a vehicle for studying the input–output response characteristics of the systems. In later chapters, more specific system techniques will be discussed. The remainder of this chapter will be devoted to illustrating the fundamental concepts of optimization.

5.2 MOTIVATION AND FREEDOM OF CHOICE

For optimization to be possible, the engineer must have the desire or at least the incentive to find the best solution. This motivation comes from the need to efficiently utilize materials and resources to improve the quality of life. The degree of success in optimization is often governed by the extent to which the designer is motivated to seek an optimum solution.

In addition, optimization is possible only if a freedom of choice exists. The problem statement may be so restrictive that there is no opportunity for choice or selection; that is, there may be only one feasible solution to the problem as stated. As the problem constraints are relaxed, the number of feasible solutions increases and the task of optimization becomes correspondingly more complex. It is important, therefore, that during the problem definition phase, the system constraints imposed must be truly representative of the goals and objectives of the problem. Unnecessary system constraints may disqualify an otherwise attainable optimum solution from further consideration.

The following series of design problems illustrate how motivation and freedom of choice can affect the solution of a particular problem.

Design Problem 5.1

Design a plane steel truss to span a distance of 120 ft. The truss must have the configuration and dimensions shown in Figure 5.1a. It supports a vertical load of 30 kips (1 kip = 1000 lb) at joints 2 and 3. A fixed support is to be used at joint 4 and a roller support at joint 1. The steel should have a Young's modulus of elasticity of $E = 30,000$ kips/in.^{2}. Members in tension must not be subjected to stressed greater than 20 kips/in.^{2} and members in compression must not be subjected to stresses greater than 10 kips/in.^{2}. The maximum joint deflection must be less than 1.3 in. and the total weight must not exceed 2.2 tons. Steel members are available in the following cross-sectional areas and costs 2, 3, 4, 5, and 6 in.^{2} at a cost of $210, $220, $240, $260, $280, and $300 per member, respectively. The design must result in a minimum-cost structure.

It is apparent that the primary task of the design engineer in this problem is to determine the lowest-cost combination of cross-sectional areas for the members. The shape of the truss, the linear dimensions of the members, and the type of steel are all specified. Neither inventiveness nor originality is being required of the engineer, even though there may be several combinations of cross-sectional areas that satisfy the design constraints.

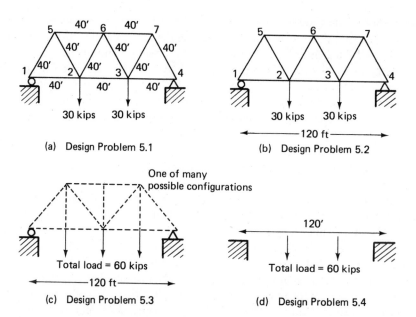

Figure 5.1 Influence of constraints on optimization problems.

Design Problems 5.2, 5.3, and 5.4 show a progressive relaxation of the design constraints in the truss problem above.

Design Problem 5.2

Design a plane truss to span a distance of 120 ft. The truss must have the configuration shown in Figure 5.1b with a vertical load of 30 kips at joints 2 and 3. A fixed support is to be used at joint 4 and a roller support at joint 1. The steel should have a Young's modulus of elasticity of $E = 30,000$ kips/in.[2]. Members in tension must not be subjected to stresses greater than 20 kips/in.[2] and members in compression must not be stressed greater than 10 kips/in.[2]. The maximum joint deflection must be less than 1.3 in. and the total weight must not exceed 2.2 tons. The cost of the structure is to be kept to a minimum. The following lengths and areas of members are available:

Member Length (ft)	Member Area (in.[2])	Number of Members
72.1	3, 4, 5, 6, 7	2 of each area
67.1	3, 4, 5, 6, 7	2 of each area
60	3, 4, 5, 6, 7	2 of each area
50	2, 3, 4, 5, 6	6 of each area
40	2, 3, 4, 5, 6	12 of each area
30	2, 3, 4, 5, 6	6 of each area

The costs of these members are as follows:

Length (ft)	Area (in.²)					
	2	3	4	5	6	7
72.1	$650	$670	$690	$720	$760	$800
67.1	450	465	490	510	535	650
60	380	390	400	415	430	450
50	300	310	320	340	360	380
40	210	220	240	260	280	300
30	110	120	140	160	180	200

In this problem, the shape of the truss and type of steel are still specified, but the designer now has more freedom in choosing the dimensions of the members.

Design Problem 5.3

Design a plane steel truss to span a distance of 120 ft. The truss can be of any configuration and dimensions. A total vertical load of 60 kips is to be evenly distributed at the joints in the same level as the two end supports. The truss must meet all the other design constraints stated in Design Problem 5.2 and at minimum cost (see Figure 5.1c).

The shape of the truss and the exact location of the loadings are no longer specified. The engineer can now choose any truss configuration and member sizes as long as they satisfy the constraints on axial stress, weight, and vertical deflection.

Design Problem 5.4

Design a minimum-cost structure to span a distance of 120 ft and support a total vertical load of 60 kips such that the maximum deflection in the plane of the supports is not greater than 1.3 in. (see Figure 5.1d).

This problem statement presents the design engineer with complete freedom to choose the type of structure, the type of materials, and the distribution of the vertical loads.

It should be obvious that the optimum design for Design Problem 5.1 is not necessarily the optimum solution for Design Problem 5.4, or vice versa. It is also apparent that the level of competence required of the designer increases with the degree of freedom in the problem.

5.3 GOALS, OBJECTIVES, AND CRITERIA

The purpose of the optimization process is to determine the system that allows the decision maker to most nearly achieve the stated goals and objectives. The degree to which the goals and objectives are achieved must be measured by a specific set of criteria. Therefore, it is absolutely essential that the goals, objectives, and criteria of the problem be clearly defined during problem definition.

The goals must truly reflect the ultimate purposes of those who have valid direct or indirect interests in the problem. For example, consider the problem of designing a high-rise building.

High-rise buildings are very common in large cities because they satisfy the social demand for sheltered space and meet the constraints imposed by the local limited availability of land and its high cost. The high-rise building reflects the impact of technology and professional creative concepts in its height, form, and efficient use of materials (Anonymous, 1972). In addition, it must provide the environment and services necessary to satisfy and attract tenants and customers and must meet financial and investment requirements.

Many complex and interacting problems exist in all phases of the building process, from its initiation by a group of investors to its rental occupancy by tenants. For a given ground area, the volume, and hence the available and potential rental floor space, increases with building heights. However, increased building height raises structural, mechanical services, and construction costs, so that eventually increased height meets financial and rental constraints from the owners and tenants.

Providing essential services, such as elevators, stairways, and heating and air conditioning consumes floor space and is directly related to the height, volume, and surface area of the high-rise building. Maximizing floor space requires improving the insulation and thermal properties of the building shell and contributes to both the cost and structural and architectural problems. Minimizing elevator service frees floor space but reduces service to tenants and transient customers, thus encountering a consumer constraint. A psychological problem is involved in excessively high buildings, which is aggravated by economies in the number and speed of elevators, as well as by structural vibrations and deflections of the building under wind loads. Finally, the overall proportions of the building and its space allocation must be viewed in the light of the owners' expectancy of the life earning capacity of the building in relation to its initial and maintenance costs.

In the illustrative sample above, many trade-offs and decisions must be made to resolve the conflicting and interacting demands made on the building. The problems raise issues involving more than technology and can be divided into two broad areas:

1. *Internal:* those associated with demands on the limited internal space of the building and the associated service and environmental requirements
2. *External:* those associated with the demands created by the building on the local political, social, and economic environment in which the building is embedded

Internally, the problem is to define, design, and provide space that fulfills the needs of those who will utilize the facilities. The allocation of space raises broad questions of whether certain spaces are required and to what extent and economic scale. It requires resolving a number of technological questions and/or a large number of conflicting demands on the space.

The construction of a high-rise building requires the blending of a number of professional disciplines, with each exercising its unique contributions to the solution. Foundation, structural, mechanical, and electrical engineers are involved, as well as architects and building contractors.

The external problems are generated by the need for space, its recognition, and attempts to provide for its satisfaction. Problems arise in sensing and evaluating the demand and a market analyses may have to be made to estimate this demand. A significant issue involves determining the nature and extent of the constraints such as building codes and zoning laws that will be imposed on the building by local authorities. An assessment must be made of the impact of the building on the various city services (water, sewerage, power, transportation, etc.). The constraints imposed represent to some degree an attempt to resolve interfacial problems equitably, so that the surrounding community is not called on to unfairly shoulder service provision costs for the high-rise building. Procedural problems arise from the legal, political, financial, and construction processes that are called into play during the creative act of producing the building. Finally, the economic feasibility of the entire process regarding investment and the life-use returns to the entrepreneur-owner must be resolved.

As the engineer approaches the planning and design of a high-rise building, the freedom to conceive and resolve alternatives may be constrained. This fact is due to the nature of the problem and the interactions with other components or elements that are involved in creating a building. The engineer achieves a better design when the broader problems and issues are considered. Thus in any problem such as the high-rise building, it is imperative that there be knowledge of the comprehensive elements that control design, the issues associated with each of these elements, and the disciplines or groups responsible if the engineer is to take an active role in the actual design and decision-making process rather than respond passively with purely technical information.

Among the many interested parties, the most important ones are the owner, the tenants, and the general public, which is represented by the city administration. The primary goal of the owner may be to maximize the return on investment or to obtain a profit large enough to justify the venture. Thus as far as the internal problems are concerned, the owner is faced with a conflict. On the one hand, there is the objective of maximizing the available useful and rental space, and on the other hand, there is the objective of minimizing the services needed to attract tenants. The owner is interested in floor space in proportion to total volume models; service space models, including garage space and transportation; architectural decor models; and floor layout models. In this situation, there is a need to resolve conflicts among the various objectives to best meet the owner's goal. To succeed, the owner must establish a value scale (criterion) with which to measure objective achievement, develop models representing the various properties of the proposed building, and solve the models to determine the optimal choices.

The tenants may be interested primarily in acquiring an attractive living or business environment for a reasonable cost. Thus the tenant's goal may be to maximize the return on the operation of a business, in which case the building may be approached from the point of view of minimizing the outlay for space consistent with satisfying the space needed for operating a business. From this point of view the tenant must also establish goals, objectives, criteria, and technological models.

The city administration may be interested primarily in improving the economic and social well-being of the city. These goals are not mutually independent. The tenants may or may not already have been identified during the design of the high-rise building, but it is apparent that if the goals of the owners are to be achieved, the interests of the prospective tenants must be taken into serious consideration. Similarly, the well-being of the city strongly affects the lives of tenants and therefore the future prospects of the high-rise building.

Although the high-rise building is used as a specific example, it portrays the typical characteristics of an engineering systems problem, as indicated by the following:

1. The engineer is involved with a project where a number of disciplines are required in order to address the variety of design aspects that must be considered. Thus the engineering discipline contribution cannot be considered in a totally isolated context.

2. Although engineering technologies play an important role in planning the building, the final design must take into account social, economic, and political aspects of the problem. These factors may provide the major rationale for decisions other than those that are made purely on technical grounds.

3. The rational solution of the planning and design problems requires a great deal of discipline interaction because of the interactive nature of the various aspects of the problem itself.

4. An organizational framework is required to resolve disciplinary conflicts and structure the manner in which the conflicts are resolved. The organizational framework must depict the peculiar nature and purpose of the high-rise building and establish the hierarchical responsibilities of the various disciplines involved.

5. Since the various technologies and disciplines permit a large number of feasible solutions involving form and spatial arrangements, a significant effort is required to formulate the building purpose and function. The purpose and function of the facility serve as the basis for guiding the entire planning and design effort.

6. Finally, the high-rise building has a large number of components of all types that can be manipulated to achieve a solution that best fulfills the requirements of the facility.

These features and considerations will occur in different degrees in any problem that the engineer confronts, and they must be evaluated and resolved early in the problem formulation.

The problem objectives specify which characteristics of the system are to be optimized in order to achieve the goals. Objectives are commonly related to such factors as cost, profit, weight, speed, time, artistic features, color, shapes, forms, and efficiency. To achieve the owner's goals of obtaining a large profit, some obvious objectives are:

1. To minimize the capital, maintenance, and operational cost
2. To maximize the total income from the high-rise building

For the tenants' goals, the objectives may be:

1. To maximize the attractiveness of the living and/or business environment, both within and around the high-rise building
2. To maximize the safety of the building
3. To maximize the convenience of entering and leaving the building

For the city administration, the objectives may be to maximize the economic gain of the city while minimizing financial and operational burdens imposed on the city by the high-rise building. Again, these goals are not mutually exclusive, but conflict with, as well as complement, one another.

Finally, the criteria are the set of parameters used to measure how optimum a solution is with respect to the goals and objectives. For the owner's objectives of minimizing cost, the criteria may be expressed in terms of the capital investment in dollars, the interest rate for loans, inflation rate, city taxes, maintenance cost, and operational cost. For maximizing income, the criteria may include rental rates in dollars, occupancy rate, tenants' income levels, and expected life of the high-rise building. Possible criteria for other objectives in this example are listed in Table 5.1.

Delineating the goals, objectives, and criteria, as shown in Table 5.1, is the necessary first step toward achieving an optimum design. For any given engineering problem, whether it be simple or complex, it may be extremely difficult to identify and state all the relevant goals, objectives, and criteria at the very beginning of the design process. However, as the problem is being studied and alternative designs are being developed, new understanding of the problem may result in new criteria, new objectives, or even new goals.

In addition, since the goals, objectives, and criteria inevitably conflict, the design engineer must also carefully rank their relative importance for the pur-

TABLE 5.1 Possible Goals, Objectives, and Criteria
for a High-Rise-Building Project

Goal	Objectives	Criteria
1. Maximize profit from investment	1. Minimize cost	Capital investment in dollars; operational cost; maintenance cost; interest rate; city taxes
	2. Maximize income	Rental rates; occupancy rate; tenants' income levels; life of the high-rise building
2. Maximize tenant satisfaction	3. Maximize safety	Structural safety factor against natural hazards such as storms, tornadoes, eathquakes, etc.; fire hazards and fire escapes; city building codes; design structural loading; emergency procedures
	4. Provide attractive living and/or business environment for tenants	Functional needs; scenic views; shopping and recreational facilities; personal security; transportation of goods and people; noise; proximity to employment; playgrounds and parks; churches and schools; prestige and pride; police protection
	5. High-rise building to be architecturally attractive	Conformity to surrounding buildings and environments; modern, traditional, or classic; degree of inventiveness; color; shape
3. Contribute to the well-being of the city	6. Maximize economic gain for the city	Creation of new industry and employment; improve urban housing environment
	7. Minimize financial and operational burden on the city	Increased demands for water and power supply, sewage disposal and processing facilities; impact on neighborhood traffic; demands for parks and recreational facilities; impact on local schools

pose of trade-offs in the optimization process. The goals and objectives in Table 5.1 are listed in the order of importance from the viewpoint of the owners. Where conflicts arise, the design engineer must decide on the solution that best resolves the conflicting issues. In order to do this, the ranked value system may be utilized to establish trade-off equations among the conflicting elements. However, the establishment of trade-off equations requires actual quantification of the relative importance of the goals, objectives, and criteria.

In general, the success of optimization depends on the ability of the design engineer to:

1. Structure the hierarchial orders of goals, objectives, and criteria
2. Define a value system that establishes priority and preference

On this basis, the engineer will undoubtedly be faced with a spectrum of problems in which there will be considerable variation in order and preference. The methods or techniques that are utilized for determining optimum solutions must recognize this aspect of the problem definition.

As a second example, consider the problem that deals with the design and location of a flood control dam on a river. A decision has been made to build a dam in a particular location; thus the goal is set. It must be understood that the decision to build the dam in the first place is part of a larger scope and involvement.

Two objectives that immediately arise are:

1. Maximize the control of floodwaters of the river
2. Minimize the damage to the environment of the area

The initial examination of these two objectives reveals that they are in direct conflict and both cannot be fulfilled.

The criteria for measuring the objectives also present a contrasting situation. In order to measure the flood control, a dollar value could be assigned to losses or potential losses from flooding. In essence, a measure of flood control is achieved. However, the damage to the environment is not so easily defined and, in some cases, it may be difficult to assess damage and consequences over a period of time. Nevertheless, they are important considerations in terms of the project.

Some type of trade-off must be established if a compromise is to be reached and the dam is to be ultimately built. In essence, increased cost of flood damage must be accepted in order to preserve the natural habitat of the area. Unless this trade-off can be achieved, the goal cannot realistically be attained and another solution to the larger problem must be sought. A choice must be made in such a problem, and the choice is an important part of the optimization problem.

5.4 OPTIMUM

An optimum solution is usually defined as the technically best solution that is achieved without compromising any goals and objectives. It represents the idealistic solution that allows all the goals and objectives to be attained. In reality, such a solution rarely exists. The nature of engineering problems invariably involves conflicting interests. Due to the state of the art in science and technology, many design factors and system constraints are not well understood; and many value criteria, such as social relevance, quality of life, ecology, and attractiveness, cannot be easily quantified or defined. The design and planning of engineering systems are always conducted within the constraints of limited manpower, equipment, financial resources, and time. Therefore, the optimization process produces only an "optimal" solution, that is, the best achievable within the design and technological constraints.

5.5 SUBOPTIMIZATION

Idealistically, all components of a system must be optimized with respect to the goals, objectives, and criteria of the total system. To do so will require a perfect understanding of the behavior of each component within the system structure, as well as its effects on the response characteristics of the total system. This requirement, in turn, calls for a single designer who has perfect understanding of every component within the system. A team of designers from various disciplines cannot perform the same function of a know-everything designer because of the difficulty in communicating technical information among different disciplines. Obviously, such requirements can never be met in complex engineering problems. Therefore, in practice, a design problem is often subdivided into component parts under the responsibility of several design teams. The quality and performance of these component parts are then controlled by sets of specifications that define the goals and objectives for each component. The process of optimizing a component of a system according to a subset of goals, objectives, and criteria is called *suboptimization*.

For example, in the high-rise building problem, the work tasks may be divided according to disciplines as follows: architectural design, structural design, mechanical services, plumbing and heating, transportation systems, construction, business management, and the like. Each work task is controlled by specifications to assure that the goals and objectives for the total building system can be at least partially attained. Indeed, each of these tasks usually involve many subtasks. For example, although the overall supervision of the construction may be conducted by a firm of consultant engineers whose sole responsibility is to serve the owners, the actual construction may be performed by a large group of subcontractors for excavation and foundation, erection of steel frame, exterior walls, interior decoration, and so forth. As the optimization processes become fragmented in large projects, the efficiency of project organization and management become vitally important to the success of the project.

The problems and methodologies associated with project management are discussed in Chapter 13.

5.6 METHODS OF OPTIMIZATION

The optimization methods that may be utilized for a given problem depend on the degree of order that can be established in the problem. The order reflects how well the system structure and behavior characteristics are known, and how well the system constraints and criteria can be quantified. The optimization methods can be broadly grouped into three approaches:

1. Analytical
2. Combinatorial
3. Subjective

5.6.1 Analytical Approach

This approach is applicable only to a totally ordered system problem. A total order implies complete knowledge of structure and behavior and either completely automates the optimization process or enables both the problem structure and the criterion preference function to be analytically stated. In such a case, the optimization process can be supported by a well-defined set of criteria that may be used to judge the quality of the design. These criteria should be included in the statements of systems objectives and should define the technical, political, and operational goals of the system.

The system variables are quantified and each variable is assigned a system value according to a common measure, such as a monetary value. Thus, if $x_1, x_2, x_3, \ldots$, and x_n are n design variables in the problem and a_i is the value of one unit of x_i to the total system, the total value of the system may be expressed in a mathematical function as follows:

$$C = \sum_{i=1}^{n} a_i x_i \qquad (5.1)$$

The objective of optimization, then, is to find the set of x_i's such that the total system value C is maximized or minimized.

Equation 5.1 represents the simplest form of a criterion function. It is a linear equation involving only first-degree terms of the design variables.

In general cases, the total system value C may appear in any function form depending on the nature of the problem; that is,

$$C = F(x_1, x_2, x_3, \ldots, x_n) \qquad (5.2)$$

There may also be more than one criterion function for a given system problem. Thus, if C_1, C_2, and C_3 are three independent measures of a system, then

$$C_1 = F_1(x_1, x_2, x_3, \ldots, x_n)$$
$$C_2 = F_2(x_1, x_2, x_3, \ldots, x_n)$$
$$C_3 = F_3(x_1, x_2, x_3, \ldots, x_n)$$

and the system design must be optimized with respect to all three criterion functions. If it is not possible to optimize simultaneously all three criterion functions, an order of priorities must be specified to resolve the conflicts.

Thus the optimization model for a totally ordered system problem takes the following general format:

Maximize or minimize the following criterion functions:

$$C_1 = F_1(x_1, x_2, x_3, \ldots, x_n)$$
$$C_2 = F_2(x_1, x_2, x_3, \ldots, x_n)$$
$$\cdot \quad \cdot$$
$$\cdot \quad \cdot$$
$$\cdot \quad \cdot$$
$$C_k = F_k(x_1, x_2, x_3, \ldots, x_n)$$

subject to the following system constraints:

Lower bounds:

$$G_1(x_1, x_2, x_3, \ldots, x_n) \geq a_1$$
$$G_2(x_1, x_2, x_3, \ldots, x_n) \geq a_2$$
$$\cdot$$
$$\cdot$$
$$\cdot$$
$$G_m(x_1, x_2, x_3, \ldots, x_n) \geq a_m$$

Upper bounds:

$$H_1(x_1, x_2, x_3, \ldots, x_n) \leq b_1$$
$$H_2(x_1, x_2, x_3, \ldots, x_n) \leq b_2$$
$$\cdot$$
$$\cdot$$
$$\cdot$$
$$H_r(x_1, x_2, x_3, \ldots, x_n) \leq b_r$$

Equalities:

$$P_1(x_1, x_2, x_3, \ldots, x_n) = c_1$$
$$P_2(x_1, x_2, x_3, \ldots, x_n) = c_2$$
$$\cdot$$
$$\cdot$$
$$P_s(x_1, x_2, x_3, \ldots, x_n) = c_s$$

Approximations:

$$Q_1(x_1, x_2, x_3, \ldots, x_n) \simeq d_1$$
$$Q_2(x_1, x_2, x_3, \ldots, x_n) \simeq d_2$$
$$\cdot$$
$$\cdot$$
$$Q_t(x_1, x_2, x_3, \ldots, x_n) \simeq d_t$$

Such an analytical approach is called *mathematical programming*. The solution of this system of mathematical expressions yields the optimum set of values for the design variables. The application of this method is discussed further in Chapters 8, 9, and 10.

In some cases, the problem structure can best be described using a graphical concept, while still keeping an analytical form for the criterion preference function. The criterion function is then used to process and evaluate the relevant graphical properties exposed by the problem. This approach is illustrated in Chapter 7.

5.6.2 Combinatorial Approach

As the degree of order possessed in the problem is weakened, the optimization process may reduce to purely evaluating a multiple selection of alternatives and selecting the best. Suppose that there are n design variables in the problem and that they are denoted as $x_1, x_2, x_3, \ldots$, and x_n. The approach is to identify the most probable range of values for each of these variables and then analyze all the possible combinations. If there are 10 discrete probable values for each parameter, a complete combinatorial analyses will involve 10^n cases.

It is obvious that a strict application of such an approach becomes impracticable even for simple systems involving only a few variables. For example, for $n = 4$, there will already be about 10^4 cases for analysis. However, technical procedures together with past experience can often be used to eliminate a large percentage of the feasible combinations and leave only a few for detailed study and analysis.

This approach is used extensively in decision analysis. All the possibilities are analyzed and compared to arrive at an optimum decision policy. The graphical methods of paths and cut-sets analysis used in Section 7.3 for analyzing the road network improvement problem is an example of this approach. Chapters 7, 9, 10, 11, 12, and 13 also illustrate to varying degrees this combinatorial approach to optimization.

5.6.3 Subjective Approach

In complex problem situations, it may be difficult or even impossible to establish specific models or system orders, in which case purely subjective methods become necessary. The subjective approach is the most important and often is the deciding method in optimization. Intangible factors such as social values, political influences, and psychological effects are extremely difficult to quantify and measure. Yet, as all practicing engineers soon learn, these factors often control the acceptance or rejection of a design.

For example, a safe highway cannot be designed without considering the response characteristics and behavior of the drivers. Nor is it possible to find an optimum location for a reservoir without paying due regard to the political forces and ecological and social consequences involved. Whether it be in choosing an alternative design or in predicting the present and future occurrence of uncertain events, sound engineering judgments based on past experience and foresight play an indispensable role in optimization. Chapters 13 and 14 introduce the reader to issues and problem situations requiring subjective approaches to problem optimization and decision making. In some cases the subjective element can be quantified and incorporated into an analytical or combinatorial approach.

5.7 SUMMARY

It is important to recognize that the identification of a best solution is contingent upon the problem formulation. The best solution is a function of the scope of the system under consideration and the components that are included in the analysis. The definition of problem constraints is particularly important in optimization because constraints may serve to limit the nature of the problem.

In some studies, goals and objectives are stated in rather vague terms. To be useful in analyzing a problem, goals and objectives should be defined in terms which can give guidance to the selection of a preferred solution. Relating goals and objectives to a hierarchical system's framework can serve to give a meaningful definition to goals and objectives.

The selection of the analysis method is a function of how well the interactions between the various system components can be quantified. In seeking a preferred solution to a problem, the engineer must attempt to coordinate the optimization method with the nature of the problem being evaluated. Later chapters contain discussions of specific analysis techniques which may be applied in the determination of a preferred solution to a problem.

PROBLEMS

P5.1. Discuss the optimization concepts implicit in the following problems:

(a) The total cost of a building project is the sum of the project costs (labor, material, etc.) and the indirect costs (organizational overhead, profit, etc.). Indirect costs, $\$_I$, amount to \$4000 per project day. Direct costs, $\$_D$, are a function of the rate of use of applied resources and vary between \$100,000 for a project duration T of 100 days to about \$200,000 for a duration of 70 days and can be expressed as

$$\$_D = \frac{\$10^9}{T^2}$$

where T is the project duration in days. Determine the project duration with minimum total cost.

(b) A manufacturing company is making household appliances and its profit is a function of the number of units produced. The total profit, $\$_P$, can be expressed as

$$\$_P = 20x - 10^{-6}x^2$$

where x is the number of units produced. This expression can be interpreted to mean that profit will increase as the number of units produced increases until the market becomes saturated, and then profit will decrease. Determine the number of units to produce such that the company maximizes its profit.

(c) The company in part (b) has only enough plant capacity to produce 8 million units. How does this constraint affect the results obtained in (b)?

(d) A city council has just opened the contractors' bids on the construction of a new city parking lot. The bids are in unit prices; however, the entire contract will be awarded for the bid with the lowest total cost. The contractors' unit price bids and quantities required for each item are as follows:

	Clearing	Cubic Yard of Subbase	Square Yard of Paving	Installation of Meters
Contractor *A*	$3000	$4.25	$26	$25 each
Contractor *B*	4000	4.00	25	30 each
Contractor *C*	2000	3.75	24	35 each
Required	1	1×10^4	7×10^4	500

Which contract should the city council accept?

P5.2. Public facilities (such as fire stations) have many different characteristics that contribute to their value or acceptance by society and prevent the formulation of unique criteria for selecting optimal facility designs. In these cases it may be possible to establish no more than a subjective preference for one complex design over another in a series of binary decisions. This situation leads to the development of preference graphs (see Kaufman, 1968) in which each facility design, as it incorporates more features, is modeled and labeled by additional design space nodes. Preference for one design over another is then modeled by a directed arrow from the less preferable to the more preferable facility.

What factors do you think should be considered in locating and designing a city fire station? Develop a preference graph for a number of fire station situations that have various levels of desirable features.

P5.3. For the hierarchical systems that were defined in P2.3, define the goals, objectives, and possible criteria for each component and system level. As the system levels increase in scope, why is it more difficult to reach a truly optimum solution? How does this situation affect the optimization method or technique that is utilized?

P5.4. The absolute and relative locations of physical facilities in a production plant are simple ways of describing or modeling the plant. How good or efficient the plant layout becomes depends on the goals, objectives, and criteria that are used to evaluate the plant.

(a) Structure the following goals and objectives relevant to a particular plant with which you are familiar.

1. Minimize product handling
2. Minimize inventory levels within the process line
3. Minimize floor space
4. Minimize cost
5. Minimize work force

6. Maximize mobility to change process lines
7. Maximize use of existing capital equipment
8. Maximize profit
9. Maximize output

(b) The investigation and modeling of a given plant will focus on a specific set of plant characteristics. The characteristics selected and hence the modeling abstraction are directly related to the goal, objective, and criteria under consideration.

Select two different objectives and show how different modeling rationales develop.

P5.5. The cost of environmental services such as heating, air conditioning, lights, and so on, may constitute as much as one-half of the cost of a building. The following represents various levels of optimization of the cost of the environmental services of a high-rise building:

1. The number of floors, the amount of floor space, the use of the building, and the building material are specified and the objective is to minimize the cost of the environmental services.

2. The number of floors, the amount of floor space, and the use of the building are specified and the objective is to minimize the cost of the building material and environmental services.

3. The amount of floor space and use is specified and the objective is to minimize the cost of the building.

4. The owner wants to invest in a high-rise building that will allow maximization of profit and does not care how large it is or what it is used for.

Describe the problems that arise at each level of optimization. How could trade-offs be developed at each level of optimization? Explain why you think it is easier or more difficult to optimize as the constraints are removed.

P5.6. Large-scale water resource projects are usually development that are built with public money for public use and benefit.
  (a) What might be some goals and objectives for public investment in water resources development? What criteria might you use for measuring the effectiveness of the projects in achieving the objectives you listed? How are goals and objectives for public investment determined?
  (b) As part of the water resources development of a region, a decision has been made to build a water supply reservoir. What might be some of the objectives that the engineer would try to meet in designing the reservoir? How are these related to the goals and objectives listed in part (a)? What criteria might be used? How might trade-offs be developed among conflicting objectives? (*Hint:* See Hall and Dracup, 1970).

P5.7. A river bisects a large city and is crossed by a number of bridges in the city transportation system. The city engineer is asked to locate a new and additional bridge over the river in a section between two existing bridges.

What criteria may develop that the engineer can use to evaluate a specific bridge location and its influence on the city transportation system? What political, social, economical, and technical constraints might affect the bridge location?

P5.8. In many design and planning situations the decision-making and optimization processes are hampered by the difficulty and delay in formulating system goals and objectives (see Hitch, 1961).

Discuss this statement in reference to a local problem with which you are familiar.

System
Evaluation

6

6.1 INTRODUCTION

In the process of addressing a problem, engineers may be required to assess the feasibility of a proposed system or evaluate alternatives that are under consideration. Basically, this involves an examination of the costs, benefits, and consequences of the system or the alternatives over a period of time. This period of time is known as the *planning horizon*.

In Chapter 5, optimization methods were discussed. One might consider some of the optimization methods as techniques by which a preferred alternative may be selected. In addition, however, there are other methods that have considerable application in engineering planning and design. For example, the concepts and methods associated with engineering economic analysis have frequent use in the evaluation of a proposed system or alternatives. These methods represent additional tools that may be applied to aid a decision maker in choosing the best course of action.

It is important to recognize that the manner in which the system will be evaluated should be considered at the time the problem is formulated and the approach to the solution is planned. In essence, the determination of the evaluation methods should be an integral part of the plan for addressing and solving the problem that has been defined.

The development of goals, objectives, and criteria was discussed in Chapter 2. As indicated in that chapter, the goals reflect the function that the system is to fulfill, and criteria are the measures of system performance. Evaluation methods should recognize and incorporate these measures. For example, traffic safety problems often utilize frequency of accident occurences as a measure of performance, or the failure of system performance in this case. In evaluating proposed improvements to a roadway system or even comparing alternative improvements, it would be necessary to use an evaluation technique that would accommodate accidents as a measure of performance.

6.2 FEASIBILITY ASSESSMENT

The assessment of feasibility involves the determination of whether the solution to a problem is suitable, acceptable, and attainable. This aspect of evaluation is extremely important because implementation decisions generally are related to the feasibility of a proposed system or project.

For most engineering problems, a proposed system will usually be required to pass tests for engineering, economic, financial, political, environmental, and social feasibility. Depending on the nature of the situation, it may not be necessary to make an assessment in terms of all of these areas. The areas that will be addressed should be established after careful consideration of the needs and the concerns of the owner of the project. This consideration should include the determination of the feasibility of a system or project from the broader viewpoints of society.

Engineering feasibility requires that the proposed system be capable of performing its intended function. Design analysis procedures as described in standard engineering texts can be used to indicate the ability of a proposed system to perform its intended function. In addition, the construction or implementation of the system must be possible.

A proposed system is *economically feasible* if the total value of the benefits that result from the system exceed the costs that result from the system. Economic feasibility depends on engineering feasibility because a system must be capable of producing the required output in order to produce benefits.

The owner must have sufficient funds to pay for system installation and operation before the proposed system is considered to be *financially feasible*. Financial feasibility may or may not be related to economic feasibility. An owner may be able and willing to pay for a system in order to fulfill noneconomic goals. It may also be that an economically feasible project is financially infeasible because the owner is not able to obtain enough money to implement the system.

Political and *social feasibility* is assured if the required political approval can be obtained and if the potential users of the system will respond favorably to system implementation. Every system is subject to review at different stages of planning. A private company has executive officers or a board of directors that reviews proposed systems. Public systems are subject to public hearings and review by committees of elected representatives. Usually, political support is gained after evidence of engineering and economic feasibility has been presented. However, political pressure may be quite strong for a specific system even if it is economically infeasible. Conversely, groups that feel that they are adversely affected often oppose economically feasible systems because noneconomic factors have not received sufficient emphasis. Political and social feasibility can best be attained by active participation of representatives from all interested groups in planning and designing a proposed system.

Environmental feasibility involves the assessment of the environmental consequences of the proposed system. Because of the increased societal concern about potential short- and long-term influences on the environment, the development and implementation of most engineering systems of any magnitude require formal study of the expected environmental consequences if the project is implemented. This study results in what is known as an *environmental impact statement*.

6.3 PLANNING HORIZON

The planning horizon is the most distant future time considered in the engineering economic study. The planning horizon to be used for a specific study depends on the purpose of the planning and the scope or areal extent for which the planning is done.

If the purpose of the planning is to develop a framework plan that is to serve as a guideline for development, a time 50 years or more in the future may be chosen as the planning horizon. The framework plan will indicate the types of systems needed to achieve the desired results from the development. A framework plan for a river basin region might include the number of reservoirs and other types of water resources development, along with the purposes that each of these elements should serve in order to sustain the projected economic and population growth of the region for the next 50 years.

A planning horizon of 20 to 30 years in the future may be chosen for a study concerning investment decisions. The size and type of water treatment plant that a city will invest in may be based on existing water treatment methods and the projected economic and population growth of the city for the next 20 years. Many systems may have even shorter planning horizons. The construction of a large dam may require that a concrete mixing plant be constructed at the dam site to provide concrete for the dam. The size and type of mixing plant to be used may be determined from a study with a planning horizon that coincides with the projected completion date of the dam, some 1 to 5 years in the future. In some cases, the planning horizon for a major plant straddles several projects, in which case the salvage value of the plant at the end of a particular project may influence both its economic and financial feasibility.

The scope or areal extent of a proposed system will also be a factor in determining the planning horizon to be used in a study. The planning horizon for a rural farm-to-market road will usually be shorter than for an interstate highway system.

Traditionally, three different periods have been considered in planning studies: the physical life of the system, the economic life of the system, and the period of analysis. The *physical life* of a system ends when it can no longer physically perform its intended function. The physical life of a building does not end if the building is converted from a hotel to a museum. Its physical life ends when it can no longer provide shelter or support the loads sustained in the use of the building.

The *economic life* of a system ends when the incremental benefit from continuing operation of the system one more time period no longer exceeds the incremental costs of continuing operation one more time period. This point usually occurs when the annual operation, maintenance, and repair (OMR) costs equal or exceed the annual benefits from the system. Since a program of regular maintenance and periodic replacement of worn parts may extend the physical life of a system almost indefinitely, the economic life is usually shorter than the physical life.

The period of time over which the system consequences are considered to affect the system benefits and costs is referred to as the *period of analysis*. The uncertainty associated with future benefits and costs increases as the length of the period of analysis increases. In actual problem solving, it is necessary to confine studies to a certain extension in space and time. This confinement within specific limitations is required because of the uncertainty associated with future benefits and costs and because of our inability to trace and evaluate the consequences of the system beyond the immediate vicinity of the system. Regardless of the period of analysis used, the salvage value of the system should be considered in the analysis. The *salvage value* is the worth of the system at the end of the period of analysis. For some systems, this worth will be positive, indicating that the system still has value or can be sold and thus can be thought of as a benefit that occurs at the end of the period of analysis. For other systems, this worth may be negative, indicating that it is a liability that must be disposed of and thus can be thought of as a cost that occurs at the end of the period of analysis.

When alternative schemes of development are being considered for the same purpose, all alternatives must be evaluated for the same period of analysis. If some component requires periodic replacement, its cost is usually assumed to be repeated at the end of each component life until the total period of analysis is completed. However, this assumption should not be made without considering the effects of inflation, the development of new production techniques through technological advance, and the changing nature of demand for the system outputs with time. Uncertainty about any of these factors tends to favor alternatives with short lives.

Sunk costs are costs that have been paid or are committed to be paid due to past actions. Sunk costs can be ignored in performing economic analysis of alternatives. For example, suppose that a company purchased a piece of machinery 2 years ago on time payment, and agreed to pay the vendor $10,000 per year for 5 years. There remain three annual payments of $10,000 each. Now the company is considering replacing that machine with a new and more productive one. In particular, it is considering three different machines, each has different cost, life expectancy, and OMR cost. Since it has to make the remaining payments on the old machine regardless of which new machine it will buy, there is no need to even consider the cost of the remaining payments in the analysis. Such costs are called sunk costs.

6.4 TIME VALUE OF MONEY

In the analysis of any system where costs and income or revenue occur at different times during a study period, it is important to recognize the time value of money. The value of the costs and income are considered to vary with time because of the ability of money to grow if profitably invested or to earn interest if placed in a savings account. A dollar today is equivalent to $1.08 one year from now if an investment opportunity is available that will earn 8 percent interest per year. If a person is given the choice of receiving $1 today or $1 one year from today, he or she would choose the $1 today because the $1 could be invested and earn interest. This concept of *equivalence* is important in economic studies that justify proposed systems because it allows values that occur at different points in time to be compared on a common time basis.

Narrowly defined, in a borrowing–lending situation, interest may be defined as money paid for the use of borrowed money. The *rate of interest* is the ratio, expressed as a percentage, of the interest payable at the end of a period of time, usually a year or less, and the money owed at the beginning of the year. Thus, if $8 is payable annually on a debt of $100, the interest is $8 and the interest rate is $8/$100 = 0.08, or 8 percent per year.

Although the interest rate may be loosely defined as any expression of the time value of money, a more precise definition distinguishes between interest rate and discount rate. Interest is the fee for borrowing money. A *discount rate* is the expression of the time value of money used in the equivalence computations. The discount rate should reflect the *opportunity cost*, which is equal to the return that would have been realized had the money been invested in other alternatives available to the person. It may or may not be equal to the interest rate. If the person can earn 15 percent interest on the money from some other source if the system is not constructed, then the person should use 15 percent for the discount rate even though he or she can borrow money from a bank at 12 percent.

Interest is said to be compounded if interest is earned on interest. If interest is based only on the original amount for each period, the interest is said to be simple interest. Most economic studies are performed at compound interest, wherein values are adjusted forward or backward in time using a compounding growth or discount factor.

The following example illustrates a situation where it would be necessary to consider the time value of money in the analysis of the problem. It should be noted that the problem is analyzed further later in this chapter.

The River Town City Council is faced with the problems of increasing its water supply during drought periods, increasing the availability of electric power that it now supplies from a city-owned steam plant, increasing the amount of recreational facilities available to its residents, and finding a way to decrease the amount of flooding that occurs in the section of the city along the river front. One alternative under consideration is a multiple-purpose reservoir that can be used for recreation, hydropower generation, water supply, and flood control. It is estimated that the reservoir will cost $26.5 million to construct; will have annual operation, maintenance, and repair costs of $0.5 million; and annual benefits of $0.9 million for recreation, $0.7 million for power, $0.4 million for water supply, and $1.0 million for flood control. The reservoir is assumed to have a 50-year life and the benefits begin the first year. It is further assumed that an 8 percent discount rate is appropriate. The situation is summarized in Figure 6.1, in which the arrows above the horizontal line represent benefits and the arrows below the line represent costs. The benefits for the four purposes total $3.0 million per year. The usual convention is to assume that the construction cost occurs at the beginning of the first year and that the annual costs and benefits occur at the end of each year.

Because the reservoir has a long life, it becomes necessary to consider the annual operating costs and annual benefits when evaluating the system. Thus there is a need to compare costs and benefits that occur at different times.

Figure 6.1 represents a model of the cash flow for this particular project. The development of such a figure requires the engineer to have a full understanding of the points in time when costs and benefits occur. This type of model is of

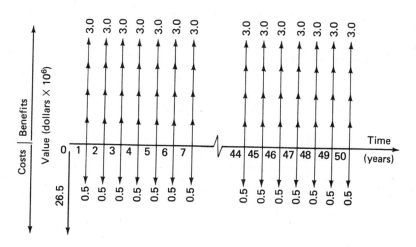

Figure 6.1   Time profile of benefits and costs for reservoir example.

great benefit when analyzing a problem of this nature, and the value of the representation increases with the complexity of the cash flow that is associated with the project.

6.4.1 Compound Interest Computations

In order to compensate for the value of expenditures or income at different times in an analysis period, an equivalent monetary value is computed for some common point in time. These calculations are accomplished through the use of compound interest or *discount factors*. There are six basic factors that are commonly used:

1. Single-payment compound amount
2. Single-payment present worth
3. Uniform series compound amount
4. Uniform series present worth
5. Sinking fund
6. Capital recovery

Using the *equivalence concept* for compound interest introduced above, note that the value today of $1 to be received one year from now is only $1/(1.08); this is the amount that would be worth $1 one year from now if invested at 8 percent interest. The value today of $1 to be received 2 years from now can be determined by computing its value one year from now, which is $1/(1.08); this is the value that must be invested at 8 percent interest one year from now to be worth $1 one year later, which is 2 years from now. The value of $1/(1.08) today is then $1/[(1.08)(1.08)] = $1/(1.08)^2. This can be generalized to

$$PWSP = \frac{F}{(1 + i)^n} \tag{6.1}$$

where $PWSP$ is the present worth of a single payment F that is n years from now when the discount rate is i.

The *present worth* of a series of equal annual payments, or receipts, such as the annual operation, maintenance, and repair costs of the reservoir, is the sum of the present worth of each cost. Therefore,

$$PWUS = \frac{A}{1 + i} + \frac{A}{(1 + i)^2} + \cdots + \frac{F}{(1 + i)^n} \tag{6.2}$$

where $PWUS$ is the present worth of a uniform series of A per time period, i is the discount rate, and n is the number of time periods. Equation 6.2 can be simplified to

$$PWUS = A \left[\frac{(1 + i)^n - 1}{i(1 + i)^n} \right] \tag{6.3}$$

The term in brackets in Equation 6.3 is known as the *present worth uniform series factor* and can be denoted as $(PWUS, i, n)$. Its value has been tabulated for various combinations of i and n in Appendix B.

The present worth uniform series factor is equal to 12.2335 for $i = 0.08$ and $n = 50$. Therefore, the present worth of the operation, maintenance, and repair costs, $PWOMR$, of the reservoir is computed as

$$PWOMR = (\$0.5 \times 10^6)(12.2335) = \$6.12 \times 10^6$$

Similarly, the present worth of the benefits, PWB, is computed as

$$PWB = (\$0.9 + \$0.7 + \$0.4 + \$1.0)(10^6)(12.2335) = \$36.70 \times 10^6$$

Hence the present worth of the total costs is $\$26.5 \times 10^6 + \$6.12 \times 10^6 = \$32.62 \times 10^6$ and the present worth of the benefits is $\$36.70 \times 10^6$. The net present worth is then equal to $\$36.70 \times 10^6 - \$37.62 \times 10^6$ and is equal to $\$4.08 \times 10^6$. This is sometimes referred to as the present value of the net benefits.

Instead of computing equivalent values of costs and benefits in terms of the present worth, another common approach is to convert all amounts to an annualized basis. For example, if Equation 6.3 is solved for A, we have

$$A = PWUS\left[\frac{i(1 + i)^n}{(1 + i)^n - 1}\right] \tag{6.4}$$

where the term in brackets is known as the *capital recovery factor* and is used to convert a single value at the present time into an equivalent series of equal annual values. The capital recovery factor is equal to 0.08174 when $i = 0.08$ and $n = 50$. Therefore, the \$26.5 million construction cost of the reservoir is equivalent to a series of annual costs C, computed as

$$C = (\$26.5 \times 10^6)(0.08174) = \$2.17 \times 10^6$$

The total annual cost is then the sum of the annual operation, maintenance, and repair costs and the annual costs for construction, which is $\$0.5 \times 10^6 + \$2.17 \times 10^6 = \$2.67 \times 10^6$. Hence the annual net benefits are determined by $\$3.0 \times 10^6 - \$2.67 \times 10^6$ and are equal to $\$0.03 \times 10^6$.

The single-payment compound amount factor is used to compute the equivalent value of a single payment for some point in the future. Also, the *sinking fund factor* annualizes a monetary amount that occurs at the end of an analysis period. The equations for all of these factors as well as the numerical values for selected interest and time periods are shown in Appendix B.

6.4.2 Period of Compounding

In banking, interests are usually compounded, not on an annual basis, but at a more frequent rate. For example, many banks offer savings accounts that pay interest compounded daily. There are also savings accounts that pay interest compounded quarterly. In such cases, distinction must be made between the stated annual interest rate (i) and the effective annual yield (i_e). Let c be the number of times per year interest is to be compounded. Then the interest rate paid at each compounding period is i/c. If $\$P$ is invested now in an account that earns at a rate i compounded c times per year, the total amount, $\$F$, available at the end of n years can be computed as follows:

$$\$F = \$P\left(1 + \frac{i}{c}\right)^{cn} \tag{6.5}$$

The amount $\$F$ can also be computed using the effective annual yield, as follows:

$$\$F = \$P(1 + i_e)^n \tag{6.6}$$

Equating the two expressions above yields the following:

$$\$P\left(1 + \frac{i}{c}\right)^{cn} = \$P(1 + i_e)^n$$

and

$$i_e = \left(1 + \frac{i}{c}\right)^c - 1 \tag{6.7}$$

For example, for $i = 0.12$, i_e equals 0.1255 for $c = 4$ (compounded quarterly), and equals 0.1275 for $c = 365$ (compounded daily).

6.5 PRICE CHANGES

In addition to the changes in monetary values due to the time value of money, it may also be necessary to consider variations in the actual price of products and services. These changes in price can be attributed to several causes: First, the value of money may decrease (inflation) or increase (deflation) with time. Second, variations in the supply and demand for specific commodities with time may cause their value to change relative to overall price levels. Finally, advancements in technology and production methods may lead to price changes for specific items.

For evaluation studies, all monetary amounts should be in terms of commensurable value units. The most satisfactory value unit is for these amounts to be expressed in constant dollars. This means that dollars spent at one point in time can be transformed to constant dollars at another date even though there may have been a price change. This transformation is accomplished through the use of a price index.

A *price index* is the cost of a preselected group of items called a bundle of goods expressed as a percentage of the cost of the same items at a base date. *Engineering News-Record (ENR) Construction Cost Index* is one of the most widely used indexes. This index is computed from the average cost of heavy construction consisting of fixed quantities of common labor, cement, steel, and lumber for 20 cities in the United States. The index was 100 for the base year of 1913. For 1960, 1970, 1980, and 1981 it equaled 824, 1385, 3237, and 3533, respectively. Other price indexes are also available and current values for most cost indexes associated with construction projects are published four times a year in the Quarterly Cost Roundup of the *Engineering News-Record*.

The use of an index assumes that cost varies with time in the same manner as the value of the items on which the index is based. Before a price index is used, the cost items under study should be compared with the cost items on which the index is based to determine whether their respective patterns of cost change are similar.

In estimating future costs for economic studies, prices under normal or average conditions should be used. This normalized price is obtained by averaging the price in constant dollars over a number of recent time periods to prevent project feasibility from depending on short-term market abnormalities.

The current normalized price in constant dollars needs adjusting only when the prices of goods or services are expected to change relative to the general price level. This adjustment involves multiplying the future cost by the ratio of the present to the future value of money. The future cost is determined by multiplying current cost by the ratio of the future to present index for the item. If the present value of the item is $100 and the value of money changes from 110 to 120 (as determined by a general price index) and the value of the item under consideration changes from 115 to 160 (as determined from a cost index for this item), the future cost in constant dollars is (160/115)(110/120)(100), or $127, which is the value that should be used in the economic study. Uncertainty in predicting future changes in indexes usually limits this type of adjustment to less than 10 years into the future.

Construction cost indexes are particularly useful for estimating current construction costs of system components from historical data of completed systems. These estimates can then be used in the economic analysis.

6.6 ECONOMIC ANALYSIS METHODS

There are several economic analysis methods that can be used to compare and rank alternatives where the benefits and costs are expressed in monetary terms. Common methods that are likely to be used in connection with engineering projects are (1) present worth, (2) annual equivalent value, (3) benefit/cost ratio, and (4) internal rate of return. Although the underlying concept for each method may differ, the resulting selection of the preferred alternative or the ranking of the alternatives should be the same regardless of the method. It should be noted that any book that deals specifically with economic analysis will probably contain additional methods.

6.6.1 Present Worth

The net present worth is an economic criterion that can be used to measure the effectiveness of alternatives. Calculating net present worth from a time profile of costs and benefits is straightforward. The following summarizes the procedure:

1. Discount each present worth to the same base year because sums of money at different times are different economic goods.

2. Use the same discount rate for all present values.

3. Base the present worth of each alternative on the same period of analysis. This may be done by evaluating the cost of extending the shorter-lived alternative to the entire period of analysis.

The net present worth of the River Town reservoir was shown to be 4.08×10^6.

If the choice of alternatives is simply how large to make the reservoir, the maximum net present worth is obtained when the incremental present worth of the benefits is exactly equal to the incremental present worth of the costs. This occurs when the present worth of increasing the size of the reservoir one unit of size is just equal to the present worth of the costs associated with this unit increase in size.

6.6.2 Annual Equivalent Value

The basic idea with regard to the use of the annual equivalent value is to reduce the net present worth to a period (or annualized) basis. This, of course, shifts focus from the size of the system to the length of its expected life. The method for converting the net present worth value to an annual equivalent value is to multiply the net present worth value by a capital-recovery factor, as illustrated above for the River Town reservoir.

Now alternatives with different life spans can be evaluated and the equivalent annual values may serve as a criterion for ranking the alternatives. If the choice of alternatives is simply how large to make the reservoir, then either the present worth or equivalent annual values may be used as the criterion.

6.6.3 Benefit/Cost Ratio

The *benefit/cost ratio* is the ratio of the present value of the benefits to the present value of the costs. Annual values can be used without affecting the ratio. For the River Town reservoir, the benefit/cost ratio is

$$B/C = \frac{\$32.62 \times 10^6}{\$36.7 \times 10^6} = \frac{\$3.0 \times 10^6}{\$2.67 \times 10^6} = 1.12$$

When comparing alternatives, the same period of analysis must be used if the benefit/cost ratio is computed using present values. If the annual values are used, the periods of analysis do not have to be the same for different alternatives because the annual values are based on the assumption that the shorter-lived alternatives would be extended to the entire period of analysis.

6.6.4 Internal Rate of Return

The *internal rate of return* is the rate of return used for discounting such that the present value of all benefits exactly equals the present value of all costs. This value cannot be determined directly, but must be solved for by trial and error; that is, different interest rates are tried until one is found that causes the present value of benefits to exactly equal the present value of costs. In addition to being computationally cumbersome, there may be more than one solution for certain cases.

For example, suppose that a high-rise building is estimated to cost $100 million, and that the net annual return or benefits from the project is estimated to be $12.5 million for a period of 50 years. Let i be the discount rate. Then the equivalent annual cost of the initial construction cost can be computed using the uniform series capital recovery factor (CRF) as follows:

$$\text{Equivalent annual cost} = \$100,000,000 \times (CRF, i, 50)$$

Equating the equivalent annual cost with the annual benefit yields the following expression:

$$100,000,000 \times (CRF, i, 50) = \$12,500,000$$

or

$$(CRF, i, 50) = \frac{12,500,000}{100,000,000} = 0.12500$$

By interpolation of the discount tables in Appendix B, we obtain

$$i = 0.12 + \frac{0.12500 - 0.12042}{0.14020 - 0.12042} \times 2$$

$$= 0.1246$$

The internal rate of return of the project is then said to be 12.46 percent.

6.7 EVALUATION USING NONCOMMENSURATE VALUES

The economic analysis methods described in the preceding section require that the costs and benefits be expressed in commensurate terms. This means that all values are measured in the same way, which in this case are monetary amounts. For instance, the example of the River Town water supply indicated that the benefits from providing recreation facilities, hydroelectric power, water supply, and flood control were all measured in dollar amounts. There are situations, however, where there may be no way of objectively assigning dollar amounts to some items. Although the River Town example indicated a value for recreation, the expression of the benefits in such terms may require a subjective evaluation. The same comment might apply to problems that involved measures related to accidents, user-time costs, and the consequences of pollution.

When encountering a situation where noncommensurate values are included in the analysis, there are several ways in which these values can be incorporated into the evaluation process. One procedure is to assign a value to the item based on the value judgment of the owner. This can be accomplished by choosing a dollar value for each unit of the item, as was the case with recreation in the example. For situations involving accidents, organizations frequently assign a monetary amount to an accident.

A second procedure for incorporating value judgments is for the owner to specify some minimum quantity of a particular item. This minimum quantity then becomes a constraint that must be satisfied. It is not unusual for the owner to specify that a certain quantity of water supply must be provided by the reservoir. The optimum reservoir must then be determined based on maximizing the present worth of the costs and benefits subject to this constraint.

Additional procedures involve the use of cost/effectiveness analysis or the development of a goals-achievement matrix (Hill, 1968).

6.7.1 Cost/Effectiveness Analysis

The underlying concept of cost/effectiveness analysis is that the cost per unit of benefit is determined. For this analysis, the costs are normally expressed in monetary amounts, and benefits are measured in terms of a noncommensurate value. The preferred alternative would be the one with the least cost per unit of benefit.

As an example, consider a situation where several intersections are being considered for safety improvements. The costs would be the construction, operating, and maintenance costs of making the improvements at each of the intersections. The benefits for this example would be expected reduction in accidents at each of the intersections. Although it may be possible to assign a monetary value to each accident, cost/effectiveness analysis permits evaluation of the problem without having to make this subjective judgment.

Assume that four intersections are being evaluated. The annualized cost of the improvement and the expected annual reduction in accidents for each of the intersections is required for the analysis. The annualized costs are determined using the compound interest calculations. Based on these data, the cost per accident that is reduced is computed. The following represents this information for the four intersections:

Intersection	Annualized Cost of Improvement	Expected Accident Reduction	Cost per Accident Reduced
A	$35,000	10	$3,500
B	12,000	4	3,000
C	5,000	2	2,500
D	16,000	4	4,000

TABLE 6.1 Typical Goals-Achievement Matrix for River Town City Council

Alternative*		Objective			
		Water Supply	Hydroelectric Power	Flood Control	Recreation Facilities
A	Costs	$0.4 × 10^6 and 500 acres of natural woodland will be removed from public access	$0.6 × 10^6 and will remove 1000 acres of natural woodland from public access	$0.9 × 10^6 and will remove 2000 acres of natural woodland from public access	$0.8 × 10^6 and 5000 acres of a wildlife refuge area will be converted to recreational use
A	Benefits	$0.4 × 10^6 from sale of water and will support continued community growth for next 30 years	$0.7 × 10^6 in power revenue	$1 × 10^6 reduction in flood damages each year	Provides facilities to be used by 1.2 × 10^6 visitors per year
B	Costs	$0.2 × 10^6	$0.9 × 10^6	$0.6 × 10^6 and two streets along river will have to be closed	$0.9 × 10^6 and remove 10 miles of river from use by those who like to canoe
B	Benefits	$0.5 × 10^6 from sale of water and will support continued community growth for next 40 years	$0.7 × 10^6 in power revenue	$1.1 × 10^6 reduction in flood damages each year	Provides facilities to be used by 2.1 × 10^6 visitors per year in water-based recreation

*Dollar values are annual equivalent values. All values are the portion of total values for the alternative that is assigned to that objective.

If an engineer wishes to rank the alternatives based on the least cost per accident reduced, the ranking of the four intersections would be *C*, *B*, *A*, and *D*.

6.7.2 The Goals-Achievement Matrix

Table 6.1 is a typical goals-achievement matrix that might be prepared for the River Town City Council. Where possible, the benefits and costs associated with each objective of each alternative have been expressed in terms of dollars. It is almost impossible to express the value of some items, such as natural woodlands or a wildlife refuge, in dollar terms. However, the impact of the alternative on these items is shown in the matrix in terms of units that are more easily expressed and understood, and the city council must now directly consider this impact when selecting the alternative to be implemented. If there were some objective way to assign a relative weight to the cost and benefit associated with each item, they could all be combined into one criterion function that could be maximized. Since this cannot be done at present, the decision maker must subjectively make the decision based on subjective values for the contributions that each alternative makes toward fulfilling the stated objectives.

The goals-achievement matrix, as with most evaluation procedures, was designed to help compare and rank alternatives rather than to test their absolute value.

6.8 FINANCIAL ANALYSIS

A system may be economically feasible but not financially feasible. In addition to the requirement that the benefits to the owner must be greater than the cost, there must be a plan for paying for the system. Few large systems are ever paid for from cash on hand. Most large systems are financed by loans that have to be repaid.

6.8.1 Bonds

Loans are often obtained through the sale of bonds. A *bond* is an instrument setting forth the conditions under which money is loaned. It consists of a pledge by a borrower to repay a specified principal at a stated time and to pay a stated interest rate on the principal in the meantime.

The interest rate that a borrower must pay on a particular bond issue is determined by competitive bidding. Interested buyers, usually large banks or related financial institutions, bid the amount they are willing to pay to secure the fixed sum of money on a fixed schedule. The borrower accepts the highest bid. The interest rate is calculated as the rate of return obtained by the investors on the face value of the bonds because the high bid hardly ever equals the face value of the issue.

Assume that the River Town City Council has decided to construct the reservoir discussed at the beginning of Section 6.4 and that the construction will be financed by the sale of bonds. The annual operation, maintenance, and repair costs can be paid out of yearly income from the reservoir. Therefore, bonds will have to be sold only to finance the construction of the reservoir. Since the construction cost of the reservoir is estimated to cost 26.5×10^6, the council decides to sell 30×10^6 worth of bonds to make sure that they have enough money in case the cost estimate is low. They state that the 30×10^6 will be repaid at the end of 50 years and that 8 percent interest will be paid each year.

The highest bid they receive for the bonds is 28×10^6, which means they can have the use of 28×10^6 in exchange for a payment of $(0.08)(30 \times 10^6)$, or 2.4×10^6, at the end of each of the next 50 years and a principal payment of 30×10^6 at the end of the 50-year period. The rate of interest that the city must pay on the amount of money borrowed (28×10^6) can be determined on the basis of the present worth of the two time profiles of money; that is, the present value of 28×10^6 must be equal to the present value of 50 annual payments of 2.4×10^6 plus the present value of a 30×10^6 payment 50 years from now.

This condition may be written

$$\$28 \times 10^6 = \$2.4 \times 10^6(PWUS, i, 50) + \$30 \times 10^6(PWSP, i, 50)$$

which must be solved by trial and error. The present worth of the amount received for the bonds is 28×10^6. To determine the present worth of the cost to the city, try $i = 0.08$. Then

$$\text{Present worth of cost} = \$2.4 \times 10^6(12.2335) + \$30 \times 10^6(0.02132)$$

$$= \$30 \times 10^6$$

Try $i = 0.10$. Then

$$\text{Present worth of cost} = \$2.4 \times 10^6(9.9148) + \$30 \times 10^6(0.00852)$$

$$= 24.05$$

The value of i falls between $i = 0.08$ and $i = 0.10$. Thus, by interpolation for i,

$$i = 0.08 + 0.02 \times \frac{28 - 30}{24.05 - 30} = 0.08 + 0.02 \times \frac{-2}{-6.95}$$

$$= 0.0857$$

Therefore, the city is actually paying 8.57 percent interest to borrow money.

6.8.2 Repaying the Loan

The city must provide a means of repaying the loan. The money for repayment may come partly from revenues received as a result of the sale of hydroelectric power and water and partly from tax revenues. Because these revenues are collected annually, the city needs to determine what the annual revenues must be in order to make the yearly payment of 2.4×10^6 for interest and still have enough left at the end of the 50 years to pay the principal of 30×10^6. To accumulate sufficient funds to redeem the bonds, annual payments must be made to a fund that, when invested at compound interest, will produce the required amount of principal when it becomes due. The time profile of the annual payments, A, as shown in Figure 6.2, must be equivalent to a final payment, F, at the end of the period.

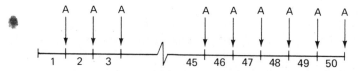

Figure 6.2 Time profile of payments to sinking fund.

The value of the last annual payment at the end of year 50 is A. The value of the payment made at the end of year 49 has increased in value to $A(1 + i)$ at the end of the 50-year period. Similarly, the payment made at the end of year 48 has increased to $A(1 + i)^2$. Since the final sum, F, is equal to the future value of all annual payments,

$$F = A + A(1 + i) + A(1 + i)^2 + \cdots + A(1 + i)^{n-1} \qquad (6.8)$$

where n is the number of time periods. This equation can be simplified to

$$F = A\left[\frac{(1 + i)^n - 1}{i}\right] \qquad (6.9)$$

where the term in brackets is called the uniform series compound amount factor. Equation 6.9 can be solved for A to yield

$$A = F\left[\frac{i}{(1 + i)^n - 1}\right] \qquad (6.10)$$

where the term in brackets is called the sinking fund payment factor.

Using Equation 6.10, the annual payment to a sinking fund is

$$A = \$30 \times 10^6(0.00174) = \$0.052 \times 10^6$$

if the sinking fund money can be invested at 8 percent interest. Therefore, the income for the city must include at least 0.052×10^6 per year for the sinking fund plus 0.5×10^6 for OMR costs, plus 2.4×10^6 for the annual interest payment. If the annual income from the reservoir is not at least 2.952×10^6, the city must provide money from some other source, such as taxes.

6.9 SUMMARY

The manner in which a system will be evaluated should be considered at the time of the formulation of the problem. Evaluation measures should reflect the goals, objectives, and criteria that are established.

All proposed engineered systems usually have to pass engineering, economic, financial, political, and social feasibility tests. These tests are applied to ensure that the system will be able to perform its intended function, that the benefits to be gained from the system are greater than the cost of implementing and operating the system, that the owner will be able to pay for the system, and that the individuals who will be affected by the system will respond favorably.

The purpose and scope of the evaluation effort are major factors in determining the planning horizon to be used in a study. There is considerable uncertainty associated with many of the benefits and costs of any system under study. The longer the period of analysis used, the more uncertain many of the costs and benefits become. Sometimes techniques, such as simulation (Chapter 12), are used to gain an understanding of the possible time profile of cash flows associated with the system.

The value of money varies with time because of its ability to grow if properly invested or to earn interest if placed in a savings account. Therefore, discounting factors must be used to convert costs and benefits to the same time period so that they can be compared. Along with the comparison of benefits and costs comes the desire to rank alternatives according to their desirability.

If all benefits and costs associated with the system can be expressed in terms of dollars, criteria such as the net present worth, annual equivalent value, benefit/cost ratio, and internal rate of return can be used to rank the desirability of alternatives. The particular criterion to be used will depend on the objective of the owner. If the costs can be expressed in terms of dollars, but the benefits cannot, it may be possible to rank alternatives according to the cost required to provide the same level of benefits. This criterion is usually satisfactory only for single-purpose systems.

The economic ranking criteria were originally developed for single-purpose systems that were independent of one another, and their use further assumes an unlimited capital available for investment. The ranking of interrelated systems under conditions of limited capital becomes a much more complex problem, as shown by Masse (1962).

When the benefits and costs cannot all be expressed in terms of dollars or any other common units, the final selection must be made on the basis of the decision maker's subjective judgment. This is usually the type of problem encountered in reality. Hence we see that evaluation techniques are helpful in ranking alternatives, but the final selection depends on the values and objectives of the decision maker, who is often the owner (individual, company, or government agency). A financial analysis determines if money can be made available for system implementation after the owner has determined which alternative is preferred.

If the proposed system passes the feasibility tests, the engineer must begin to plan the sequence and duration of activities necessary to construct the system. In addition, the engineer must develop support organizations to perform the construction. These are discussed in Chapter 13.

PROBLEMS

P6.1. (a) Equivalence calculations

1. If a company puts $5000 in a savings account that pays 6 percent compounded annually, how much will it have at the end of 10 years?

2. A company sets aside $5000 a year of its profits for future expansion of the company. The money is put into a savings account that pays interest at 6 percent compounded annually. How much money will be available to the company for expansion 8 years after it starts this practice?

3. A concrete company will have to replace its mixing facilities at the end of the next 7 years. It is estimated that a new mixing facility will cost $40,000. How much must be put into a savings account each year in order to have enough money to pay cash for the facility if the money will earn 8 percent interest compounded annually?

4. If the company does not pay cash, it must borrow the $40,000 at 10 percent compounded annually. What would its annual payment be in order to repay the loan in seven equal annual payments?

(b) Economic analysis calculations

The ABC Corporation purchased an electronic computer 3 years ago for $30,000, and the payment for the purchase was to be paid in five equal annual installments of $7000 each. At this time, two installments remain to be paid. The corporation is now considering expanding the computing capability by either buying a new computer or adding components to the existing one.

Additional components can be purchased for the present computer at a cost of $10,000. It is estimated that the improved computer will have a life of 5 years with an annual OMR cost of $2500. It is not expected to have any salvage value at the end of the period.

One option is to buy a new YORK 310 computer at a cost of $35,000. This unit will have an expected life of 15 years, an annual OMR cost of $1500, and a salvage value of $5000.

Another option is to purchase a new XYZ 205 computer for $27,000. For this computer, the expected life is 10 years with an OMR cost of $2000 and a salvage value of $3000.

The existing computer has a trade-in value of $5000 if either of the new computers is purchased. All three computer systems have the same computational capability.

Assuming an interest rate of 10 percent, which alternative should the corporation choose?

P6.2. A company has $1,200,000 to invest. One alternative is to invest the money in bonds that will pay 8 percent compounded annually. Another alternative is to invest the money in a power generating station. The generating equipment will last 50 years. The net income (income in excess of operating expenses) is estimated to be $10,000 the first year, $20,000 the second, $45,000 the third, $65,000 the fourth, $95,000 the fifth, $110,000 the sixth, $115,000 the seventh, $120,000 for the eighth, $125,000 for the ninth, and $130,000 for each remaining year of life of the station. Should the company invest in the power station? What would be the internal rate of return on this project?

P6.3. A section of roadway pavement cost $5000 a year to maintain. What immediate expenditure for a new replacement for the existing pavement is justified if no maintenance will be required for the first 5 years, $1100 per year for the next 10 years, and $5000 a year thereafter? Assume money to cost 6 percent.

P6.4. You are considering the improvement of some manufacturing facilities by purchasing one of three possible different machines, each with the same production capacity.

Machine A costs $30,000, has a life of 40 years, annual maintenance costs of $1500, and salvage value of $5000. Machine B costs $20,000, has a life of 20 years, annual maintenance of $2000, and salvage value of $3000. Machine C costs $10,000, has a life of 10 years, annual maintenance of $4000, and no salvage value.

Use an annual discount rate of 6 percent. Assume that initial costs, annual maintenance, and discount rates are constant throughout any period of time you desire. Determine the most economical choice.

P6.5. A firm is considering building a high-rise building that it will finance through the sale of bonds. It issues 30-year bonds with a face value of $4,000,000 bearing interest of 5 percent payable annually. The firm finds, however, that it can sell these bonds for only $3,900,000. What is the yearly rate of interest that the firm must actually pay for the funds received?

P6.6. A dam is to be constructed to provide a flood storage reservoir for a certain flood control project. Following are benefit/cost data for reservoirs of several sizes at the proposed dam site.

Reservoir Volume (acre-feet)	Initial Construction Cost ($\times 10^6$)	Average Annual OMR Cost ($\times 10^6$)	Average Annual Benefits ($\times 10^6$)
50,000	$ 4.5	$0.032	$0.317
100,000	5.0	0.079	0.761
150,000	8.0	0.127	1.078
200,000	14.0	0.143	1.269
250,000	22.0	0.190	1.332

Determine the project size (i.e., reservoir size) that yields:
(a) The maximum value of net benefits
(b) The maximum benefit/cost (B/C) ratio
Assume that the project life is 50 years and that the appropriate discount rate is 6 percent.

P6.7. A tollroad authority has decided to build a new toll bridge across a river at a cost of $15 million. The bridge is to be paid for by selling bonds that are to be retired at the end of 40 years. The bonds are to pay 5 percent interest, payable annually. A sinking fund is to be established to pay interest on the bonds and to retire them at the end of 40 years. The sinking fund is to be invested at 6 percent interest, compounded annually, and is to be established in 40 equal installments, one year apart, the first installment one year after the bonds are issued. Assume that the bonds are sold at face value. How much should the toll per vehicle be if it is estimated that 5 million vehicles per year will use the bridge? The annual operating and maintenance costs for the bridge are $500,000.

Now assume that the bridge has been in operation for 5 years. The actual traffic volume that uses the bridge is only 85 percent of the projected volume; hence the authority has not been able to make its total payment each year into the sinking fund. The authority must increase its revenue or extend the length of the bond issue. For each 10 percent increase in toll charge, the volume of traffic will decrease by 1 percent. If the bonds have to be reissued, the interest rate will be 6 percent instead of the current 5 percent rate.

(a) What toll would have to be charged in order to make up the back payments and pay off the bonds on time?

(b) What would be the value and length of a new bond issue that would replace the old bonds without a change in toll charges?

(c) Discuss the relative merits of the two approaches and a combination of the two.

P6.8. What factors would you consider in evaluating the feasibility of locating and constructing a new highway through your city? How would you develop a goals-achievement matrix to evaluate the effects of the new highway? (*Hint:* See Hill, 1968.)

Linear
Graph Analysis

7

7.1 THE GRAPHICAL MODELING AND ANALYSIS PROCESS

The use of simple linear graph concepts for the modeling of network and systems problems has been discussed in Chapter 4. A common feature of the modeling of these decision and organizational problems is that they are viewed in such a way that inherently logical or procedural aspects associated with each problem are used as a means of identifying components and structuring them together into a network system. That is, linear graph modeling concepts are used to capture, and portray, essential aspects of the problem under consideration. Unnecessary problem detail is eliminated and attention is focused on the graphical framework and logical structure of the problem. Provided that the relevant graphical properties can be formulated and used, linear graph models can be used in the modeling and analysis of network and systems problems.

The process of applying linear graphs to systems problems might be described as follows:

1. As with other systems problems, the first step is to identify the problem and the system components, structures, and attributes that are involved.

2. The problem must now be represented as a model according to some representational form and convention. In linear graph modeling, the model is limited to the use of nodes and branches as the representational forms. In addition, it is usually necessary to assign attributes to the branches and nodes.

3. The problem must be posed in terms of a graphical structural property that exists in the model.
4. A linear graph analysis procedure must be established that will enable the graphical property to be processed through the linear graph model.
5. The graphical solution must be transferred back into the system structure of the actual problem.

To apply this linear graph process successfully to the analysis of a particular problem, the relevant graphical properties must be defined and incorporated into the solution process. Graphical properties relate to the many different structural relationships that can be imagined as existing between the various nodes and branches of a linear graph. Thus the simple sequencing of network branches or a path structure may be relevant to a straightforward network problem. In other cases more complex graphical properties may need to be developed, such as cutting sections that divide the original graph into portions, or transformation rules or algorithms that convert an existing graph model in one more meaningful or amenable to a solution process. In this respect a large class of engineering networks can be viewed as capacitated flow problems in which one or more commodities (e.g., traffic, water, information, cash, etc.) can be considered as flowing through the network whose branches have various constraints and flow capacities. The development of these types of graphical properties permits graph theory to be applied to many engineering problems.

The modeling of a system as a linear graph was introduced in Chapter 4. Several of the examples that are contained in that chapter are analyzed in the sections that follow. In addition, the analysis of linear graphs is applied to several problem areas whereby the results enable an engineer to select a preferred course of action.

Finally, the development and use of linear graph topological matrices is presented as an indication of the manner in which graphical properties can be numerically expressed and made compatible with computer processing methods.

7.2 LINEAR GRAPH ANALYSIS: A PATH PROBLEM

Consider again the linear graph model of the construction plan for a house shown previously in Figure 4.3, reproduced here as Figure 7.1. It is a directed graph since technological and organizational conditions establish the various conditions before each specific work task can start. All the work tasks must be completed in specific directed sequences in order to complete the construction project. This construction requirement implies that in the graph it is necessary to *traverse all* the seven paths joining the start node A and the end node G. Consequently, if the problem is to determine the minimum project duration, the relevant graphical structural property forming the basis of the analysis is the path or the sequence of activities.

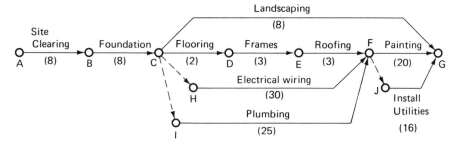

Figure 7.1   Linear graph model of a construction plan.

The total time to complete each path is as follows:

Path	Time Required for Completion (hours)
A-B-C-G	24
A-B-C-D-E-F-G	44
A-B-C-D-E-F-J-G	40
A-B-C-H-F-G	66
A-B-C-H-F-J-G	62
A-B-C-I-F-G	61
A-B-C-I-F-J-G	57

The longest path takes 66 hours. This is the minimum time required to complete the project. If there is a delay in completing any of the work tasks along this critical path, the entire project will be delayed. On the other hand, if the duration of the project is to be shortened, one or more of the work tasks along this critical path must be shortened by the adoption of a different construction method or the application of more resources, such as a larger crew. Therefore, this critical path controls the duration of the project.

A similar method of analysis can be used for the contractor problem (see Example 4.3). As indicated in Figure 7.2, the contractor has the choice of three possible decision paths:

1. Move equipment
2. Do not move equipment, but build a protective platform
3. Do not move equipment and build no protective platform

The consequence of the first path is a definite loss of $1800. However, the consequences of the other two paths depend on chance. Without definite knowledge of future events, the contractor must estimate what he can expect to gain or lose in taking path 2 or 3. These estimates provide a single measure of value for the paths with uncertain outcomes. Taking into consideration his

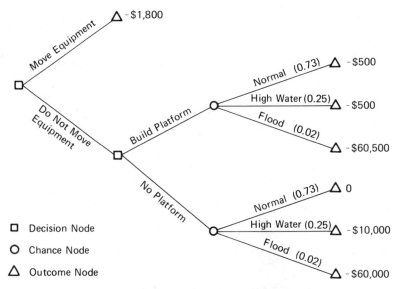

Figure 7.2 Linear graph model of a contractor's decision problem.

available capital and willingness to assume risk, he must decide on which path he will take. The methods of decision analysis are discussed in Chapter 11.

7.3 LINEAR GRAPH ANALYSIS:
A NETWORK FLOW PROBLEM

A large class of engineering networks can be viewed as capacitated flow problems in which one or more commodities (e.g., traffic, water, information, cash, etc.) can be considered as flowing through the network whose branches have various constraints and flow capacities.

Consider, for example, a two-lane, two-way highway network connecting a metropolitan area with a residential suburban town, as shown in Figure 7.3a. Along each road branch is a vector containing the relevant attributes of the road it represents: road identification number, mileage, and flow capacity in one direction in vehicles per hour (vph). Thus the road between nodes 1 and 3 is identified as road 1, which is 15 miles long and has a maximum capacity of 700 vph in the direction from node 1 to node 3. The two-way capacity would be twice this value.

The engineer is interested in determining the maximum flow, or capacity, of the network during the morning and evening peak traffic hours. Assuming that the morning and evening flows follow identical characteristics, but in opposite directions, the study can be confined to either the morning or evening

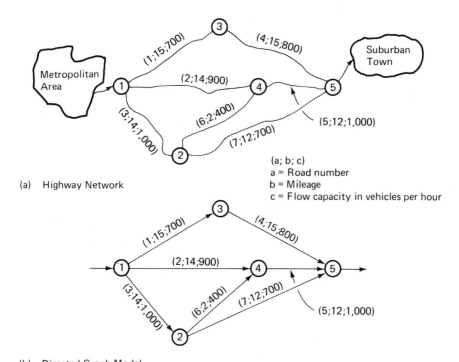

(a)   Highway Network

(a; b; c)
a = Road number
b = Mileage
c = Flow capacity in vehicles per hour

(b)   Directed Graph Model

Figure 7.3   Linear graph model of a highway network.

conditions; for example, only the flow from node 5 to node 1 or the flow from node 1 to node 5 needs to be studied, not both.

Consider the evening peak traffic. Each path joining node 1 to node 5 represents one possible route of travel. The first step is to assign directions to the flow in as many branches as possible. The direction of flow in roads 1, 2, 3, 4, 5, and 7 must be from node 1 to node 5 and can, therefore, be assigned. However, the flow in road 6 can be from node 2 to node 4 or from node 4 to node 2, and there may be no a priori way of knowing the correct direction to assign to the flow in this branch. Therefore, the problem must be analyzed for both possible conditions. For the first case, assume that the flow is from node 2 to node 4; the direction of flow in each branch is shown in the directed graph of Figure 7.3b.

If the engineer is interested in determining the flow of vehicles between two intersections that are connected by a highway segment, he or she can place a pneumatic tube across the roadway and connect it to a counting device. The road tube may be placed anywhere along the highway segment, the only require-ment being that the path between the two intersections is traversed. Figure 7.4a depicts this situation in terms of a linear graph. The intersections are repre-

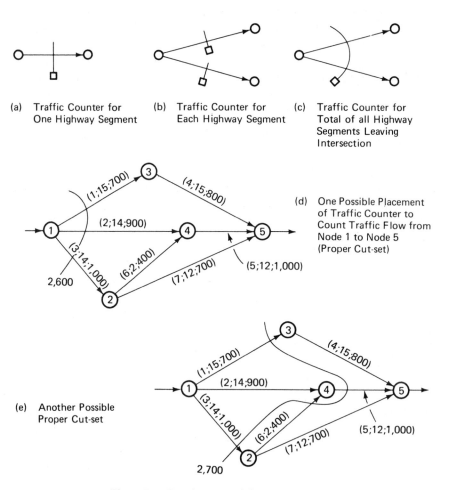

(a)   Traffic Counter for   (b)   Traffic Counter for   (c)   Traffic Counter for
      One Highway Segment   Each Highway Segment   Total of all Highway
                                                                            Segments Leaving
                                                                            Intersection

(d)   One Possible Placement
      of Traffic Counter to
      Count Traffic Flow from
      Node 1 to Node 5
      (Proper Cut-set)

(e)   Another Possible
      Proper Cut-set

Figure 7.4   Development of the cut-set concept.

sented as nodes, the roadway as a branch, and the counting tube as cutting the
branch between the nodes. If roadway traffic is flowing at its maximum rate,
the value obtained from the count represents the capacity of flow between the
two nodes.

Furthermore, if two roads are involved, two counters could be utilized
that would determine the flow on each. Thus the total flow on both roadways
could be ascertained by adding the values obtained from each counter. This
situation is shown in Figure 7.4b, in which each of the two branches are cut by
the counting device. Again, the important consideration is that counting devices
cut across the paths of travel or the branches.

The engineer is interested in obtaining the value of total maximum flow;
therefore, instead of two counting devices and two tubes, the engineer really

wants to determine the flow that could be measured by one counter and a tube that crosses both roadways. Figure 7.4c represents this concept and shows that the tube across both branches would yield the total flow on those branches.

In the case of the network shown in Figure 7.3, determining maximum flow from node 1 to node 5 would require that a tube be placed across the roadways such that node 1 is on one side of the tube and node 5 is on the other. Figure 7.4d shows one possible placement of the tube. Since roads 1, 2, and 3 all carry traffic flow from node 1 (source) side of the tube to node 5 (sink) side, the total flow capacity from the source side to the sink side is thus the sum of the flow capacity of these three roads; that is, $700 + 900 + 1000 = 2600$ vph. In the terminology of linear graph, branches 1, 2, and 3 constitute a proper cut-set. A *proper cut-set* is defined as a minimal set of branches in a connected graph that, if removed, would break the graph into only two subgraphs. Thus the proper cut-set in Figure 7.4d breaks the graph into two subgraphs: one containing only node 1; and one containing nodes 2, 3, 4, and 5 and branches 4, 5, 6, and 7. Branches 1, 2, 3, and 6 constitute a cut-set; that is, their removal separates the graph into two subgraphs. However, they do not constitute a minimal set because only branches 1, 2, and 3 need be removed to form two subgraphs.

The proper cut-set above simply indicates that the maximum network flow capacity cannot be greater than 2600 vph. At this point, the engineer is not certain that the proper cut-set has been found that indicates the maximum network capacity flow. For example, if branches 4, 5, and 7, are cut the total potential flow would only be 2500 vph. This proper cut-set further indicates that the network capacity cannot be greater than 2500 vph.

In computing the values of the proper cut-sets, consideration must be given to the direction of the branches included in the cut-set. For example, the cut-set in Figure 7.4e has branches 1, 3, and 5 going from the source side to the sink side, and branch 6 going from the sink side to the source side. The maximum capacity flow from the source side of the cut-set to the sink side should be $700 + 1000 + 0 + 1000 = 2700$ vph. That is, the branch that crosses the cut-set from the sink side to the source side has zero value. In the physical sense, if the pneumatic tube were placed in that position to measure the flow capacity from the source to sink, the tube should not count the cars traveling from node 2 to node 4.

For total network analysis, all the possible proper cut-sets that separate the source from the sink must be considered. Figure 7.5 depicts all the possible proper cut-sets for this particular road network. As can be seen, there are eight possible proper cut-sets that separate the source node 1 from the sink node 5. The components and capacities of the cut-sets are also tabulated in Table 7.1. An entry of $+1$ denotes a positive flow (from source side to sink side); an entry of -1 denotes a negative flow (from sink side to source side); and a blank entry (or zero) indicates that a branch is not included in the cut-set. The maximum network flow capacity is thus determined by the proper cut-set that has the

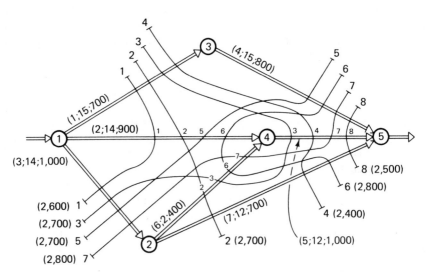

Figure 7.5   Road network proper cut-sets.

TABLE 7.1   Road Network Proper Cut-Set Capacities

Highway		Cut-Set Number							
Number	Capacity	1	2	3	4	5	6	7	8
1	700	1	1	1	1				
2	900	1	1			1	1		
3	1000	1		1		1		1	
4	800					1	1	1	1
5	1000			1	1			1	1
6	400		1	−1			1	−1	
7	700		1		1		1		1
Cut-set capacity		2600	2700	2700	2400	2700	2800	2800	2500

smallest total capacity; in this case, it is cut-set 4, which has a capacity of 2400 vph. Therefore, the maximum network flow capacity between nodes 1 and 5 is 2400 vph.

7.4 MAXIMUM FLOW–MINIMUM CUT THEOREM

The analysis approach of the preceding section illustrates the maximum flow–minimum cut theorem, which states: For any source–sink connected network *the maximum flow capacity from source to sink is equal to the minimum of the proper cut-set capacities for all the proper cut-sets separating the source from the sink.*

In the preceding example a solution was obtained by finding and evaluating all the different proper cut-sets of the graph. In large networks, this approach is both laborious and uncertain because of the difficulty of ensuring that all proper cut-sets have been found. The maximum flow–minimum cut theorem can be used to provide upper and lower limits to the network flow capacity, as described in the following paragraphs.

Let C be the network flow capacity. Then the initial cut-set 1, which has a value of 2600 vph, immediately establishes the relationship

$$C \leq 2600 \qquad (7.1)$$

Again, with cut-set 8 evaluated at 2500 vph, there results

$$\begin{array}{l} C \leq 2600 \\ C \leq 2500 \end{array} \quad \text{(i.e., } C \leq 2500) \qquad (7.2)$$

As each cut-set is found and evaluated, it plays a part in determining whether the current tentative value of the upper limit for the network capacity must be lowered.

Suppose now that instead of seeking to evaluate all the remaining proper cut-sets, an attempt is made to establish actual path flows through the network. Introducing 700 vph along the path produced by highways 1 and 4 establishes a lower bound for the network capacity as

$$700 \leq C \qquad (7.3)$$

The addition of 900 vph along the path produced by roads 2 and 5 and of 700 vph along the path produced by roads 3 and 7 establishes for all three path flows

$$2300 \leq C \qquad (7.4)$$

Hence the network capacity can be bounded from both sides:

$$2300 \leq C \leq 2500 \qquad (7.5)$$

Therefore, if an actual flow can be found equal to the minimum value of the several cut-sets found at this stage, it is not necessary to locate and evaluate the remaining proper cuts because they cannot have capacities less than the flow already established. In determining the lower limit on the flow capacity a branch can be in only one path.

In order to determine if 2400 vph is the maximum flow in the network, the flow from node 4 to node 2 in road 6 must be analyzed. Figure 7.6a shows the directed graph for this condition. The cut-set analysis for this case is shown in Figure 7.6b. In this case the maximum flow capacity is only 2300 vph. Thus *the maximum flow will occur in the network with the flow from node 2 to node 4 on road 6.*

For flow networks that have many branches for which the flow directions cannot be assigned a priori, the graphical cut-set analysis becomes very long and tedious.

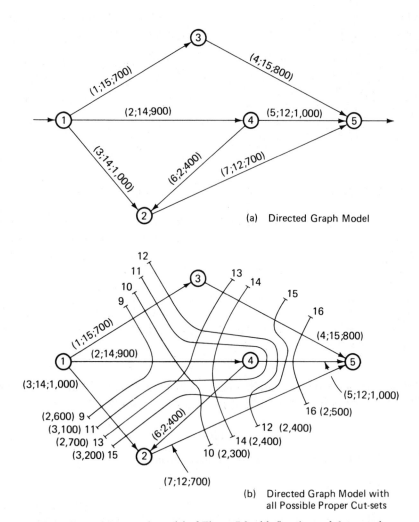

Figure 7.6 Linear graph model of Figure 7.3 with flow in road 6 reversed.

Once the road network flow capacity has been established and the constraining cut-set identified, an analysis can be made for flow capacity improvement. Because the network flow capacity is limited by the capacity of the roads forming cut-set 4, road improvement elsewhere is not effective at this stage. Therefore, at least one of the cut-set 4 roads must be increased in capacity for any overall network improvement.

Road 1 can be improved alone, but only optimally for the additional capacity of 100 vph, because the path flow through it then matches the road 4 capacity of 800 vph.

Either road 5 or 7 can be improved by an amount of 200 vph, at which state the new cut-set 4 capacity equals the cut-set 1 capacity of 2600 vph.

Which of the two roads should be improved will depend on other considerations, such as length of road and costs. A typical problem will be considered later (see Sections 7.6 and 10.1).

The use of the maximum flow–minimum cut theorem for this problem is interesting because it illustrates a problem solution based on the interaction of two different linear graph analysis methods. Although both methods used the same graph, method one focuses on cut-sets and the other on paths. This illustrates that when the problem is clearly understood, relevant graphical properties can be selected or formulated for the specific problem.

7.5 LINEAR GRAPH TRANSFORMATIONS: PRIMAL–DUAL GRAPHS

In some cases, a problem formulation requires an analysis based on a graph property that is inconvenient for an analytical, logical, or computational reason. Unless another formulation is possible, it may be necessary to transform synthetically the original graph and graph property into an equivalent problem using a different graph model and graph property. In essence, the information contained in the linear graph is transformed into a more convenient and usable analysis format.

As an example, consider again the network flow problem and cut-set analysis approach. Suppose that the cut-sets of Figure 7.5 are redrawn as shown in Figure 7.7a in order to develop a concept for readily identifying and evaluating the cut-sets. In every case, the cut-sets mingle at focal points inside the graph. This suggests that the various cut-sets are obtained by linking these

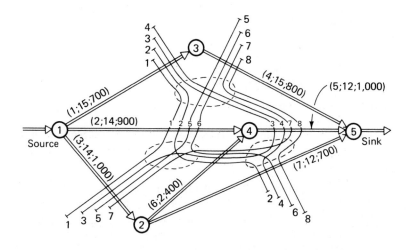

(a)   Road Network Cut-set Bundles

Figure 7.7   Transformation of primal graph to dual graph.

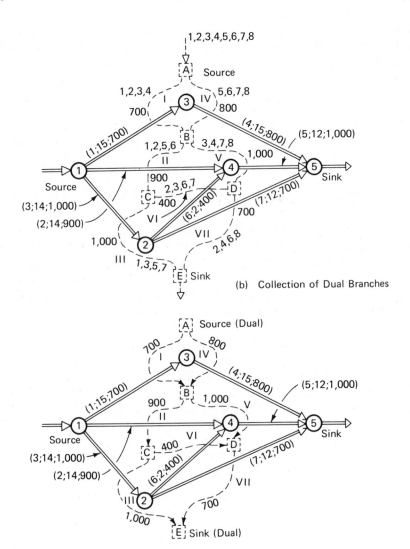

(b)   Collection of Dual Branches

(c)   Road Network: Primal and Dual Graphs

Figure 7.7   Continued.

internal focal points in different ways and raises immediately a chain concept for the cut-set representation. Figure 7.7b shows this concept developed to the stage where the focal points of Figure 7.7a are replaced by square nodes and the bundles of cut-sets are replaced by dashed branches. All the cut-sets in the original or *primal* graph can be identified as different chains through the new dashed *dual* graph. Furthermore, each branch in the primal graph is intersected by a dashed branch in the dual. The graphical property of the primal is, in fact, the basis of the dual; and the cut-sets in the primal are transformed into chains in the dual. The transformation, however, is not complete unless the directions in the primal are translated to the dual. A simple sign convention can be easily derived from the cut-set principle. The following procedure can be followed:

1. Because there are two nodes in the dual graph that are outside the confines of the primal, designate one arbitrarily as the source node and the other as the sink node of the dual. It is generally more convenient to designate the top node as the source node in the dual graph.

2. Because any chain in the dual is a cut-set in the primal, if a branch in the primal crosses this cut-set from the *source* side to the *sink* side, then the corresponding branch in the dual should point along the direction of travel in the dual chain from the source in the dual to the sink in the dual.

3. If a branch in the primal crosses the cut-set from the *sink* side to the *source* side, the corresponding branch in the dual should point against the direction of travel in the dual chain, from the dual source to the dual sink.

For example, for the dual graph in Figure 7.7b, let node A be the dual source and node E be the sink. Since branch 1 crosses the chain (I-II-III) from the source side of the primal to the sink side of the primal, the corresponding branch I in the dual should be directed along the direction of travel from source A to E, that is, from A to B in the chain (I-II-III). On the other hand, since branch 6 crosses the chain (IV-V-VI-III) from the sink side of the primal to the source side, the corresponding branch VI in the dual should be pointed against the direction of travel from A to E, that is, from C to D in the chain (IV-V-VI-III). Using this sign convention, a unique direction is assigned to each branch in the dual, as shown in Figure 7.7c.

7.5.1 Network Analysis with Dual Graphs

The method for constructing the dual described above implies that the analysis of cut-sets in the primal can be replaced by a chain analysis in the dual. It is now necessary to identify the primal branch attribute of flow capacity with a corresponding dual branch attribute. The primal cut-set evaluation of capacity required the summation of individual branch capacities for each branch member of the cut-set. The corresponding evaluation in the dual requires the summation of the dual branch attributes along the relevant chain through the dual graph. In summing along a chain, it is convenient to think of the attributes as length. Hence, branch I of the dual has a length of 700, which corresponds to a capacity of 700 for branch 1 of the primal. Thus, for cut-set 3, the following one-to-one correspondence exists between the two methods of analysis:

Primal (Cut-Set Analysis)		Dual (Chain Analysis)	
Branch	Capacity	Branch	Length
1	700	I	700
5	1000	V	1000
6	0	VI	0
3	1000	III	1000
Total	2700	Total	2700

The primal problem of determining the minimum cut-set capacity now becomes the dual problem of determining the minimum-length chain. The result of a complete chain analysis and its correspondence to the cut-set method is given in Table 7.2. The advantage of the chain method of analysis is that once the dual graph is constructed, all the possible chains going from the source to the sink can be easily and systematically enumerated.

TABLE 7.2   Chain Analysis*

Chain	Total Length	Corresponding Primal Cut-Set
I-II-V	$700 + 900 + 1000 = 2600$	1
I-II-VI-VII	$700 + 900 + 400 + 700 = 2700$	2
I-V-VI-III	$700 + 1000 + 0 + 1000 = 2700$	3
I-V-VII	$700 + 1000 + 700 = 2400$	4
IV-II-III	$800 + 900 + 1000 = 2700$	5
IV-II-VI-VII	$800 + 900 + 400 + 700 = 2800$	6
IV-V-VI-III	$800 + 1000 + 0 + 1000 = 2800$	7
IV-V-VII	$800 + 1000 + 700 = 2500$	8

*Maximum flow capacity = minimum chain length = 2400.

Instead of enumerating all the possible chains, the minimum path can be found by selective addition of the dual branches. Thus, commencing at A, node B can be reached via branch I of length 700 or by branch IV of length 800; therefore, select branch I. Node C can be reached from node B via branch II or via branches V and VI for total lengths of 1600 or 1700; hence, select branch II. The remaining analysis portrayed in Figure 7.8a is left to the reader.

The dual graph can also be used when the direction cannot be assigned a priori for the flow in some of the branches. Recall the original problem from Figure 7.3a. The direction of flow is known for all branches except branch 6. Therefore, draw the dual graph as shown in Figure 7.8b without an assigned direction to branch VI. The analysis proceeds the same as for Figure 7.8a until the computations are to be made for nodes C and D. Since the flow in branch VI may be either from C to D or from D to C, node C can be reached from

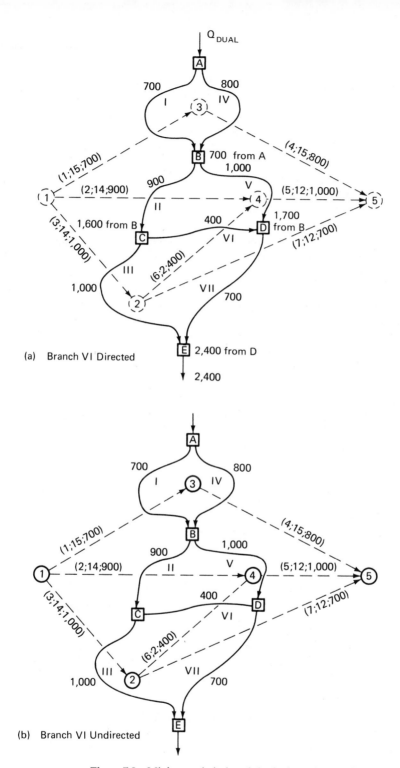

Figure 7.8 Minimum chain length in dual graph.

node *B* via branch II for a length of 1600 or via branches V and VI for a length of 2100; hence select branch II, which gives the minimum. Similarly, node *D* can be reached from node *B* via branch V for a length of 1700 or via branches II and VI for a length of 2000; hence select branch V and the analysis proceeds as for Figure 7.8a; the minimum chain length is branches I, V, and VII, as before.

However, if road 6 must cross a river by means of a ferry that has a capacity of 50 vph, the length of branch VI would be 50 and the analysis would yield the following results. Since the flow in branch VI can be either *C* to *D* or *D* to *C*, node *C* can be reached from *B* via branch II for a length of 1600 or via branches V and VI for a length of 1750; hence select branch II. Node *D* can be reached from *B* via branch V for a length of 1700 or via branches II and VI for a length of 1650; hence select branches II and VI. Proceeding to node *E*, the minimum chain has a length of 2350 and includes branches I, II, VI and VII. This indicates that the direction of branch VI must be from *C* to *D* and hence the direction of flow in road 6 must be from node 2 to 4 in order to obtain the maximum flow capacity of the network. Thus the dual graph analysis not only provides the maximum flow capacity of the network, but also provides a means of determining the direction of flow for branches on the critical chain for which flow directions cannot be assigned a priori. It is important to note that the transformation permits the use of procedures or methods that can result in a better understanding of the problem and a simplified solution process.

7.5.2 Constructing the Dual Graph

Constructing the dual graph can be formalized into the following steps, as illustrated in Figure 7.9.

1. Let the source and sink nodes of the primal subdivide the outside region into an upper and a lower region; allocate the dual source to the upper region and the dual sink to the lower region.
2. Allocate a dual node inside each bounded region in the primal graph.
3. Cut each primal branch with a short line segment.
4. Connect each of the line sections to the dual node inside the same region.
5. Assign directions to the dual branches according to the sign convention established in the preceding section.
6. Transfer branch quantity from the primal to the dual.

7.5.3 Planar and Nonplanar Graphs

The method of constructing the dual described above is applicable only to planar graphs. A *planar* graph is a graph in which the branches *can be* drawn such that no two branches cross each other. Otherwise, the graph is called *nonplanar*. A graph has a dual if and only if it is planar.

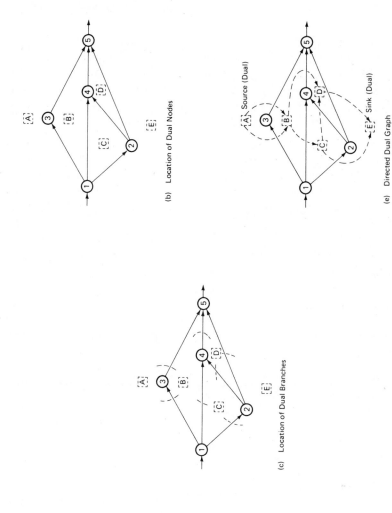

Figure 7.9 Constructing the dual graph.

(a) Location of Dual Source and Sink

(b) Location of Dual Nodes

(c) Location of Dual Branches

(d) Undirected Dual Graph

(e) Directed Dual Graph

For example, Figure 7.10a is a planar graph because it can be redrawn as in Figure 7.10b. According to the definition of a linear graph, these two graphs are identical because they express the same connectivity among the nodes and the branches. In constructing the dual of a planar graph, such as Figure 7.10a, it is always advisable first to redraw the branches so that no two cross each other. Both Figures 7.10c and 7.10d are nonplanar graphs.

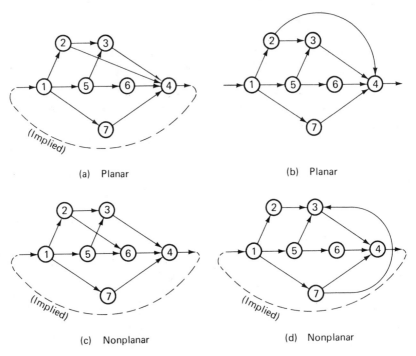

(a) Planar   (b) Planar

(c) Nonplanar   (d) Nonplanar

Figure 7.10   Planar and nonplanar graphs.

7.5.4 Multiple Source–Multiple Sink Networks

A network that has more than one source or more than one sink node can be easily converted into a one source–one sink network. For example, Figure 7.11a shows a linear graph representation of a street network connecting two separate business districts and two residential areas. The branch arrows indicate the direction of traffic flow during evening rush hours, and the number associated with each branch denotes the flow capacity in vehicles per hour (vph). Thus this graph has two source nodes and two sink nodes. To convert this network to a one source–one sink network, simply connect the existing source nodes to an artificial source node and the existing sink nodes to an artifical sink node, as shown in Figure 7.11b. Because the artifical source and sink nodes are created primarily for purposes of convenience in analysis, the branches directly connecting these nodes must be assumed to have an infinite

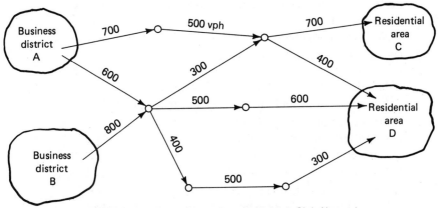

(a) Original Graph: A Multiple Source–Multiple Sink Network

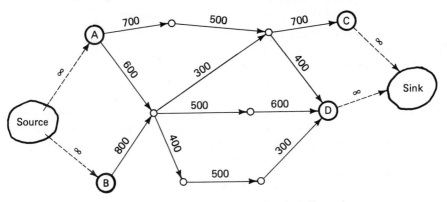

(b) Modified Graph: A One Source–One Sink Network

Figure 7.11   Multiple source–multiple sink network.

flow capacity. With this modification in the network it can be noticed that the minimum cut-set will not include any of these branches.

7.6 LINEAR GRAPH ANALYSIS: A NETWORK IMPROVEMENT PROBLEM

In order to illustrate the application of linear graph methods in network analysis, consider the problem faced by a state highway department as it prepares its request for funds for capital improvements for the network of roads between two cities.

7.6.1 General Statement of the Problem

The following are some of the more obvious issues that are posed by the problem of improving the road network system.

1. The network characteristics of the road network system
   a. As currently existing
   b. With recognized new routes
2. The individual road characteristics
   a. Length
   b. Number of lanes, and hence, traffic capacity
   c. Pavement status and, hence, level of service provided
3. Improvement objective defined as
   a. Improved network capacity
   b. Improved level of service
   c. Constant maintenance of current level of service
4. Financial constraints defined as
   a. Federal user funds for new road improvement construction
   b. State taxation funds for maintenance and road improvement
   c. Local taxation funds for road system maintenance
5. Road users' requirements
   a. Simple demand traffic flows (special purpose)
   b. Complex network traffic flows
   c. Social and political objectives

In order to proceed with the linear graph analysis, consider the following specific case. The problem concerns a transportation system between cities A and B passing through other intermediate towns, as shown in Figure 7.12. Towns and cities have been modeled as nodes and roads as branches. A feasible new road connecting towns 2 and 6 is shown as a dashed branch. Along each

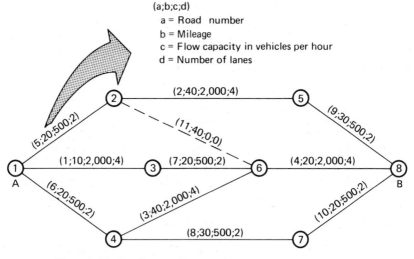

Figure 7.12   Road network system between towns A and B.

branch of the linear graph is a vector containing the relevant attributes of the road it represents: road identification number, mileage, flow capacity, and number of lanes. The existing roads are either two or four lanes. A two-lane road has a capacity flow of 500 vph in each direction and a four-lane road has a capacity flow of 2000 vph in each direction, as illustrated in Figure 7.13.

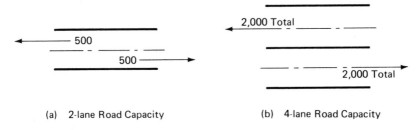

(a)   2-lane Road Capacity   (b)   4-lane Road Capacity

Figure 7.13   Road capacities.

Further investigation reveals that there is a demand for a flow of about 6000 vph from city A to city B and that capacity improvements can be made by adding two lanes to existing two-lane roads to increase their capacity from 500 to 2000 vph or by constructing a new road connecting towns 2 and 6. Road construction costs have been determined as follows:

1. The addition of two lanes to an existing two-lane road costs R per mile
2. The construction of a new two-lane road costs $1.5R$ per mile
3. The construction of a new four-lane road costs $2.3R$ per mile

Assume that the intermediate towns between A and B do not generate sufficient traffic to affect traffic flow considerations between A and B.

7.6.2 Objectives and Investment Criteria

The highway department perceives its objective to be that of increasing the flow capacity of the road network between cities A and B such that it can accommodate 6000 vph from A to B. The department has also established the following critieria for choosing among alternative methods of increasing the network capacity.

1. The alternative with the largest increase in network capacity per dollar of investment ($\Delta C/\$$) will be developed first.
2. It two or more alternatives have the same capacity increase per dollar of investment, the alternative that provides the most direct connection between A and B will be developed first.

The foregoing objective and criteria are reasonably realistic, although in some instances political influences and regional developments affecting the intermediate towns may have to be considered. There may already have been some kind of influence exerted to cause road 3 between cities 4 and 6 to be built as a four-lane road, because as will be seen later, it is of no importance for maintaining or increasing the capacity of flow from *A* to *B*.

7.6.3 Analysis Procedure

Figure 7.14 is a linear graph model of the network flow. The problem is simplified by considering only flow from city 1 to city 8, since the reverse flow is identical in capacity. In Figure 7.14, the four-lane roads (e.g., those with two lanes in each direction) are represented by double-line branches.

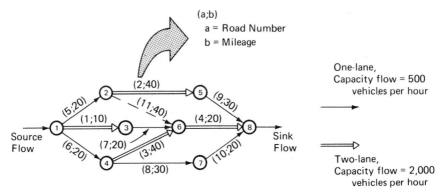

Figure 7.14   Network flow model.

A sequential procedure is followed. The roads that limit the flow capacity of the existing network are first identified from the dual graph. The feasible alternatives for increasing the capacity are outlined and ranked according to the established criteria. The best alternative is chosen and the limitation in the network with the added improvement is determined. This procedure is repeated until the total demand can be satisfied or no additional improvement is possible.

Figure 7.15a shows that the existing network has a maximum flow capacity of 1500 vph and is limited by the roads indicated by two paths through the dual graph. These two paths are a *A-B-D-F* and *A-C-B-D-F*. To expand the network, both paths must be lengthened; that is, the capacity of roads represented in both paths must be increased. Thus the following feasible alternatives exist at this stage:

Alternative	Increase in Capacity, ΔC	Total Cost	$\Delta C/\$$
1. Add two lanes to roads 5 and 9	1500	$ 50R	30/R
2. Add two lanes to road 7	1500	20R	75/R
3. Add two lanes to road 6	1500	20R	75/R
4. Add two lanes to road 5 and build four lanes at road 11	1500	112R	13.4/R

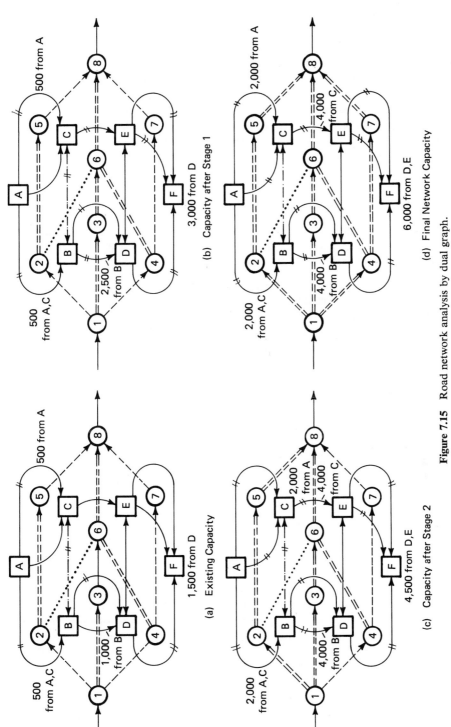

(a) Existing Capacity

500 from A

500 from A,C

1,000 from B

1,500 from D

(b) Capacity after Stage 1

500 from A

500 from A,C

2,500 from B

3,000 from D

(c) Capacity after Stage 2

2,000 from A

2,000 from A,C

4,000 from B

4,500 from D,E

2,000 from A

4,000 from C

(d) Final Network Capacity

2,000 from A

2,000 from A,C

4,000 from B

6,000 from D,E

4,000 from C

Figure 7.15 Road network analysis by dual graph.

Alternatives 2 and 3 result in the same capacity increase per dollar invest-ment. But road 7 provides a more direct connection between the two cities, 1 and 8. Thus, according to the second design criterion, the best choice at this stage is to add two lanes to road 7 at a cost of $20R.

Figure 7.15b shows all minimum-length paths after the improvement described above has been completed. This network capacity is now 3000 vph. Using the same logic as before, in order to increase the network capacity, all the minimum-length paths must be lengthened. Many feasible alternatives exist. But from Figure 7.15b, because all paths must be lengthened, it is obvious that the two best alternatives are as follows:

Alternative	Increase in Capacity, ΔC	Total Cost	$\Delta C/\$$
1. Add two lanes to roads 5 and 9	1500	$50R	30/R
2. Add two lanes to roads 6, 8, and 10	1500	70R	21.4/R

Therefore, the best choice at this stage is alternative 1.

Figure 7.15c shows the minimum-length paths after the second improve-ment has been completed. It can be easily shown that for a further expansion of the network flow capacity the best choice is to add two lanes to roads, 6, 8, and 10 at the cost of $70R, resulting in an increase in network capacity of 1500 vph and, hence, providing a total capacity of 6000 vph, as shown in Figure 7.15d.

Thus the department has developed a cost–capacity relationship for the road network, as indicated by Table 7.3. Therefore, to provide capacity for

TABLE 7.3 Cost–Capacity Relationships for Road Network

Capacity	Total Cost of Capital Improvement	Number of Lanes	Roads
1500	0		
3000	$ 20R	2	7
4500	70R	2	5, 7, 9
6000	140R	2	5, 6, 7, 8, 9, 10

6000 vph, the department should request a capital appropriation of $140R in order to add two lanes to roads 5, 6, 7, 8, 9, and 10. If the department receives a capital appropriation below its $140R request, Table 7.3 indicates how the money should be invested; for example, if an appropriation of only $70R is received, two lanes should be added to roads 5, 7, and 9. This addition increases the capacity to only 4500 vph, and an attempt must be made to secure the addi-tional $70R if the total demand is to be satisfied.

Another way of interpreting these results is that Table 7.3 shows the minimum cost of providing different network capacities. Thus the analysis would have been the same if the department had stated that it wanted to minimize the cost of providing a network capacity of 6000 vph between cities 1 and 8.

The preceding example uses the dual graph and the method of path analysis to identify the road sections that constrain the maximum network flow capacity. For a simple flow network such as the one discussed above, the method of cut-sets may also be used. Figure 7.16 shows the minimum cut-sets and the network flow capacity at each of the four stages illustrated in Figure 7.15.

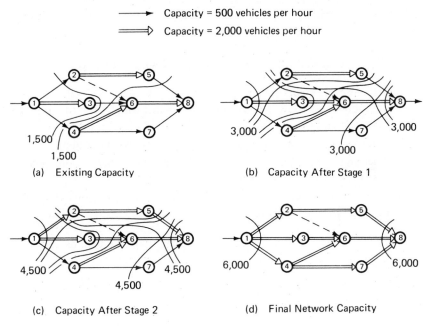

Figure 7.16   Road network analysis by cut-sets.

7.7 LINEAR GRAPH TOPOLOGICAL MATRICES

The many graphical structural properties provide the opportunity for storing information about specific linear graphs. For example, a path can be identified with the nodes and branches that are traversed along the path. A convenient way of describing graph structure is through the use of matrices. A matrix consisting of a rectangular array of elements has a unique capacity for fulfilling this function.

In some cases, the ultimate usefulness of the linear graph method of analysis depends on the ability to describe the graphs in some numerical format so that complex computations may be performed logically, either by hand or with the help of an electronic computer.

In addition, matrix algebra notation represents a system of grouping numerical quantities in such a way that a single symbol suffices to denote the whole group. In this way, an algebraic form of shorthand can be developed.

Since matrices are commonly considered as two-dimensional arrays with rows and columns, they can be used to model binary relationships. Thus a path matrix could be developed to sort the various nodes or branches into the different paths existing in the linear graph. A row could be provided and identified for each node (or branch) and a column provided for each unique path. An entry in the matrix could then signify whether or not a node (or branch) is a component of the graph structural property considered in the path. In this way, a node-path matrix or a branch-path matrix could be developed. Finally, a sign convention can be introduced to capture the additional information portrayed in directed graphs.

The development of a linear graph topological matrix therefore requires:

1. Selecting the structural property attribute to be described and the graph component attribute (node or branch) against which the structural property is to be mapped
2. Allocating the two attributes to the matrix rows and columns
3. Selecting the mapping rule that distinguishes whether the relationship exists (i.e., true $\equiv$ 1) or not (i.e., false $\equiv$ 0)
4. Selecting the sign convention to be followed for directed graphs and directed structural properties

In the following sections, several topological matrices will be developed to illustrate the general approach to the matrix algebra modeling of graph structure.

7.7.1 The Branch–Node Incidence Matrix A

The fundamental property of a linear graph is the connectivity of its branches and nodes. A linear graph is completely specified once the incidence of each branch on each node is known. This is conveniently done by specifying a matrix $\bar{\mathbf{A}}$ called the augmented branch–node incidence matrix, consisting of b rows (one per branch) and n columns (one per node). For undirected graphs, the typical element $\bar{a}_{ij} = (1, 0)$ if the ith branch (is, is not) incident on the jth node (see Figure 7.17a). For directed graphs, the typical element $\bar{a}_{ij} = (+1, -1, 0)$ if the ith branch is (positively, negatively, not) incident on the jth node (see Figure 7.17b).

Each row corresponds to a particular branch that connects two nodes, and has two nonzero elements that for directed graphs are one $+1$ and one -1. Each column corresponds to a particular node, and the number of nonzero elements in it indicates the number of branches that are incident on the node. In addition, by summing independently the number of the positive and negative entries in a column, the number of originating and terminating branches on the

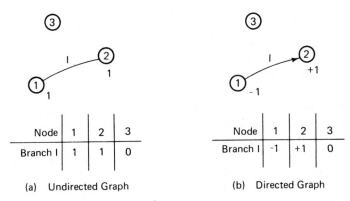

Node	1	2	3
Branch I	1	1	0

(a)   Undirected Graph

Node	1	2	3
Branch I	-1	+1	0

(b)   Directed Graph

Figure 7.17   Branch–node incidence.

node can be determined. The $2b$ nonzero entries in the augmented branch–node incidence matrix completely specify an orientated linear graph if the branch and node ordering within the rows and columns are known.

As an illustration, consider the directed linear graph shown in Figure 7.18. The following table can be identified:

		Node			
		n_0	n_1	n_2	n_3
	b_1	-1	$+1$	0	0
	b_2	-1	0	$+1$	0
Branch	b_3	0	-1	$+1$	0
	b_4	0	-1	0	$+1$
	b_5	0	0	-1	$+1$

where the rows and columns have been identified and labeled. Once the row and column labels have been prescribed, it is no longer essential to include them as descriptors on the matrix itself.

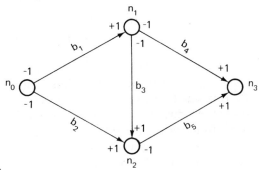

Figure 7.18   Notation for branch–node incidence matrix.

n_0 = Datum Node

It is easy to see that the $\bar{\mathbf{A}}$ matrix contains redundant information. Since each row contains exactly one $+1$ coefficient and one -1 coefficient, then independent of their column locations within the row, the sum of all coefficients in the row is zero. That is, for a connected graph, the columns are onefold linearly dependent and one nodal column (i.e., the datum node) can be suppressed without loss of information.

The reduced matrix is called the branch–node incidence matrix $\mathbf{A}$ and has b rows and $n - 1$ columns for a connected graph. The suppressed nodal (datum) column can be chosen at will from any node in the connected graph, but is usually chosen for convenience as the first or last within the nodal sequence of $\bar{\mathbf{A}}$. Practical examples of datum nodes include the starting node of a CPM network, the source node of a flow network, and the beginning node of a decision tree.

Using n_0 as the datum node, the branch–node incidence matrix $\mathbf{A}$ becomes

$$\underset{(5\times 3)}{\mathbf{A}} = \begin{bmatrix} +1 & 0 & 0 \\ 0 & +1 & 0 \\ -1 & +1 & 0 \\ -1 & 0 & +1 \\ 0 & -1 & +1 \end{bmatrix} \tag{7.6}$$

7.7.2 Cut-Set Matrix

In analyzing the road network capacity problem, it was necessary to identify the various proper cut-sets and highway branches of the linear graph model. A cut-set matrix, $\mathbf{D}$, may be derived directly from the entries in Table 7.1 for the directed graph in Figure 7.5 as follows:

$$\underset{(7\times 8)}{\mathbf{D}} = \begin{bmatrix} 1 & 1 & 1 & 1 & 0 & 0 & 0 & 0 \\ 1 & 1 & 0 & 0 & 1 & 1 & 0 & 0 \\ 1 & 0 & 1 & 0 & 1 & 0 & 1 & 0 \\ 0 & 0 & 0 & 0 & 1 & 1 & 1 & 1 \\ 0 & 0 & 1 & 1 & 0 & 0 & 1 & 1 \\ 0 & 1 & -1 & 0 & 0 & 1 & -1 & 0 \\ 0 & 1 & 0 & 1 & 0 & 1 & 0 & 1 \end{bmatrix} \tag{7.7}$$

where $d_{ij} = -1$ if the ith branch is included in the jth cut-set and flows from the sink side to the source side; $d_{ij} = 0$ if the ith branch is not included in the jth cut-set; and $d_{ij} = +1$ if the ith branch is included in the jth cut-set and flows from the source side to the sink side. The $\mathbf{D}$ matrix above adequately describes the components of all the possible proper cut-sets of the directed graph in Figure 7.5. For undirected graphs, the d_{ij} elements of the matrix may simply be either 1 or 0, denoting whether the ith branch is or is not included in the jth cut set.

Other topological matrices may be constructed to describe other graphical properties in the same manner (Busacker and Saaty, 1965; Marshall, 1971). The needs for such matrices depend on the method of analysis and the graphical properties of the network.

7.8 SUMMARY

Many engineering systems can be modeled and analyzed using linear graph models and concepts. Linear graph analysis requires that (1) the problem be expressed in graphical form, (2) the problem be identified with a graphical property, and (3) an analysis procedure be developed that focuses on this specific graphical property. The problem can frequently be identified in terms of a path or a network flow problem.

In analyzing the network flow problem, the mini-max theorem may be applied to determine the maximum flow through the network. The determination of the maximum flow can be simplified through the transformation of the graph to the dual graph. Although the transformation concept has been applied specifically to linear graph analysis in this chapter, engineers frequently may find other applications where data or models can be transformed to permit the use of different analysis methods.

The network improvement example represents a problem in which linear graph analysis can be used to assist the engineer with determining the priority of improvements. Finally, the topological matrices reflect a means of describing and analyzing linear graphs through the use of a numerical format.

PROBLEMS

P7.1. Identify a problem that might be associated with each of the situations in Problem P4.1. Develop a linear graph and describe a linear graph procedure that can be used to solve the problem.

P7.2. Figure P7.2 shows a water flow network. The arrows show the direction of flow and the numbers indicate the flow capacity in cubic feet per second (ft^3/sec).
(a) Draw a dual for the graph.
(b) Determine the network flow capacity.
(c) Given the following alternatives, which can be considered for increasing the network capacity, which has the highest benefit/cost ratio? Show the methods of analysis.

   1. Add new inlet for inflow at node B—$60,000
   2. Add new outlet for outflow at node G—$50,000
   3. Add new pipe section A-E with a flow of 20 ft^3/sec—$75,000

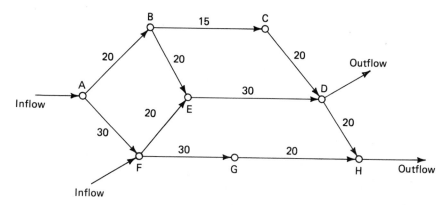

Figure P7.2

P7.3. Figure P7.3 represents the portion of a street network in a city. The direction of
travel that is permitted on each street is indicated by the arrows in the figure.
Also, the capacity of each street segment is shown in vehicles per hour.

(a) Currently, the street segment between intersections C and D is two-way
with a capacity of 500 vph in each direction. A plan is being considered to
convert this segment to one-way operation with a capacity of 1500 vph in
order to increase the flow of traffic. Would such a plan increase the flow of
traffic in the network between intersections A and G? If so, which direction
should the one way street be? Justify your answer.

(b) Based on your decision in part (a), what is the maximum flow capacity
between intersections A and G?

(c) A traffic study reveals that the parking on street segments II, IV, and IX
might be banned to increase the street capacity. For each street segment, the
parking ban will increase the capacity by 400 vph. Because of a shortage of
parking in the area, the city council has consented to ban parking from only

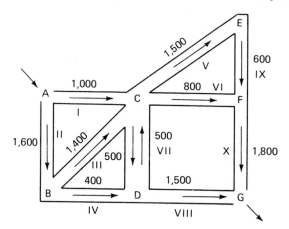

Figure P7.3

one segment. Which one should be chosen to increase the flow capacity of the street network between intersections *A* and *G*?

P7.4. Develop a flow process linear graph for manufacturing precast concrete wall panels using metal molding forms and steam curing. The panels are reinforced with steel mesh and contain utility ducting as well as both window and door frame inserts. Use the following mode symbols and attributes:

●   Operation (description)

➤   Transport (distance, mode)

▲   Storage (time period, quantity)

■   Inspection (detail, quality)

P7.5. Identify the components and component characteristics for the following systems in a high-rise building. Develop linear graph models for each system and explain how linear graph analysis procedures can aid in the design of each system.

(a) An electrical network

(b) A heating and ventilation system

(c) A hot- and cold-water distribution system

(d) A waste collection system

(e) A transportation or circulation system within the building

(f) A structural system

(g) A telephone system

P7.6. A city plans to expand its water supply by developing a well field northeast of town, as shown in Figure P7.6. The water must be pumped from the wells to the water treatment plant before distribution. The distances in feet are shown in Figure P7.6.

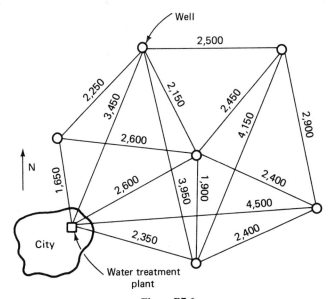

Figure P7.6

(a) In what pattern should the wells be connected to use the minimum length of pipe?

(b) In part (a) the minimum length of pipe was used as a criterion for defining the best solution. What might be another criterion to use for this problem? Can the same analysis procedure be used for both criteria?

P7.7. Figure P7.7 shows the classical Königsberg bridge problem, which was investigated by Euler using linear graph analysis. There are seven bridges connecting the two sides of the river (*B* and *C*) and two islands (*A* and *D*).

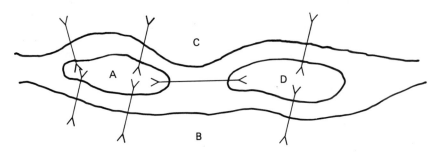

Figure P7.7

You are a bridge engineer and your office is located at *C*. You are planning an inspection trip to visit all seven bridges and return to your office at *C*.

(a) Illustrate your inspection route with a linear graph that denotes each bridge as a branch.

(b) Illustrate your inspection route with a linear graph that denotes each bridge as a node.

(c) Show that it is impossible to find a route in which you pass through each bridge only once.

P7.8. The way in which a problem is formulated will often control which analysis procedure can be used. Either a dual graph procedure or a cut-set analysis procedure can be used to solve the problem in Section 7.6. What are some possible advantages of attempting several different formulations of the problem using different graphical properties with respect to each of the following?

(a) The analysis procedure that can be used

(b) The amount and kinds of data required

(c) The kinds and detail of answers that can be obtained

Calculus Methods
for Optimization

8

8.1 THE CALCULUS METHOD

Calculus methods are very powerful in solving optimization problems when explicit functional relationships can be developed between the relevant problem decision variables and when, in addition, these functional relationships can be readily differentiated to give condition equations for maxima and minima that can be readily solved. From the classicial calculus extreme values for unconstrained functions occur at the values of the variables at which first derivatives are zero, and maxima when negative second derivatives result for these extreme values.

Application of the methods of calculus in solving optimization problems will be illustrated in this chapter through the use of examples. First, optimization models consisting of only one and two variables will be used to illustrate the method of finding minima and maxima and the meaning of an objective function. Then optimization problems involving equality and inequality constraints are introduced to illustrate the methods of direct substitution and Lagrangian multipliers.

8.2 MODEL WITH ONE DECISION VARIABLE

Suppose that a county is planning to build a reservoir for the purpose of flood control, water supply, power generation, and recreation. Detailed economic studies have been made to study the benefits and costs of building a reservoir of several different sizes. County decision makers must now decide on the optimum reservoir size. The primary objective of the county is to maximize the net benefit from the project. The only practical constraint, because of the topography and the existing land-use pattern of the area, is that the size of the reservoir cannot be larger than 5 million acre-feet.

This problem involves only one decision variable, the size of the reservoir in millions of acre-feet. Let it be denoted by X. If the net benefit is denoted by P (in millions of dollars), a mathematical expression must be established to relate P as a function of X.

Suppose that the benefits and costs of four different sizes of reservoirs have been determined, and that these are plotted as shown in Figure 8.1. An examination of this figure reveals that the total benefit varies approximately linearly with the size of the reservoir, but the cost increases at a higher rate with increasing size. The method of regression analysis may be used to find the mathematical expressions that best fit the trends shown in Figure 8.1. Suppose that the following mathematical expressions represent the curves (or relationships) depicted in Figure 8.1:

$$B = 10 + 5X \tag{8.1}$$
$$C = 5 + 2X^2 \tag{8.2}$$

where B is the total benefit and C is the total cost both expressed in millions of dollars.

The net benefit, P, can thus be expressed as follows:

$$P = B - C$$
$$= (10 + 5X) - (5 + 2X^2)$$
$$= 5 + 5X - 2X^2$$

The decision problem facing the county can now be represented by the following optimization model:

Maximize

$$P = 5 + 5X - 2X^2 \tag{8.3}$$

subject to

$$0 \le X \le 5 \tag{8.4}$$

From the calculus

$$\frac{dP}{dX} = 5 - 4X$$

and the condition for a maximum or minimum is

$$\frac{dP}{dX} = 0$$

that is,

$$5 - 4X = 0$$
$$X = 1.25 \text{ million acre-feet}$$
$$P = 5 + 5(1.25) - 2(1.25)^2 = \$8.13 \text{ million}$$

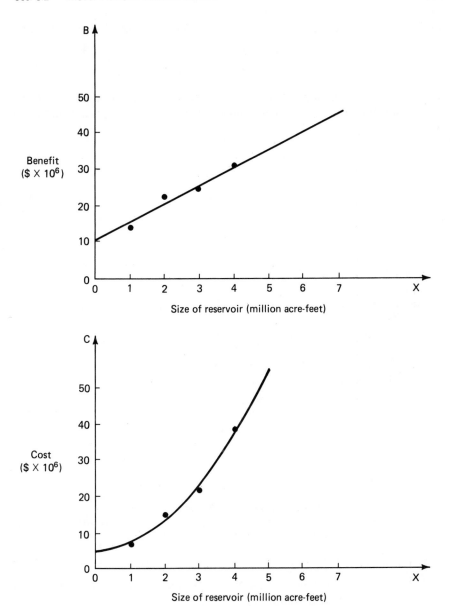

Figure 8.1   Benefit/cost curves for reservoir size.

Also, since

$$\frac{d^2P}{dX^2} = -4 \qquad \text{(i.e. negative)}$$

$X = 1.25$ represents a local maximum. To prove that it is also the maximum (i.e., global maximum) within the range between $X = 0$ and $X = 5$, the value of P must also be computed for the two extreme values of X. Therefore, for

$$X = 0 \qquad P = \$5 \text{ million}$$
$$X = 5 \qquad P = -\$20 \text{ million}$$

Hence $X = 1.25$ million acre-feet is indeed the global maximum and the optimum solution to the problem. This point is illustrated graphically in Figure 8.2.

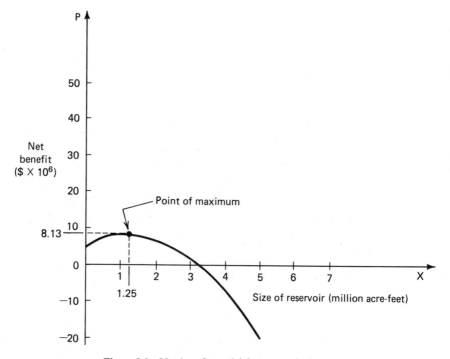

Figure 8.2   Net benefit model for reservoir size.

The procedure above can be summarized as follows:

1. Formulate an optimization model.
2. Find the extreme points (points of maximum and minimum) by the method of differentiation.
3. Check for maximum or minimum by taking second derivatives.
4. Check for global maximum or minimum by comparing points of extreme from step 2 with boundary or constraints limits.

If the objective of the county were to maximize the benefit/cost ratio, the optimization model would be as follows:

Maximize

$$R = \frac{10 + 5X}{5 + 2X^2} \tag{8.5}$$

subject to:

$$0 \leq X \leq 5 \tag{8.6}$$

The optimum solution would then be $X = 0.55$ million acre-feet and $R = 2.27$. The corresponding value of net benefit (P) will amount to only \$7.15 million, which is a decrease of about \$1 million from the previous optimum solution. This illustrates that different objectives can lead to different optimal solutions.

8.3 MODEL WITH TWO DECISION VARIABLES

The scope of the reservoir problem can be expanded so that it becomes a problem involving two variables. In addition to the size of the proposed reservoir in millions of acre-feet, suppose that the county must decide at the same time the amount of park space to be developed for recreational purposes. Let Y represent the number of thousands of acres of recreational parks to be developed. An economic study can again be conducted to develop a mathematical model relating the size of the reservoir (X) and the amount of recreational park space (Y) to the total cost and benefit. Suppose that such a study results in the following model:

$$B = 10 + 5X + XY \tag{8.7}$$

$$C = 5 + 2X^2 + Y^2 \tag{8.8}$$

where B and C again represent the total benefit and cost, respectively, in millions of dollars. If the objective were to maximize the total net benefit, the objective function may be expressed as follows:

Maximize

$$P = B - C$$
$$= (10 + 5X + XY) - (5 + 2X^2 + Y^2)$$
$$= 5 + 5X + XY - 2X^2 - Y^2$$

If the reservoir cannot exceed 5 million acre-feet and only a maximum of 4000 acres can be made available for recreational parks, the optimization model can be expressed as follows:

Maximize

$$P = 5 + 5X + XY - 2X^2 - Y^2 \tag{8.9}$$

subject to:

$$0 \leq X \leq 5 \tag{8.10}$$

$$0 \leq Y \leq 4 \tag{8.11}$$

The optimum values of X and Y can be found by the method of derivatives as follows:

$$\frac{\partial P}{\partial X} = 5 + Y - 4X = 0 \tag{8.12}$$

$$\frac{\partial P}{\partial Y} = X - 2Y \quad\quad = 0 \tag{8.13}$$

From Equation 8.13,

$$X = 2Y \tag{8.14}$$

Substituting Equation 8.14 into Equation 8.12 yields

$$5 + Y - 8Y = 0$$

or

$$Y = \frac{5}{7} = 0.71 \text{ thousand acres}$$

and

$$X = 2Y = 1.42 \text{ million acre-feet}$$

The corresponding value of P is \$8.57 million. Second derivatives are next derived to show that this is a point of maximum.

$$\frac{\partial^2 P}{\partial X^2} = -4 \quad\quad \text{negative}$$

$$\frac{\partial^2 P}{\partial Y^2} = -2 \quad\quad \text{negative}$$

$$\frac{\partial^2 P}{\partial X \partial Y} = 1$$

and

$$\left(\frac{\partial^2 P}{\partial X \partial Y}\right)^2 - \frac{\partial^2 P}{\partial X^2}\frac{\partial^2 P}{\partial Y^2} = 1 - (-4)(-2) = -7 \quad\quad \text{negative}$$

Therefore, $X = 1.42$ and $Y = 0.71$ define a point of maximum and it lies within the constraints of $0 \le X \le 5$ and $0 \le Y \le 4$. Furthermore, it can be easily shown that if $X = 0$, P is maximum at $Y = 0$ and is equal to \$5 million; and if $Y = 0$, P is maximum (\$8 million) at $X = 1.25$ million acre-feet. Thus the point $(X = 1.42, Y = 0.71)$ is also a global maximum. Figure 8.3 graphically illustrates the location of the optimum point within the feasible solution space.

A solution point would represent a point of local minimum if the following threee conditions are satisfied:

1. $\dfrac{\partial^2 P}{\partial X^2}$ is positive

2. $\dfrac{\partial^2 P}{\partial Y^2}$ is positive

3. $\left(\dfrac{\partial^2 P}{\partial X \partial Y}\right)^2 - \dfrac{\partial^2 P}{\partial X^2}\dfrac{\partial^2 P}{\partial Y^2}$ is negative

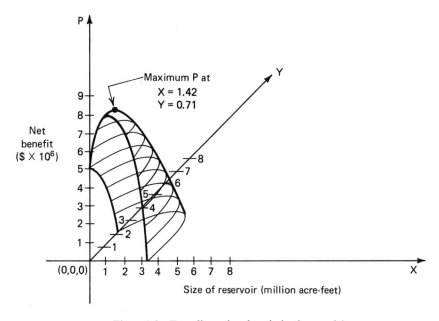

Figure 8.3   Two-dimensional optimization model.

If the term

$$\left(\frac{\partial^2 P}{\partial X \partial Y}\right)^2 - \frac{\partial^2 P}{\partial X^2}\frac{\partial^2 P}{\partial Y^2}$$

is positive, the solution represents a saddle point as shown in Figure 8.4. If this term is equal to zero, the function $P = f(X, Y)$ is indeterminant and a reexamination of the problem model becomes necessary.

It is beyond the scope of this book to discuss the conditions for local maxima and minima for functions involving more than two variables. They are, however, covered in most textbooks on calculus of more than two variables.

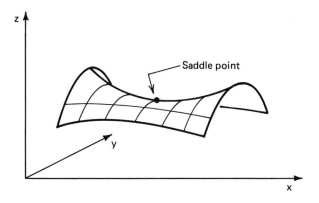

Figure 8.4   Saddle point.

8.4 MODEL WITH EQUALITY CONSTRAINT

Suppose that the county further decides that there be sufficient recreational facilities to accommodate 20,000 people. It is estimated that each 1 million acre-feet of reservoir capacity can accommodate 15,000 people in activities such as swimming, boating, and fishing; and that each 1000 acres of park space can accommodate 1000 people in activities, such as camping, picnics, and hiking. Thus this new constraint can be expressed as

$$15,000X + 1000Y = 20,000$$

Dividing throughout by 1000 yields

$$15X + Y = 20 \tag{8.15}$$

Thus the optimization model can now be written as:
Maximize

$$P = 5 + 5X + XY - 2X^2 - Y^2 \tag{8.9}$$

subject to:

$$15X + Y = 20 \tag{8.15}$$

$$0 \leq X \leq 5 \tag{8.10}$$

$$0 \leq Y \leq 4 \tag{8.11}$$

This model can be solved by either the method of direct substitution or the method of Lagrangian multipliers.

8.4.1 Solution by Direct Substitution

From Equation 8.15,

$$Y = 20 - 15X \tag{8.16}$$

Substituting Equation 8.16 into the objective function represented by Equation 8.9 yields

$$P = -395 + 625X - 242X^2 \tag{8.17}$$

Differentiating P with respect to X and equating the result to zero yields

$$\frac{dP}{dX} = 625 - 484X = 0$$

Therefore,

$$X = \frac{625}{484} = 1.29 \text{ million acre-feet}$$

$$Y = 0.65 \text{ thousand acres}$$

Then $P = \$8.54$ million. This solution represents a point of maximum because

$$\frac{d^2P}{dX^2} = -484 \qquad \text{is negative}$$

The solution also satisfies the condition that $0 \leq X \leq 5$ and $0 \leq Y \leq 4$.

8.4.2 Solution by Lagrangian Multiplier

Equation 8.15 may be rewritten as follows:

$$15X + Y - 20 = 0 \qquad (8.18)$$

Next, let λ be an artificial variable. The objective function can now be rewritten as follows:

Maximize

$$P = 5 + 5X + XY - 2X^2 - Y^2 - \lambda(15X + Y - 20) \qquad (8.19)$$

The variable λ is called the *Lagrangian multiplier*. It should be noted that the term $\lambda(15X + Y - 20)$ is equal to zero because of Equation 8.18. The reason for the negative sign before λ will be explained later in this section.

Now that the objective function includes the equality constraint expressed by Equation 8.15, the values of X, Y, and λ which will maximize P can be solved by the method of differentiation:

$$\frac{\partial P}{\partial X} = 5 + Y - 4X - 15\lambda = 0 \qquad (8.20)$$

$$\frac{\partial P}{\partial Y} = X - 2Y - \lambda = 0 \qquad (8.21)$$

$$\frac{\partial P}{\partial \lambda} = -(15X + Y - 20) = 0 \qquad (8.22)$$

From Equation 8.21

$$\lambda = X - 2Y \qquad (8.23)$$

Substituting Equation 8.23 into Equation 8.20 and collecting terms yields

$$5 - 19X + 31Y = 0 \qquad (8.24)$$

But from Equation 8.22,

$$Y = -15X + 20 \qquad (8.25)$$

Therefore,

$$5 - 19X + 31(-15X + 20) = 0$$

or

$$-484X + 625 = 0$$

and

$$X = \frac{625}{484} = 1.29 \text{ million acre-feet}$$

Then, from Equations 8.25, 8.23, and 8.9,

$$Y = 0.65 \text{ thousand acres}$$

$$\lambda = -0.01$$

$$P = \$8.54 \text{ million}$$

which is the same solution as that obtained previously by direct substitution. The proof for maximum (or minimum) in a solution using Lagrangian multipliers is considerably involved. It is beyond the scope of this book to discuss the subject here, and interested readers are referred to Hestemus (1975).

It can be shown that the Lagrangian multiplier, λ, represents the rate of change of P with respect to the constant term in the equality constraint of Equation 8.15. Let the constant term in Equation 8.15 be replaced by a variable b. Equation 8.19 can thus be written as

$$P = 5 + 5X + XY - 2X^2 - 2Y^2 - \lambda(15X + Y - b)$$

Differentiating P with respect to b yields

$$\frac{\partial P}{\partial b} = \lambda \tag{8.26}$$

In the solution above, $\lambda = -0.01$. It means that each additional unit of b (in this case, 1000 people), will result in a decrease of \$0.01 million of net profit. That is, if the constraint requires that there be sufficient recreational facilities to accommodate 21,000 people (rather than 20,000 people), the maximum net benefit from the project will be \$8.53 million (instead of \$8.54 million). Thus the Lagrangian multiplier, λ, provides a measure of the sensitivity of the optimum solution with respect to the specified constraints. It should also be noted that the partial derivative in Equation 8.26 has a positive sign for λ only because there is a negative sign preceding λ in Equation 8.19.

Finally, if the optimization model has more than one equality constraint, one Lagrangian multiplier should be introduced for each constraint and they may be represented as $\lambda_1, \lambda_2, \lambda_3$, and so on.

8.5 MODEL WITH INEQUALITY CONSTRAINT

The constraint on recreational space may be changed to require that there be sufficient space to accommodate at least 20,000 people. Then, instead of an equality as represented by Equation 8.15, the constraint would be specified as an inequality as follows:

$$15X + Y \geq 20 \tag{8.27}$$

The optimization model would then be represented as follows:

Maximize

$$P = 5 + 5X + XY - 2X^2 - Y^2$$

subject to:

$$15X + Y \geq 20$$
$$0 \leq X \leq 5$$
$$0 \leq Y \leq 4$$

This model can also be solved by either the method of direct substitution or the method of Lagrangian multiplier. The inequality constraint can first be converted to an equality by introducing a slack function, S^2. In this case,

$$15X + Y - S^2 = 20 \qquad (8.28)$$

where S^2 may be interpreted as a function that represents the number of people in excess of the specified minimum of 20,000. The same basic procedures as those outlined previously can be used to solve for the optimum values of X, Y and S.

8.6 SUMMARY

The use of calculus methods of solving optimization problems can become extremely difficult or impossible if the objective and/or constraint functions involve a large number of variables, or have nonlinear terms. Some of the most commonly encountered difficulties are:

1. Nonexistence of the partial derivatives
2. Necessity of solving a system of nonlinear equations
3. Testing for conditions of local maximum or minimum
4. Proving global maximum or minimum

Therefore, during the formulation of optimization problems, there is a strong incentive to keep the optimization model as simple as possible. However, over-simplification can result in a model that does not represent the real system and is thus useless as a planning tool. A good planner must develop the ability to make a proper trade-off between simplicity and close representation of reality in modeling.

   Special techniques have been developed for solving optimization problems. Chapter 9 discusses how models with only linear functions can be solved. Chapter 10 presents techniques that are applicable to models with nonlinear functions.

PROBLEMS

P8.1. Find the values of x and y that will result in minimizing C in the following optimization model:
Minimize

$$C = 10 + 6x + y^2$$

subject to:

$$2x + 6y = 100$$

   (a) By the method of direct substitution
   (b) By the method of Lagrangian multipliers
   Verify that your answer is indeed the optimum.

P8.2. Solve the following:
Maximize

$$Z = 2x_1 - 2x_1^2 + 6x_1x_2 - 2x_2^2$$

subject to:

$$x_1 + x_2 \leq 2$$
$$x_1 \geq 0$$
$$x_2 \geq 0$$

P8.3. A bridge is to be constructed to span a distance of 3000 ft across a river. From the preliminary study it is concluded that a series of equal span simple trusses is most suitable. It is estimated that the two end piers, A and B, will cost \$40,000 each. The remaining piers, which are to be located in water, can be constructed anywhere at the crossing at a cost of \$80,000 per pier. Each bridge truss of length l (feet) costs $8l^2$ dollars. Determine the optimum length l for the trusses.

P8.4. A manufacture of large crates has in stock a large quantity of steel bars which are used for the frames of crates. In order to make use of these bars which are 25 ft long, each bar is to be cut into three sections, x_1, x_2, and x_3, corresponding to length, width, and depth of the crate, respectively.

Determine the lengths x_1, x_2, and x_3 so that the volume of the crates will be maximum. You do not have to prove that the conditions for optimality have been achieved.

P8.5. A high-rise building being planned will house apartments, stores, and offices. Market research has shown that the potential rental incomes for the three usages are as follows:

$$R(x) = 10x$$
$$R(y) = 10y - 0.1y^2$$
$$R(z) = 16z - 0.025z^2$$

where x, y, and z represent the amount of floor space in thousands of square feet to be made available for apartments, stores, and offices, respectively; and $R(x)$, $R(y)$, and $R(z)$ are the potential rental incomes from the three types of usage. The high-rise building will have a total of 300,000 square feet of floor space.

Determine the optimum amount of floor space to be allocated for apartments, stores, and offices.

P8.6. A water company wants to pump water from its well field to the city where the water will be sold. The well field is 3 miles from the city. The pipe in place will cost \$10D/foot of length, where D is diameter of the pipe in feet. The annual pumping cost is given by \$150h per year, where h is the head loss, or drop in the hydraulic grade line, in the pipe. The head loss is given by

$$h = 0.02 \frac{L}{D} \frac{V^2}{2g}$$

where L is the length of pipe in feet, V is the average velocity of flow in feet per second, and g is the constant due to gravity. The pipe will last 15 years and the interest rate is 10 percent. The smallest diameter pipe that can be purchased is 6 in. If the company wants to pump 12,000 gallons per minute of water, what size pipe should the company use to minimize cost? Solve using the method of Lagrangian multipliers.

P8.7. In the process of preparing a bid for the construction of a high-rise building, a contractor has just completed a detailed analysis of the project schedule by the critical path method. He has determined that the direct and indirect cost can be related to the project duration by the following mathematical expressions:

Direct cost in millions of dollars $(D) = 53 - 4.75x + 0.125x^2$

Indirect cost in millions of dollars $(I) = 4 + 0.5x$

in which x is the project duration in months. He has also determined that the project cannot be completed in less than 14 months, and the owners in the project require that the project must be completed in 20 months or less.

Determine the project duration that will result in minimum total cost.

P8.8. The cross section of a tunnel has the shape of a rectangle surmounted by a semicircle as shown in Figure P8.8. To make the most effective use of the available lining material, the inside perimeter of the cross section must be equal to 60 ft. Determine the radius (r) and the height (h) that will yield the greatest cross-sectional area.

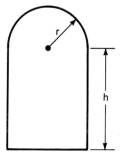

Figure P8.8

Linear
Programming
9

9.1 THE NATURE OF MATHEMATICAL PROGRAMMING

Mathematical programming is used to find the best or optimal solution to a problem that requires a decision or set of decisions about how best to use a set of limited resources to achieve a stated goal or objective. It makes three important contributions to the study of the problem.

1. The problem situation must be structured into a mathematical model that abstracts the essential elements so that a solution relevant to the decision maker's objective can be sought. This involves looking at the problem in the context of the entire system.
2. The structure of solutions must be explored and systematic procedures must be developed for obtaining the solutions.
3. It yields an optimal value of desirability for the system or has at least compared alternative courses of action by evaluating their desirability.

This chapter deals primarily with the linear programming aspects of mathematical programming and contains an introduction to the principles of integer programming. The principles of nonlinear programming will be discussed in Chapter 10.

9.2 THE LINEAR PROGRAMMING MODEL

Linear programming, the simplest and most widely used of the programming techniques, requires that all the mathematical functions in the model be linear functions. The mathematical statement of a general linear programming problem is the following: Find the values $x_1, x_2, x_3, \ldots, x_n$ (called *decision variables*) that maximize the linear function Z (i.e., the *criterion function*, commonly called the *objective function*)

$$Z = c_1 x_1 + c_2 x_2 + \cdots + c_n x_n \tag{9.1}$$

subject to the following relationships (called *constraints* or *restrictions*):

$$a_{11} x_1 + a_{12} x_2 + \cdots + a_{1n} x_n \leq b_1$$
$$a_{21} x_1 + a_{22} x_2 + \cdots + a_{2n} x_n \leq b_2$$

$$\cdot \qquad\qquad\qquad \cdot \tag{9.2}$$

$$a_{m1} x_1 + a_{m2} x_2 + \cdots + a_{mn} x_n \leq b_n$$

$$\text{all } x_j \geq 0$$

where a_{ij}, b_i, and c_j are given constants. The linear programming model can be written in more efficient notation as

Maximize

$$Z = \sum_{j=1}^{n} c_j x_j$$

subject to:

$$\sum_{j=1}^{n} a_{ij} x_j \leq b_i$$

where

$$i = 1, 2, \ldots, m \tag{9.3}$$

and

$$x_j \geq 0$$

where

$$j = 1, 2, \ldots, n$$

The decision variables, $x_1, x_2, \ldots, x_n$, represent levels of n competing activities. These decision variables represent variables in the real situation being modeled that can be freely changed in magnitude by management. If each activity is manufacturing a certain product, then x_j would be the number of units of the jth product to be produced during a given time period; Z is the overall measure of effectiveness (e.g., profit over the given time period); and c_j is the increase in the objective function (profit) that would result from each unit increase in x_j. Each of the first m linear inequalities corresponds to a restriction on the availability of one of the resources; b_i is the total amount of the ith resource available; and a_{ij} is the amount of the ith resource used for each unit of the jth product. The restriction that all x_j be greater than or equal to zero models the fact that negative quantities cannot be produced. This constraint implies that the minimum value that any decision variable can attain is zero. The "less than or equal" sign in the constraint inequalities may be replaced by "greater than or equal," "equal," or "approximately equal" signs to suit the description of the particular problem that is being modeled.

9.3 DEVELOPING LINEAR PROGRAMMING MODELS

The variety of situations to which linear programming has been applied ranges from agriculture to zinc smelting. However, it is not always easy to recognize that a decision-making problem can be solved by linear programming. A certain amount of skillful trial and error is usually required to capture the essence of a decision problem in a linear model. The decision maker must determine the objective to be considered and this objective must be described by a criterion function in terms of the decision variables, x_i, that can be controlled or decided by the decision maker. The decision variables are not always obvious at first glance. Furthermore, the decision maker is always limited by constraints on what can be done. Care must be exercised in depicting the relevant constraints to be imposed on the optimization. An analysis procedure is then followed, which leads to the selection of values for the decision variables that optimize the criterion function while satisfying all the constraints imposed on the problem. One way to gain an understanding of the types of problems to which linear programming can be applied is to study how linear models have been formulated for several problems that have been presented verbally—the form in which the decisionmaker usually meets the problem.

9.3.1 A Product-Mix Problem

General problem.   A manufacturer has fixed amounts of different resources such as raw material, labor, and equipment. These resources can be combined to produce any one of several different products. The quantity of the ith resource required to produce one unit of the jth product is known. The decision maker wishes to produce the combination of products that will maximize total income.

Example problem.   The N. Dustrious Company produces two products: I and II. The raw material requirements, space needed for storage, production rates, and selling prices for these products are given in Table 9.1. The total amount of raw material available per day for both products is 1575 lb,

TABLE 9.1   Production Data for N. Dustrious Company

	Product	
	I	II
Storage space (ft^2/unit)	4	5
Raw material (lb/unit)	5	3
Production rate (units/hr)	60	30
Selling price ($/unit)	13	11

the total storage space for all products is 1500 ft², and a maximum of 7 hours per day can be used for production. All products manufactured are shipped out of the storage area at the end of the day. Therefore, the two products must share the total raw material, storage space, and production time. The company wants to determine how many units of each product to produce per day to maximize its total income.

Formulating the problem.   The company has decided that it wants to maximize its sale income, which depends on the number of units of product I and II that it produces. Therefore, the decision variables, x_1 and x_2 can be the number of units of products I and II, respectively, produced per day. The object, then, is to maximize the equation

$$Z = 13x_1 + 11x_2$$

subject to the constraints on storage space, raw materials, and production time. Each unit of product I requires 4 ft² of storage space and each unit of product II requires 5 ft². Thus a total of $4x_1 + 5x_2$ ft² of storage space is needed each day. This space must be less than or equal to the available storage space, which is 1500 ft³. Therefore,

$$4x_1 + 5x_2 \leq 1500$$

Similarly, each unit of product I and II produced requires 5 and 3 lb, respectively, of raw material. Hence a total of $5x_1 + 3x_2$ lb of raw material is used. This must be less than or equal to the total amount of raw material available, which is 1575 lb. Therefore,

$$5x_1 + 3x_2 \leq 1575$$

The limitation on production time is a little more difficult to translate into a mathematical statement. Product I can be produced at the rate of 60 units per hour. Therefore, it must take 1 minute or 1/60 of an hour to produce 1 unit. Similarly, it can be reasoned that it requires 1/30 of an hour to produce 1 unit of product II. Hence a total of $x_1/60 + x_2/30$ hours is required for the daily production. This quantity must be less than or equal to the total production time available each day. Therefore,

$$\frac{x_1}{60} + \frac{x_2}{30} \leq 7$$

This expression can be cleared of fractions by multiplying each term by 60. The result is

$$x_1 + 2x_2 \leq 420$$

Finally, the company cannot produce a negative quantity of any product (it is impossible to produce less than nothing). This means that x_1 and x_2 must each be greater than or equal to zero.

The linear programming model for this example can be summarized as:
Maximize
$$Z = 13x_1 + 11x_2$$
subject to:

$$4x_1 + 5x_2 \leq 1500$$
$$5x_1 + 3x_2 \leq 1575 \qquad (9.4)$$
$$x_1 + 2x_2 \leq 420$$
$$x_1 \geq 0$$
$$x_2 \geq 0$$

The model above is typical of most linear programming models. A specific objective (i.e., income) has been chosen from the many possible objectives that may pertain to the situation, such as cost of production or full use of machines. In addition, the number of constraint statements that can be formulated is not rigorously prescribed, but depends on the perception of the modeler of features in the real problem. This emphasizes the creative aspects of linear programming modeling.

9.3.2 A Blending Problem

General problem.   Blending problems refer to situations in which a number of components (or commodities) are mixed together to yield one or more products. Typically, different commodities are to be purchased. Each commodity has known characteristics and costs. The problem is to determine how much of each commodity should be purchased and blended with the rest so that the characteristics of the mixture lie within specified bounds and the total cost is minimized.

Example problem.   The Sunnyflush Company has two plants located along a stream, as shown in Figure 9.1. Plant 1 is generating 20 units of pol-

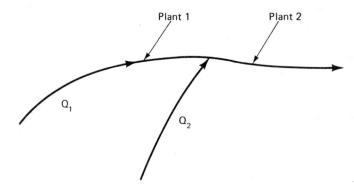

Figure 9.1   Location of Sunnyflush Company's plants.

lutants daily and plant 2 is generating 14 units. Before the wastes are discharged into the stream, part of these pollutants are removed by a waste treatment facility in each plant. The costs associated with removing a unit of pollutant are $1000 and $800 for plants 1 and 2, respectively. The rates of flow in the streams are Q_1 equal to 5 million gallons per day (mgd) and Q_2 equal to 2 mgd, and the flows contain no pollutants until they pass the plants. Stream standards require that the number of units of pollutants per million gallons of flow should not exceed 2. Twenty percent of the pollutants entering the stream at plant 1 will be removed by natural processes (i.e., oxidation by the air) before they reach plant 2. The Sunnyflush Company wants to determine the most economical operation of its waste treatment facilities that will allow it to satisfy the stream standards.

Formulating the problem.   The most economic operation of the waste treatment plants would be the operation that minimizes the total cost of removing part of the pollutants from the wastes of both plants. The decisions to be made, then, are how many units of pollutants are to be removed by each waste treatment facility. Let x_1 and x_2 be the number of units of pollutants to be removed by the waste treatment facilities of plants 1 and 2, respectively. The total cost of removing the pollutants is $1000x_1 + 800x_2$ and is the quantity to be minimized. If x_1 units of pollutants are removed from the wastes at plant 1, $20 - x_1$ units of pollutants are released into the stream. This quantity cannot be greater than the quantity of pollutants permitted by the stream standards, which is two times the flow rate Q_1 (i.e., 10 units). Thus the first constraint is

$$20 - x_1 \le 10$$

The second point of concern is at plant 2. Here the amount of pollutants released is $14 - x_2$ units. In addition, at the outlet from plant 2 there are pollutants from plant 1. Since 20 percent of the pollutants released at plant 1 are removed by natural processes, 80 percent remains by the time the flow reaches plant 2. Therefore, the total amount of pollutants in the stream at plant 2 is $0.8(20 - x_1) + (14 - x_2)$. This quantity must not be greater than the stream standards allow. Thus the second constraint is

$$0.8(20 - x_1) + (14 - x_2) \le 14$$

where the right-hand side, 14, is 2 times the combined flow rate of Q_1 and Q_2. Since the waste treatment facilities cannot remove a negative amount, x_1 and x_2 must each be zero or positive. Furthermore, the treatment facilities cannot remove more pollutants than exist in the waste, which means that x_1 must be less than or equal to 20 and x_2 must be less than or equal to 14. The linear programming model for this example can be summarized as:

Minimize

$$Z = 1000x_1 + 800x_2$$

subject to:

$$20 - x_1 \leq 10$$
$$0.8(20 - x_1) + (14 - x_2) \leq 14$$
$$x_1 \leq 20$$
$$x_2 \leq 14 \tag{9.5}$$
$$x_1 \geq 0$$
$$x_2 \geq 0$$

The constraints can further be simplified to

$$0.8x_1 + x_2 \geq 16$$
$$20 \geq x_1 \geq 10 \tag{9.6}$$
$$14 \geq x_2 \geq 0$$

9.3.3 A Production Scheduling Problem

General problem. A manufacturer knows that he must supply a given number of items of a certain product each month for the next n months. They can be produced either in regular time, subject to a maximum each month, or in overtime. The cost of producing an item during overtime is greater than during regular time. A storage cost is associated with each item not sold at the end of the month. The problem is to determine the production schedule that minimizes the sum of production and storage costs.

Example problem. The Hard Rock Company provides gravel from its quarry for concrete mixing companies. During the next 4 months its sales, costs, and available time are expected to be those given in Table 9.2. There is

TABLE 9.2 Sales, Costs, and Available Time for Hard Rock Company

	Month			
	1	2	3	4
Gravel required (1000 yd³)	1000	2500	2100	2900
Cost regular time ($/1000 yd³)	100	100	110	110
Cost overtime ($/1000 yd³)	110	116	120	124
Regular operation time of rock crushers (hours)	2400	2400	2400	2400
Overtime (hours)	990	990	990	990

no gravel in stock at the beginning of the first month. It takes 1.5 hours of production time to produce 1000 cubic yards (yd^3) of gravel. It costs \$5 to store 1000 yd^3 of gravel from one month to the next. The company wants a production schedule that does not exceed the production time limitations each month, that meets the demand requirements, and that minimizes total cost.

Formulating the problem.   Let x_1, x_2, x_3, and x_4 be the number of units of gravel produced in months 1, 2, 3, and 4, respectively, on regular time where a unit of gravel is 1000 yd^3. Also, let x_5, x_6, x_7, and x_8 be the number of units of gravel produced during months 1, 2, 3, and 4, respectively, on overtime, and let x_9, x_{10}, and x_{11} be the number of units of gravel in stock at the end of months 1, 2, and 3, respectively. This assumes that there is to be no gravel left in stock at the end of month 4. The objective is to minimize costs. Therefore, the company wants to minimize

$$Z = 100x_1 + 100x_2 + 110x_3 + 110x_4 + 110x_5 + 116x_6$$
$$+ 120x_7 + 124x_8 + 5x_9 + 5x_{10} + 5x_{11}$$

subject to demand requirements for gravel sales and production time limitations. In order to meet sales demands in the first month, the total number of units of gravel produced ($x_1 + x_5$) must be equal to the sum of the sales demand (1000 yd^3) and the number of units in storage at the end of the month, x_9. Thus the first constraint is

$$x_1 + x_5 = 1000 + x_9$$

During the second month, the quantity in stock at the beginning of the month, x_9, plus the quantity produced, $x_2 + x_6$, must be equal to the sum of the quantity on stock at the end of the month, x_{10}, and the sales demand of 2500 units. Hence the second constraint is

$$x_9 + x_2 + x_6 = 2500 + x_{10}$$

Similarly, the third and fourth constraints are

$$x_{10} + x_3 + x_7 = 2100 + x_{11}$$
$$x_{11} + x_4 + x_8 = 2900$$

In addition, the production time limitations cannot be violated. Therefore,

$$1.5x_i \leq 2400 \qquad \text{where } i = 1, 2, 3, \text{ and } 4$$
$$1.5x_i \leq 990 \qquad \text{where } i = 5, 6, 7, \text{ and } 8$$

Finally, a negative quantity of gravel cannot be produced at any time, and thus each x_i must be greater than or equal to zero.

The linear programming model for this problem can be summarized as

Minimize
$$Z = 100x_1 + 100x_2 + 110x_3 + 110x_4 + 110x_5 + 116x_6$$
$$+ 120x_7 + 124x_8 + 5x_9 + 5x_{10} + 5x_{11}$$

subject to:

$$x_1 + x_5 - x_9 = 1000$$
$$x_9 + x_2 + x_6 - x_{10} = 2500$$
$$x_{10} + x_3 + x_7 - x_{11} = 2100 \tag{9.7}$$
$$x_{11} + x_4 + x_8 = 2900$$
$$1.5x_i \leq 2400 \quad \text{where } i = 1, 2, 3, \text{ and } 4$$
$$1.5x_i \leq 990 \quad \text{where } i = 5, 6, 7, \text{ and } 8$$
$$x_i \geq 0 \quad \text{where } i = 1, 2, \ldots, 11$$

A linear graph model can be developed to illustrate the derivation of the constraint operation. In Figure 9.2, for example, the nodes represent months and the branches storage stocks. Nodal inputs and outputs model crusher production and sales. Continuity statements at nodes and capacity conditions produce the system constraints. A formal mathematical solution to Equation 9.7 is possible, as indicated later, but the reader may develop a graphical solution based on Figure 9.2.

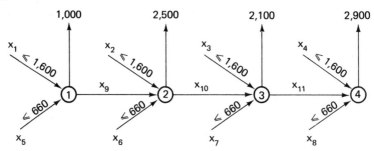

Figure 9.2   Gravel flow model.

9.3.4 A Transportation Problem

General problem. A product is to be shipped in the amounts a_1, $a_2, \ldots, a_m$ from m shipping origins and received in amounts $b_1, b_2, \ldots, b_n$ at each of n shipping destinations. The cost of shipping a unit from the ith origin to the jth destination is known for all combinations of origins and destinations. The problem is to determine the amount to be shipped from each origin to each destination such that the total cost of transportation is a minimum.

Example problem. The Cheap Concrete Company has three sand pits from which it can obtain sand for its two concrete plants. Plant 1 requires 24 units of sand, and plant 2 requires 28 units. Unfortunately, the sand pits are almost depleted. Pit 1 has only 8 units of sand left, pit 2 has 23 units, and pit 3 has 21 units. It costs $19 to haul 1 unit of sand from pit 1 to plant 1. The remaining haul costs are given in Table 9.3. The company wants to determine how much to haul from each pit to each plant to minimize its total hauling cost.

TABLE 9.3   Hauling Costs

	Plant	
Pit	1	2
1	$19	$12
2	6	9
3	7	18

Formulating the problem.   Let $x_{i,j}$ be the number of units of sand hauled from the ith pit to the jth plant. To minimize hauling costs, the objective function is

Minimize
$$Z = 19x_{1,1} + 12x_{1,2} + 6x_{2,1} + 9x_{2,2} + 7x_{3,1} + 18x_{3,2} \qquad (9.8)$$

subject to availability and demand constraints. The quantity hauled from pit 1, $x_{1,1} + x_{1,2}$, cannot be greater than the amount available at pit 1; that is,

$$x_{1,1} + x_{1,2} \leq 8$$

Similarly, for pits 2 and 3,

$$x_{2,1} + x_{2,2} \leq 23$$
$$x_{3,1} + x_{3,2} \leq 21 \qquad (9.9)$$

The total amount hauled from the three pits to plant 1, that is, $x_{1,1} + x_{2,1} + x_{3,1}$, must equal the quantity needed at plant 1, that is,

$$x_{1,1} + x_{2,1} + x_{3,1} = 24$$

Similarly, at plant 2, $\qquad (9.10)$

$$x_{1,2} + x_{2,2} + x_{3,2} = 28$$

Finally, each $x_{i,j}$ must be positive or zero since it is impossible to haul a negative quantity.

Special cases of the transportation problem include the transshipment and the assignment problems. Transshipment can occur in the distribution system of a company that has regional warehouses that ship to smaller district warehouses, which in turn ship to final destinations. The transshipment problem is a direct extension of the transportation problem and permits each source or distination to act as an intermediate point for shipments from other sources to other destinations. The assignment problem occurs when n individuals or machines must be assigned to perform n different jobs. Each individual has a rating of effectiveness on each job and the objective is to maximize the total effectiveness for all n jobs.

Although the transportation, transshipment, and assignment problems can be formulated as linear programming models, more efficient algorithms exist for solving these problems (Dantzig, 1963; Gass, 1975; Llewellyn, 1964; Wagner, 1975).

9.3.5 A Flow-Capacity Problem

General problem. One or more commodities (e.g., traffic, water, information, cash, etc.) are flowing from one point to another through a network whose branches have various constraints and flow capacities. The direction of flow in each branch and the capacity of each branch are known. The problem is to determine the maximum flow, or capacity of the network.

Example problem. Consider again the traffic flow problem of Section 7.6 as shown in Figure 7.12. The engineer can use the linear graph procedures of Sections 7.3 and 7.6 or the following model to determine the current flow capacity.

The network capacity model is developed for current flow capacity based on the following assumptions:

1. Traffic flow from A to B is independent and equal to flow from B to A.
2. Traffic flow direction on a specific road is known, as shown in Figure 7.14.
3. The traffic flow originating or terminating at the towns between A and B is insignificant.

Formulating the problem. Let

x_i = actual flow in road i (i.e., branch flow)
c_i = capacity of road i (i.e., branch capacity)
y_i = imposed flow at node i (i.e., flow originating or terminating at node i); $y_i = 0$ for $i = 2, 3, \ldots, 7$ from assumption 3 above

Therefore, the maximum flow will occur in the network when y_1 is a maximum. The value of y_1 is limited, however, by two constraints:

1. The capacity of each road cannot be exceeded.
2. The flow at each node must be in equilibrium.

The capacity constraints on the existing roads can be determined from Figure 7.12. They are:

$$\text{for existing roads} \begin{cases} \text{four-lane} \begin{cases} 0 \leq x_1 \leq 2000 \\ 0 \leq x_2 \leq 2000 \\ 0 \leq x_3 \leq 2000 \\ 0 \leq x_4 \leq 2000 \end{cases} \\ \text{two-lane} \begin{cases} 0 \leq x_5 \leq 500 \\ 0 \leq x_6 \leq 500 \\ 0 \leq x_7 \leq 500 \\ 0 \leq x_8 \leq 500 \\ 0 \leq x_9 \leq 500 \\ 0 \leq x_{10} \leq 500 \end{cases} \end{cases} \quad (9.11)$$

In matrix form,

$$\mathbf{0}_{(10 \times 1)} \leq \mathbf{X}_{(10 \times 1)} \leq \mathbf{C}_{(10 \times 1)}$$

where $\mathbf{X}$ = column matrix of road traffic flow

$\mathbf{C}$ = column matrix of road flow capacities and represents a physical behavior property of the road components

The second constraint means that the actual road flows, x_i, at the nodes must balance at road junctions because of continuity requirements.

If the flow into a node is considered as positive, then at source A, node 1, the continuity equation must relate the network nodal flow, y_1, to the road flows x_1, x_5, and x_6. Furthermore, the direction of traffic flows is known; hence

$$y_1 - x_1 - x_5 - x_6 = 0$$

or

$$x_1 + x_5 + x_6 = y_1 \quad (9.12)$$

Similarly, for

Node 2:	$-x_2 + x_5$	$= 0$	
Node 3:	$+x_1 - x_7$	$= 0$	
Node 4:	$-x_3 + x_6 - x_8$	$= 0$	
Node 5:	$+x_2 - x_9$	$= 0$	(9.13)
Node 6:	$+x_3 - x_4 + x_7$	$= 0$	
Node 7:	$+x_8 - x_{10}$	$= 0$	
Node 8:	$+x_4 + x_9 + x_{10}$	$= y_8$	

Equations 9.13 can be expressed in matrix form as

$$\mathbf{A}^T_{(7 \times 10)} \mathbf{X}_{(10 \times 1)} = \mathbf{Y}_{(7 \times 1)}$$

where $\mathbf{Y}$ = column matrix of imposed nodal flows exclusive of y_1

$\mathbf{A}$ = branch–node incidence matrix for Figure 7.14.

The flow-capacity model can be summarized in matrix form as:

Maximize

$$y_1 = x_1 + x_5 + x_6 \qquad (9.14)$$

subject to:

$$\mathbf{0}_{(10 \times 1)} \leq \mathbf{X}_{(10 \times 1)} \leq \mathbf{C}_{(10 \times 1)} \qquad (9.15)$$

$$\mathbf{A}^T_{(7 \times 10)}\mathbf{X}_{(10 \times 1)} = \mathbf{Y}_{(7 \times 1)} \qquad (9.16)$$

Equations 9.14, 9.15, and 9.16 constitute a linear programming model with 10 flow variables $x_1, \ldots, x_{10}$. The values of these variables must be determined to obtain the solution. Computer programs are available on most computer systems that can solve these equations in seconds of computer processor time. One of these algorithms is explained below.

9.4 GRAPHICAL SOLUTION
TO LINEAR PROGRAMMING PROBLEM

In this day of the high-speed digital computer, the graphical method of solution is still useful for small problems involving two decision variables, and it also provides clues to a systematic method of solving linear programming problems that have more than two decision variables.

Consider the following linear programming model, which was formulated in Section 9.3.1 as a product-mix problem.

Maximize

$$Z = 13x_1 + 11x_2$$

subject to:

$$
\begin{aligned}
4x_1 + 5x_2 &\leq 1500 \\
5x_1 + 3x_2 &\leq 1575 \\
x_1 + 2x_2 &\leq 420 \\
x_1 \geq 0, \qquad x_2 &\geq 0
\end{aligned}
\qquad (9.17)
$$

Any pair of values of x_1 and x_2 that satisfies the set of inequalities is a feasible solution to the problem. An optimal solution is a feasible solution that also maximizes the objective function.

Since the problem consists of only two variables, x_1 and x_2, a graphical method may be used to obtain the solution. An equation of the form $4x_1 + 5x_2 = 1500$ defines a straight line in the x_1, x_2 plane. An inequality defines an area bounded by a straight line. Therefore, the region below and including the line $4x_1 + 5x_2 = 1500$ in Figure 9.3 represents the region defined by $4x_1 + 5x_2 \leq 1500$. The region below and including the line $5x_1 + 3x_2 = 1575$ in Figure 9.3 represents the region defined by $5x_1 + 3x_2 \leq 1575$. Similarly, the region defined by $x_1 + 2x_2 \leq 420$ is represented in Figure 9.3 by the region below and including the line $x_1 + 2x_2 = 420$. The shaded area of Figure 9.3 comprises

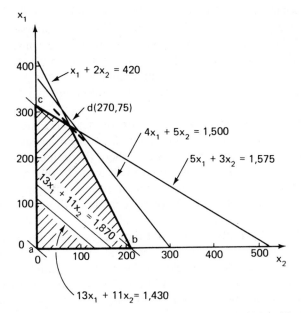

Figure 9.3   Graphical solution for linear programming problem.

the area common to all the regions defined by the constraints and contains all pairs of x_1 and x_2 that are feasible solutions to the problem. This area is known as the *feasible region* or *feasible solution space*. The optimal solution must lie within this region.

There are various pairs of x_1 and x_2 that satisfy the constraints. One pair, $x_1 = 0$ and $x_2 = 0$, satisfies the constraints and provides an income equal to 0. The pair of variables in a solution can be written in vector form.

$$X = \begin{bmatrix} x_1 \\ x_2 \end{bmatrix} = \begin{bmatrix} 0 \\ 0 \end{bmatrix}$$

Various pairs of x_1 and x_2 lead to the same income. An income of \$1430, for instance, would be obtained for all pairs of values of x_1 and x_2 satisfying the equation $13x_1 + 11x_2 = 1430$, that is, lying on the line representing this equation in Figure 9.3. The line for any other value of the income would be a line parallel to $13x_1 + 11x_2 = 1430$, such as $13x_1 + 11x_2 = 1870$. The line representing income would be further from the origin for a larger income, but would be nearer the origin for a smaller income. The maximum value of the objective function that satisfies all the constraints is determined by finding the point that, although within the feasible region, lies on a line parallel to the line $13x_1 + 11x_2 = 1430$ and is as far from the origin as possible. From Figure 9.3 it can be seen that point *d* is such a point.

The optimal solution from Figure 9.3 is $X = \begin{bmatrix} 270 \\ 75 \end{bmatrix}$ and means that a maximum income of \$4335 is obtained by producing 270 units of product I and 75 units of product II. In this solution, all the raw material and available time are used, because the optimal point lies on the two constraint lines for these resources. However, $1500 - [4(270) + 5(75)]$, or 45 ft² of storage space, is not used. Thus the storage space is not a constraint on the optimal solution; that is, more products could be produced before the company ran out of storage space. Thus this constraint is said to be *slack*.

The graphical concepts associated with the feasible space of a problem gives some insight into why linear programming problems are generally over- or underdetermined sets of inequations (i.e., where the number of variables are more than or less than the number of inequations). In this case a constraint inequation is produced for the three relationships that can be determined between x_1 and x_2; two of these effectively assist in defining the feasible space, one is redundant.

It may be that a problem does not have a feasible solution space. If the problem above had another constraint such that $2x_1 + x_2 \geq 800$, there would be no pair of x_1 and x_2, which would satisfy all the constraints.

If there is a feasible region for a linear programming problem, it may be represented graphically by a polygon. Notice in Figure 9.3, that a linear objective function must attain its maximum value at an extreme point of such a region. The feasible region of Figure 9.3 is convex; that is, all internal angles are less than 180 degrees. The graphical method of solution can also be applied to non-convex regions; that is, regions which have at least one interior angle that is greater than 180 degrees, but the simplex method described below is only applicable to convex regions.

If the objective function happens to be parallel to one of the edges of the feasible region, any point along this edge between the two extreme points may be an optimal solution that maximizes the objective function. When this occurs, there is no unique solution, but there is an infinite number of optimal solutions.

The graphical method of solution may be extended to a case in which there are three variables. In this case, each constraint is represented by a plane in three dimensions, and the feasible region bounded by these planes is a polyhedron. As in the two-variable case, the objective function attains its maximum value at an extreme point of the three-dimensional feasible region. Although the graphical method of solution is of no practical value when the number of decision variables exceeds three, it does provide clues to the simplex method.

9.5 THE SIMPLEX METHOD

The simplex method is not used to examine all the feasible solutions. It deals only with a small and unique set of feasible solutions, the set of vertex points (i.e., extreme points) of the convex feasible space that contains the optimal solution. It considers the members of this set one at a time, and ends when the optimum has been achieved. The simplex method involves the following steps:

1. Locate an extreme point of the feasible region.
2. Examine each boundary edge intersecting at this point to see whether movement along any edge increases the value of the objective function.
3. If the value of the objective function increases along any edge, move along this edge to the adjacent extreme point. If several edges indicate improvement, the edge providing the greatest rate of increase is selected.
4. Repeat steps 2 and 3 until movement along any edge no longer increases the value of the objective function.

9.5.1 The Simplex Method Applied

Again, consider the product-mix problem represented in Section 9.3.

Step 1.   The first step in the simplex solution is to convert all the inequality constraints into equalities by the use of slack variables. Let

$$S_1 = \text{unused storage space}$$
$$S_2 = \text{unused raw materials}$$
$$S_3 = \text{unused production time}$$

Introducing these slack variables into the inequality constraints and rewriting the objective function such that all variables are on the left-hand side of the equation, Equation 9.4 can be expressed as

$$
\begin{array}{lllll}
Z - 13x_1 - 11x_2 & & & = & 0 & \text{(A1)} \\
4x_1 + 5x_2 + S_1 & & & = & 1500 & \text{(B1)} \\
5x_1 + 3x_2 & + S_2 & & = & 1575 & \text{(C1)} \\
x_1 + 2x_2 & & + S_3 & = & 420 & \text{(D1)} \\
\end{array}
\qquad (9.18)
$$

$$x_i \geq 0, \quad i = 1, 2$$
$$S_j \geq 0, \quad j = 1, 2, 3$$

From the equations above, it is obvious that one feasible solution that satisfies all the constraints is: $x_1 = 0$, $x_2 = 0$, $S_1 = 1500$, $S_2 = 1575$, $S_3 = 420$, and $Z = 0$.

It is also obvious from Equation A1 that this is not the optimum solution. Since the coefficients of x_1 and x_2 in Equation A1 are both negative, the value of Z can be increased by giving either x_1 or x_2 some positive value in the solution. Since x_1 has a smaller negative coefficient (i.e., larger income value), it is intuitively obvious that a reasonable strategy would be to give x_1 as large a positive value as possible in the next trail solution.

In Equation B1, if $x_2 = S_1 = 0$, then $x_1 = 1500/4 = 375$. That is, there is only sufficient storage space to produce 375 units at product I. Similarly, from

Equation C1, there is only sufficient raw materials to produce $1575/5 = 315$ units of product I; and from Equation D1, there is only sufficient time to produce $420/1 = 420$ units of product I. Therefore, considering all three constraints, there is sufficient resource to produce only 315 units of x_1.

Thus the maximum value of x_1 is limited by Equation C1. It is obvious from this equation and from the analysis above that another feasible solution to the optimization problem is $x_1 = 315$, $x_2 = 0$, and $S_2 = 0$. Is this the optimum solution? What are the values for S_1, S_3, and Z in this solution? To answer these questions, some further manipulation of Equation 9.18 is necessary.

Step 2. From Equation C1, which limits the maximum value of x_1,

$$x_1 = -\frac{3}{5}x_2 - \frac{1}{5}S_2 + 315 \tag{9.19}$$

Substituting this equation into Equation 9.18 yields the following new formulation of the model:

$$
\begin{aligned}
Z - \frac{16}{5}x_2 \quad && + \frac{13}{5}S_2 \quad && = 4095 \quad (A2) \\
+ \frac{13}{5}x_2 + S_1 && - \frac{4}{5}S_2 \quad && = 240 \quad (B2) \\
x_1 + \frac{3}{5}x_2 \quad && + \frac{1}{5}S_2 \quad && = 315 \quad (C2) \\
\frac{7}{5}x_2 \quad && - \frac{1}{5}S_2 + S_3 && = 105 \quad (D2)
\end{aligned}
\tag{9.20}
$$

It is now obvious from these equations that the new feasible solution is: $x_1 = 315$, $x_2 = 0$, $S_1 = 240$, $S_2 = 0$, $S_3 = 105$, and $Z = 4095$. It is also obvious from Equation A2 that it is also not the optimum solution. The coefficient of x_2 in the objective function represented by A2 is negative $(-16/5)$, which means that the value of Z can be further increased by giving x_2 some positive value.

Again, following the same analysis procedure used in step 1, it is clear from constraint Equation B2 that x_2 can take on the value of $(5/13)(240) = 92.3$ if $S_1 = S_3 = 0$. Similarly, from constraint Equation C2, x_2 can take on the value $(5/3)(315) = 525$ if $x_1 = S_2 = 0$. And from Equation D2, x_2 can take on the value $(5/7)(105) = 75$ if $S_2 = S_3 = 0$. Therefore, constraint $D2$ limits the maximum value of x_2 to 75.

Thus a new feasible solution includes $x_2 = 75$, $S_2 = S_3 = 0$. Again, to check the optimality of the new solution, x_2 is first expressed as function S_2 and S_3 from Equation D2 and then substituted into the remaining equations.

Step 3. From Equation D2,

$$x_2 = \frac{1}{7}S_2 - \frac{5}{7}S_3 + 75 \tag{9.21}$$

Substituting this expression into Equation 9.20 yields

$$Z \qquad + \frac{15}{7}S_2 + \frac{16}{7}S_3 = 4335 \quad (A\,3)$$

$$S_1 - \frac{3}{7}S_2 - \frac{13}{7}S_3 = \quad 45 \quad (B\,3)$$

$$x_1 \qquad + \frac{2}{7}S_2 - \frac{3}{7}S_3 = \quad 270 \quad (C\,3) \qquad (9.22)$$

$$x_2 \qquad - \frac{1}{7}S_2 + \frac{5}{7}S_3 = \quad 75 \quad (D\,3)$$

From these equations, the new feasible solution is readily found to be: $x_1 = 270$, $x_2 = 75$, $S_1 = 45$, $S_2 = 0$, $S_3 = 0$, $Z = 4335$.

Because the coefficients in the objective function represented by Equation A3 are all positive, this new solution is also the optimum solution. If either S_2 or S_3 takes on a positive value, the value of Z would decrease.

Referring to the graphical solution in Figure 9.3, it is seen that the solution procedure above initially identifies point a in the graph as a feasible solution. The feasible solution in step 2 is actually point c, and the solution in step 3 is point d. Thus the procedure moves from extreme point to extreme point until an optimum solution is achieved.

9.5.2 Simplex Tableau for Maximization Problem

Because the solution procedure of the simplex method is repetitive in nature, the calculations can be performed in a table, called the simplex tableau. Table 9.4 illustrates a solution of the product-mix problem using the simplex tableau.

The solution procedure using the tableau may be summarized as follows.

Step 1.   Set up the initial tableau using Equation 9.18. Notice that a column is specified for basic variables. In any iteration, a variable that has a nonzero value in the solution is called a *basic variable*. On the other hand, a variable that has a zero value in a solution is called a *nonbasic variable*. There are usually as many basic variables as there are constraints in the problem model. In the present problem, there are three constraints involving five variables $(x_1, x_2, S_1, S_2, \text{and } S_3)$. Thus a feasible solution will include three basic variables and two nonbasic variables.

Step 2.   Identify the variable that will be assigned a nonzero value in the next iteration so as to increase the value of the objective function. That is, identify the variable that will enter the next solution as a basic variable. This variable is called the *entering variable*. It is that nonbasic variable which is associated with the smallest negative coefficient in the objective function. If

TABLE 9.4　Summary of Simplex Tableau Procedure for Maximization Problem

Row Number	Basic Variable	Coefficients of:						Right-Hand Side	Upper Bound on Entering Variable
		Z	x_1	x_2	S_1	S_2	S_3		
Initial tableau									
A1	Z	1	-13	-11	0	0	0	0	
B1	S_1	0	4	5	1	0	0	1500	375
C1	S_2	0	5	3	0	1	0	1575	315
D1	S_3	0	1	2	0	0	1	420	420
Second tableau at end of first iteration									
A2	Z	1	0	$-\frac{16}{5}$	0	$+\frac{13}{5}$	0	4095	
B2	S_1	0	0	$\frac{13}{5}$	1	$-\frac{4}{5}$	0	240	92.3
C2	x_1	0	1	$\frac{3}{5}$	0	$\frac{1}{5}$	0	315	525
D2	S_3	0	0	$\frac{7}{5}$	0	$-\frac{1}{5}$	1	105	75
Third tableau at end of second and final iteration									
A3	Z	1	0	0	0	$+\frac{15}{7}$	$+\frac{16}{7}$	4335	
B3	S_1	0	0	0	1	$-\frac{3}{7}$	$-\frac{13}{7}$	45	
C3	x_1	0	1	0	0	$\frac{2}{7}$	$-\frac{3}{7}$	270	
D3	x_2	0	0	1	0	$-\frac{1}{7}$	$\frac{5}{7}$	75	

two or more nonbasic variables are tied with the smallest coefficients, select one of these arbitrarily and continue.

In the initial tableau of Table 9.4, x_1 has a coefficient of -13 and x_2 has a coefficient of -11. Thus x_1 is the entering variable in the next iteration. It is helpful to circle the column of the entering variable as shown in Table 9.4.

Step 3.　Identify the variable, called the *leaving variable*, which will be changed from a nonzero to a zero value in the next solution, that is, the variable that will be changed from being a basic variable to a nonbasic variable. The upper bound of the entering variable (x_1) imposed by each constraint is computed and shown in the last column of the initial tableau. The upper bound for row B1 is $1500/4 = 375$; for row C1, it is $1575/5 = 315$; and for row D1, it is $420/1 = 420$. Since 315 is the smallest positive upper bound, row C1 is the limiting constraint and S_2 is the leaving variable. Again, it is helpful to identify row C1 by circling it.

In computing the upper bounds, consider those constraint equations in which the coefficients of the entering variable are positive. If all the coefficients in the column of the entering variable are negative, the problem has no feasible solution. A tie between two or more leaving variables indicates a degenerate solution (see Section 9.6).

Step 4.   Enter the basic variables for the second tableau. The row sequence of the previous tableau should be maintained, with the leaving variable being replaced by the entering variable. Thus S_1 is in row B2, x_1 takes the place of S_2 in row C2, and S_3 is in row D2.

Step 5.   Compute the coefficients for the second tableau. A sequence of operations will be performed so that at the end the x_1 column in the second tableau will have the following coefficients:

	x_1
Z	0
S_1	0
x_1	1
S_3	0

That is, it will have a coefficient of 1 in the x_1 row and a coefficient of 0 elsewhere. Remember that x_1 is the entering variable. The row operations proceed as follows:

1. The coefficients in row C2 are obtained by dividing the corresponding coefficients in row C1 by 5.

2. The coefficients in row A2 are obtained by multiplying the coefficients of row C2 by 13 and adding the products to the corresponding coefficients in row A1.

3. The coefficients in row B2 are obtained by multiplying the coefficients of row C2 by -4 and adding the products to the corresponding coefficients in row B1.

4. The coefficients in row D2 are obtained by multiplying the coefficients of row C2 by -1 and adding the products to the corresponding coefficients in row D1.

The coefficients in the column for upper bound are not included in the row operations above. Furthermore, it can be seen that the coefficients in the second tableau correspond exactly with the coefficients in Equation 9.20. Thus the row operations above serve exactly the same purpose as the operations that were used to transform Equation 9.18 into Equation 9.20.

The second tableau yields the following feasible solution:

$$x_1 = 315, \quad x_2 = 0, \quad S_1 = 240, \quad S_2 = 0, \quad S_3 = 105, \quad \text{and} \quad Z = 4095$$

Step 6. Check for optimality. The second feasible solution is also not optimal, because the objective function (row A2) contains a negative coefficient. Another iteration beginning with step 2 is necessary. In the third tableau in Table 9.4 all the coefficients in the objective function (row A3) are positive. Thus an optimal solution has been reached and it is as follows:

$$x_1 = 270, \quad x_2 = 75, \quad S_1 = 45, \quad S_2 = 0, \quad S_3 = 0, \quad \text{and} \quad Z = 4335$$

9.5.3 Marginal Values of Additional Resources

The simplex solution described above yields the optimum production program for N. Dustrious Company. The company can maximize its sale income to $4335 by producing 270 units of product I and 75 units of product II. There will be no surplus of raw materials or production time. But there will be 45 units of unused storage space. Assuming that the company is making an acceptable profit from its sale income of $4335, its managers are interested to know if it is worthwhile to increase its production by purchasing additional units of raw materials and by either expanding its production facilities or working overtime. The critical questions are: (1) What is the income value (or marginal value) of each additional unit of each type of resources? and (2) What is the maximum cost (or marginal cost) that they should be willing to pay for each additional unit of resources? Answers to these questions can be obtained from the objective function in the last tableau of the simplex solution.

From the third tableau in Table 9.4, the following expression is obtained for the objective function:

$$Z + \frac{15}{7}S_2 + \frac{16}{7}S_3 = \$4335$$

that is,

$$Z = \$4335 - \frac{15}{7}S_2 - \frac{16}{7}S_3$$

Because S_1, S_2, and S_3 represent surplus resources, the negatives of these variables (i.e., $-S_1$, $-S_2$, $-S_3$) represent additional units of these resources that can be made available. The income values (or marginal values of additional units of these resources can be obtained by taking the partial derivatives of Z with respect to $-S_1$, $-S_2$, and $-S_3$. Therefore, the marginal value of one additional unit of

$$\text{Storage space} = \frac{\partial Z}{\partial(-S_1)} = \$0$$

$$\text{Raw materials} = \frac{\partial Z}{\partial(-S_2)} = \$\frac{15}{7}$$

$$\text{Production time} = \frac{\partial Z}{\partial(-S_3)} = \$\frac{16}{7}$$

Thus, the marginal values of additional units of resources can be obtained directly from the coefficients of the objective function in the last tableau of a simplex solution.

The N. Dustrious Company should be willing to pay up to $15/7 for an additional unit of raw materials and $16/7 for an additional unit of production time. If the actual cost of an additional unit (i.e., marginal cost) of these resources are smaller than the marginal value, the company should be able to increase its income by increasing production.

To illustrate the validity of the marginal values computed above, consider, for example, the resources required to produce one additional unit of product I (i.e., x_1). It needs 4 units of storage space (which are already available), 5 additional units of raw materials, and 1 additional unit of production time. The total increase in income can be computed from the marginal values as $5 \times 15/7 + 1 \times 16/7 = \$13$, which is exactly equal to the selling price of 1 unit of product I.

The marginal values above are valid, however, only as long as there is surplus storage space available. Once the production is increased to the stage where all storage space is used up, the marginal value of storage space will no longer be zero and the marginal values of other resources will also change.

9.5.4 Sensitivity Analysis

Additional analysis can be performed to test the sensitivity of the optimum solution with respect to changes of the coefficients in the objective function, coefficients in the constraints inequalities, or the constant terms in the constraints. For example, on the production-mix problem of N. Dustrious Company, the actual selling prices (or market values) of the two products may vary from time to time. Over what ranges can these prices change without affecting the optimality of the present solution? Will the present solution remain the optimum solution if the amount of raw materials, production time, or storage space is suddenly changed because of shortages, machine failures, or other events. It is also possible that the amount of each type of resources needed to produce one unit of each type of product can be either increased or decreased slightly. Will such changes affect the optimal solution? There are relatively simple techniques that can be used to find the answers to these questions. It is beyond the scope of this book to discuss in detail such techniques on sensitivity analysis. But coverage on this topic can be found in most textbooks on linear programming, such as Zionts (1974).

9.6 COMPLICATIONS IN APPLYING THE SIMPLEX METHOD

The solution procedure of the simplex method is based on the standard form of the linear programming model, which is represented by Equation 9.3. The simplex method will require modifications for use with forms other than the standard form, although the basic concept of the simplex method will remain the same. Complications that may be encountered include:

1. An objective function to be minimized instead of maximized.
2. Greater-than-or-equal-to constraints.
3. Equalities instead of inequalities for constraints.
4. Decision variables unrestricted in signs.
5. Zero constants on the right-hand side of one or more constraints.
6. Some or all decision variables must be integers.
7. Nonpositive constants on the right-hand side of the constraints.
8. More than one optimal solution, that is, multiple solutions such that there is no unique optimal solution.
9. The constraints are such that no feasible solution exists.
10. The constraints are such that one or more of the variables can increase without limit and never violate a constraint (i.e., the solution is unbounded).
11. Some or all of the coefficients and right-hand-side terms are given by a probability distribution rather than a single value.

The modification necessary for overcoming complications 1, 2, 3, and 4 are quite simple and will be discussed in this section. The problems associated with complications 5 and 6 will also be discussed briefly. Detailed treatment of modifications necessary in each of these and other cases of complications can be found in most textbooks on linear programming, such as Dantzig (1963), Gass (1975), Greenberg (1971), and Zionts (1974) and in books on operations research, such as Wagner (1975).

9.6.1 Minimization Problem

In a minimization problem, the objective is to find the values of $x_1, x_2,$ $x_3, \ldots,$ and x_n which minimize instead of maximize a linear function Z; that is,

Minimize

$$Z = c_1 x_1 + c_2 x_2 + c_3 x_3 + \cdots + c_n x_n \qquad (9.23)$$

This objective function can be converted to the standard form of maximization by a simple procedure. Let $Z' = -Z$, that is,

$$Z' = -c_1 x_1 - c_2 x_2 - c_3 x_3 - \cdots - c_n x_n$$

Then, since maximum $Z' =$ minimum (Z), the objective function of Equation 9.23 can then be replaced by the following:

Maximize

$$Z' = -c_1 x_1 - c_2 x_2 - c_3 x_3 - \cdots - c_n x_n \qquad (9.24)$$

After the value Z' has been found by the standard simplex procedure, Z is computed from the relationship

$$Z = -Z'$$

Another approach to solving a minimization problem is by making simple modifications on the simplex procedure. It is recalled that in a maximization problem, an optimum solution is reached when all the nonbasic variables have nonnegative coefficients in row 1 (objective function) of the simplex tableau. Furthermore, the entering variable is that which has the smallest negative cofficient. In the case of a minimization problem, an optimum solution is reached when all the nonbasic variables have nonpositive coefficients in row 1 of the simplex tableau; and the entering variable is that which has the largest positive coefficient in row 1. All the other operations in the simplex method remain unchanged. The justification for these modifications is intuitively obvious.

An example of the solution of a minimization problem will be presented later in this section.

9.6.2 Greater-Than-Or-Equal-To Constraints

In the standard form of the linear programming model, the constraints are all expressed as less than or equal to ($\leq$) a certain amount, that is,

$$\sum_{j=1}^{n} c_{ij} \leq b_i$$

The constant terms b_i may be interpreted as the maximum amount of resource that is available. In many occasions, however, the constraints must specify the lower bounds rather than the upper bounds. The general form of a constraint on lower bounds is

$$\sum_{j=1}^{n} c_{ij} \geq b_i$$

which involves the inequality "greater than or equal to."

When a linear programming model includes one or more greater-than-or-equal-to constraints, some modification of the basic simplex procedure must be made. These modifications can best be illustrated by a numerical example. Consider the following linear programming model:

Minimize
$$P = 1500y_1 + 1575y_2 + 420y_3$$

subject to:
$$4y_1 + 5y_2 + \ y_3 \geq 13$$
$$5y_1 + 3y_2 + 2y_3 \geq 11$$
$$\text{all } y_1 \geq 0$$

To start the solution, slack variables must first be assigned to convert all inequalities to equalities. Let S_1 and S_2 be slack variables. Furthermore, rearrange the objective function so that all the variables are on the left-hand side of the equation. Equation 9.25 can then be transformed to the following form:

$$P - 1500y_1 - 1575y_2 - 420y_3 = 0$$
$$4y_1 + 5y_2 + y_3 - S_1 = 13$$
$$5y_1 + 3y_2 + 2y_3 - S_2 = 11 \qquad (9.26)$$
$$\text{all } y_i \geq 0, \text{ all } S_i \geq 0$$

Because the constraint inequalities are greater-than-or-equal-to ($\geq$), the slack variables must be subtracted form the left-hand side of the constraints. The negative signs for S_1 and S_2 make it no longer feasible to set all the decision variables (i.e., y_1, y_2, y_3) equal to zero as the initial solution. If $y_1 = y_2 = y_3 = 0$, $S_1 = -13$ and $S_2 = -11$, which violates the nonnegativity requirements.

To assure a starting feasible solution, artificial variables can be added to the greater-than-or-equal-to constraints. Let W_1 and W_2 be two artificial variables. The constraints in Equation 9.26 can be rewritten as follows:

$$4y_1 + 5y_2 + y_3 - S_1 + W_1 = 13$$
$$5y_1 + 3y_2 + 2y_3 - S_2 + W_2 = 11 \qquad (9.27)$$

Now, a starting feasible solution can be easily derived from Equation 9.27 as follows:

$$y_1 = y_2 = y_3 = S_1 = S_2 = 0, \quad W_1 = 13, \quad \text{and} \quad W_2 = 11$$

The variables W_1 and W_2 are called *artificial variables* because they have no significant meaning in the problem model. Their only purpose is to provide an easy means of obtaining a feasible solution. Because of this fact, some additional modifications must be made to make sure that all the artificial variables will have zero values in the optimum solution. This is easily achieved by assigning a very large penalty to these variables in the objective function. For example, since the present problem is to minimize the objective function, the penalty is imposed by assigning a significantly large positive coefficient to these variables in the original form of the objective function. The objective function in Equation 9.25 then becomes

Minimize
$$P = 1500y_1 + 1575y_2 + 420y_3 + 5000W_1 + 5000W_2 \qquad (9.28)$$

Because the coefficients of W_1 and W_2 are much larger than the coefficients of y_1, y_2, and y_3, the optimum will favor y_1, y_2, and y_3 and force W_1 and W_2 to have zero values. For a maximum problem, the artificial variables would be assigned a large negative coefficient in the objective function.

Another algebraic step can be taken before entering the simplex tableau. From Equation 9.27,

$$W_1 = 13 - 4y_1 - 5y_2 - y_3 + S_1$$
$$W_2 = 11 - 5y_1 - 3y_2 - 2y_3 + S_2$$

Substituting these expressions into Equation 9.28 and collecting terms yields the following new expression for the objective function:

$$P = -43{,}500y_1 - 38{,}425y_2 - 14{,}580y_3 + 5000S_1 + 5000S_2 + 120{,}000$$

The objective function may now be combined with Equation 9.27 to express the problem model as follows:

$$P + 43{,}500y_1 + 38{,}425y_2 + 14{,}580y_3 - 5000S_1 - 5000S_2 = 120{,}000$$
$$4y_1 + 5y_2 + y_3 - S_1 + W_1 = 13 \tag{9.29}$$
$$5y_1 + 3y_2 + 2y_3 - S_2 + W_2 = 11$$

The coefficients and constants in Equation 9.29 can now be arranged in the initial tableau, as shown in Table 9.5. Notice that in Equation 9.29, the objective function does not include any W_1 or W_2 terms. Because W_1 and W_2 are to be assigned as basic variables in the initial solution, the coefficients of these variables must be zero before a check for optimization can be made.

The complete solution for the minimization problem is given in Table 9.5. It differs from the previous solution for a maximization problem (Table 9.4) only in the criteria used for deciding on the optimality of a feasible solution and for selecting an entering variable. Instead of selecting an entering variable to increase the value of the objective function as in the maximization problem, the new entering variable is selected as the variable that causes the greatest rate of decrease in the objective function. This is accomplished by selecting the variable in row 1 with the largest positive value as the entering variable. The optimal solution is reached when all nonbasic variables have nonpositive coefficients in the objective function row. The other steps in the simplex method remain unchanged.

9.6.3 Equality Constraint

An equality constraint has the following general form:

$$c_1x_1 + c_2x_2 + c_3x_3 + \cdots + c_nx_n = b \tag{9.30}$$

As in the use of a greater-than-or-equal-to constraint, an artificial variable must be assigned to each equality constraint to start the simplex solution. Otherwise, the constraint would be violated when all the decision variables are assumed to be zero.

TABLE 9.5 Summary of Simplex Tableau Procedure for Minimization Problem

Row Number	Basic Variable	P	y_1	y_2	y_3	S_1	S_2	W_1	W_2	Right-Hand Side	Upper Bound on Entering Variable
Initial tableau											
A1	P	1	43,500	38,425	14,580	$-5,000$	$-5,000$	0	0	120,000	
B1	y_6	0	4	5	1	-1	0	1	0	13	$\frac{13}{4}$
C1	y_7	0	5	3	2	0	-1	0	1	11	$\frac{11}{5}$
Second tableau											
A2	P	1	0	12,325	$-2,820$	$-5,000$	3,700	0	$-8,700$	24,300	
B2	y_6	0	0	$\frac{13}{5}$	$-\frac{3}{5}$	-1	$\frac{4}{5}$	1	$-\frac{4}{5}$	$\frac{21}{5}$	$\frac{21}{13}$
C2	y_1	0	1	$\frac{3}{5}$	$\frac{2}{5}$	0	$-\frac{1}{5}$	0	$\frac{1}{5}$	$\frac{11}{5}$	$\frac{11}{3}$
Third tableau											
A3	P	1	0	0	24.23	-259.62	-92.31	$-4,740.38$	$-4,907.69$	4,390.38	
B3	y_2	0	0	1	$-\frac{3}{13}$	$-\frac{5}{13}$	$\frac{4}{13}$	$\frac{5}{13}$	$-\frac{4}{13}$	$\frac{21}{13}$	
C3	y_1	0	1	0	$\frac{7}{13}$	$\frac{3}{13}$	$-\frac{5}{13}$	$-\frac{3}{13}$	$\frac{5}{13}$	$\frac{16}{13}$	$\frac{16}{7}$
Fourth tableau											
A4	P	1	-45	0	0	-270	-75	$-4,730$	$-4,925$	4,335	
B4	y_2	0	$\frac{3}{7}$	1	0	$-\frac{2}{7}$	$\frac{1}{7}$	$\frac{2}{7}$	$-\frac{1}{7}$	$\frac{15}{7}$	
C4	y_3	0	$\frac{13}{7}$	0	1	$\frac{3}{7}$	$-\frac{3}{7}$	$-\frac{3}{7}$	$\frac{5}{7}$	$\frac{16}{7}$	

9.6.4 Variables Unrestricted in Signs

Sometimes a decision variable may take on either negative or positive values. It x_i is unrestricted in sign, replace it throughout the model by the difference of two new nonnegative variables:

$$x_j = x'_j - x''_j \qquad \text{where } x'_j \geq 0, \quad x''_j \geq 0$$

because x'_j and x''_j can have any nonnegative values their difference $(x'_j - x''_j)$ can have any value (positive or negative). After substitution the simplex method can proceed with just nonnegative variables. Hillier and Lieberman (1980) and Wagner (1975) discuss other procedures for handling variables unrestricted in sign.

9.6.5 Degenerate Solution

If the number of basic variables is fewer than the number of constraints in a solution, the solution is said to be degenerate. A zero constant term for one or more basic variables in any iteration of the simplex solution would be a clear indication of a degenerate solution. The normal simplex procedure cannot solve a degenerate problem. However, there are methods that can ensure that all linear programming problems, whether degenerate or not, are effectively non-degenerate. References on such methods can be found in most textbooks on linear programming, such as Dantzig (1963) and Zionts (1974).

It is interesting to note here, although without presenting the proof, that there can be no tie for determining the leaving variable as long as the problem is nondegenerate.

9.6.6 Integer and Mixed-Integer Problems

A linear programming problem in which all the decision variables must have integer values is called an *integer programming problem*. A problem in which only some of the decision variables must have integer values is called a *mixed-integer programming problem*. Sometimes, some (or all) of the decision variables must have the value of either 0 or 1. Such problems are then called *zero–one mixed-integer* (or *integer*) *programming problems*.

Occasionally, the simplex method may yield the optimal solution to an integer or mixed-integer problem. However, in general, the simplex method cannot always be depended upon to yield the optimal solution to such problems. In fact, although effective methods have been developed for solving integer and mixed-integer problems, there is not yet available an algorithm that can assuredly always lead to an optimal solution. Special-purpose algorithms are often developed to solve large problems which involve several hundreds of variables by taking advantage of the special structure of the problems. In spite of these shortcomings, integer and mixed-integer programming play a major role in the field of optimization and systems engineering. Further references on the topic can be found in Greenberg (1971), Zionts (1974), and Salkin (1975).

9.7 DUALITY

Associated with every linear programming problem is another linear programming problem which is called the *dual* of the original (or the *primal*) problem. The interrelationship between the dual and the primal problems is of significant importance in both problem interpretation and in the development of efficient computing algorithms for linear programming.

9.7.1 Formulating the Dual Problem

Consider again the production mix problem of N. Dustrious Company. Suppose that the company is considering leasing out the entire production facility to another company, and it must decide on the minimum daily rental price that will be acceptable. This decision problem can also be formulated as a linear programming problem.

Let y_1, y_2, and y_3 represent the unit price of each unit of storage space, raw materials, and production time, respectively. The unit prices are in fact the income values of each unit of resource to the N. Dustrious Company. There are available 1500 ft^2 of storage space, 1575 lb of raw materials, and 420 minutes of production time per day. Thus the total income value (P) of all the available resources may be expressed as follows:

$$P = 1500y_1 + 1575y_2 + 420y_3$$

The objective of the problem is to minimize P subject to the condition that the N. Dustrious Company will earn at least as much income as when it operates the production facility itself. Since the market value (or selling price) of 1 unit of product I is $13 and it requires 4 ft^2 of storage space, 5 lb of raw materials, and 1 minute of production time, the following constraint must be satisfied:

$$4y_1 + 5y_2 + y_3 \geq 13$$

Similarly, it requires 5 ft^2 of storage space, 3 lb of raw materials, and 2 minutes of production time to produce 1 unit of product II, which has a selling price of $11. Therefore, the following constraint must also be satisfied:

$$5y_1 + 3y_2 + 2y_3 \geq 11$$

In addition, the unit prices y_1, y_2, and y_3 must all be greater than or equal to zero. The new linear programming problem may now be summarized as follows:

Minimize

$$P = 1500y_1 + 1575y_2 + 420y_3$$

subject to:

$$4y_1 + 5y_2 + y_3 \geq 13 \qquad (9.31)$$
$$5y_1 + 3y_2 + 2y_3 \geq 11$$
$$\text{all } y_i \geq 0$$

The following interesting observations can now be made:

1. It is intuitively obvious that the minimum daily rental price (P) in this problem must equal the maximum daily sale income (Z) when the com-

pany operates the production facility itself. Therefore, from the solution of the previous maximization problem in Table 9.4, $P = Z = \$4335$. Furthermore, the unit selling prices (y_1, y_2, and y_3) of the resources must be the same as the marginal values for the corresponding resources, which have been computed in Section 9.5. That is, $y_1 = \$0$, $y_2 = \$15/7$, and $y_3 = \$16/7$.

2. The linear programming problem in Equation 9.31 is identical to the example problem in Equation 9.25, the solution of which has already been presented in Table 9.5. From Table 9.5 it is seen that the optimum solution is $P = \$4335$, $y_1 = \$0$, $y_2 = \$15/7$, and $y_3 = \$16/7$, which verifies the observation above. Furthermore, the coefficients (ignoring the negative signs) of the slack variables (S_1 and S_2) in the objective function row of the optimum tableau in Table 9.5 are 270 and 75 for S_1 and S_2, respectively. These numbers, in fact, correspond to the optimum solution for x_1 and x_2 in Table 9.4. Consequently, it can be concluded that the solution to the new minimization problem can be obtained directly from the solution to the previous maximization problem, and vice versa.

3. By comparing the minimization problem in Equation 9.31 with the maximization problem in Equation 9.4, it can be seen that there is a direct correspondence between the coefficients of the two problems. The coefficients in the objective function of the minimization problem are the constant terms in the constraints of the maximization problem, and vice versa. The coefficients in the first constraint of the minimization problem are the coefficients of the first decision variable (x_1) in the maximization problem. The coefficients in the second constraints of the minimization problem are the coefficients of the second decision variable (x_2) in the maximization problem. The reverse is also true.

9.7.2 The Primal–Dual Relationship

If the maximization problem of Equation 9.4 is called the primal problem, then the minimization problem of Equation 9.31 is called the dual problem. The dual can be derived directly from the primal problem as indicated by the following general formulation:

Primal Problem	*Dual Problem*
Maximize	Minimize
$Z = c_1 x_1 + c_2 x_2$	$P = b_1 y_1 + b_2 y_2 + b_3 y_3$
subject to:	subject to:
$k_{11} x_1 + k_{12} x_2 \leq b_1$	$k_{11} y_1 + k_{21} y_2 + k_{31} y_3 \geq c_1$
$k_{21} x_1 + k_{22} x_2 \leq b_2$	$k_{12} y_2 + k_{22} y_2 + k_{32} y_3 \geq c_2$
$k_{31} x_1 + k_{32} x_2 \leq b_3$	all $y_i \geq 0$
all $x_i \geq 0$	

Every linear programming model has a dual associated with it. The dual of the dual problem is the primal. Hence the interpretation of the dual varies from one application to another. The value of the basic variables in the dual solution is equal to the coefficients of the nonbasic variables in row 1 of the primal solution. Also, the optimal values of the basic variables in the primal are the coefficients of the nonbasic variables in row 1 in the dual solution with their signs changed, and the optimal value of Z is equal to the optimal value of P. Hence it is not necessary to solve both the primal and the dual since the solution to one contains the solution to the other. Which method to use is a matter of convenience or computational efficiency since the dual and the primal problems are not always equally easy to solve. Usually, an additional constraint requires more computational effort than an additional variable. Thus, if the primal should have a large number of constraints and relatively few variables, its dual will probably require less computation since the number of variables and constraints are interchanged.

9.7.3 Complete Regularization of the Primal Problem

For the purpose of deriving the dual problem from a primal problem, it is convenient first to completely regularize the primal to the standard form of Equation 9.3. That is, all the constraints should be of the less-than-or-equal-to type for a maximization problem. For example, consider the following primal problem:

Maximize

$$Z = 12x_1 + 4x_2$$

subject to:

$$4x_1 + 7x_2 \leq 56$$
$$2x_1 + 5x_2 \geq 20 \qquad (9.32)$$
$$5x_1 + 4x_2 = 40$$
$$x_1 \geq 0$$
$$x_2 \geq 0$$

The first inequality requires no modification. The second inequality can be changed to the less-than-or-equal-to type by multiplying both sides of the inequality by -1 and reversing the direction of the inequality; that is,

$$-2x_1 - 5x_2 \leq -20 \qquad (9.33)$$

The equality constraint can be replaced by the following two inequality constraints:

$$5x_1 + 4x_2 \leq 40 \qquad (9.34)$$
$$5x_1 + 4x_2 \geq 40 \qquad (9.35)$$

If both of these inequality constraints are satisfied, the original equality constraint is also satisfied. Furthermore, multiplying both sides of the inequality in Equation 9.35 by -1 and reversing direction of the inequality yields

$$-5x_1 - 4x_2 \leq -40 \tag{9.36}$$

Thus the primal problem of Equation 9.32 can be regularized to the following standard form:

Maximize

$$Z = 12x_1 + 4x_2$$

subject to:

$$
\begin{aligned}
4x_1 + 7x_2 &\leq 56 \\
-2x_1 - 5x_2 &\leq -20 \\
5x_1 + 4x_2 &\leq 40 \\
-5x_1 - 4x_2 &\leq -40 \\
x_1 &\geq 0 \\
x_2 &\geq 0
\end{aligned}
\tag{9.37}
$$

The dual of this problem can now be obtained by observation as follows:

Minimize

$$P = 56y_1 - 20y_2 + 40y_3 - 40y_4$$

subject to:

$$
\begin{aligned}
4y_1 - 2y_2 + 5y_3 - 5y_4 &\geq 12 \\
7y_1 - 5y_2 + 4y_3 - 4y_4 &\geq 4 \\
\text{all } y_i &\geq 0
\end{aligned}
\tag{9.38}
$$

9.8 SUMMARY

Linear programming is a technique for optimizing the allocation of resources subject to constraints. The linear programming models developed in this chapter illustrate the variety of problems for which it is suited.

The development of the criterion function and the constraint equations constitute the professional aspect of problem solving. Once the model has been formulated there are algorithms, such as the simplex method, that are available for solving the model. Computer programs are also widely available for solving linear programming problems.

In practice, it is often difficult to formulate a decision problem into an exact numerical model, such as those presented in this chapter. A high degree of generality and simplification is often required to formulate the objective function and all the constraints into linear functions. The coefficients of the constraints as well as the objective function often are chosen to represent average or most probable conditions. Most decision problems also involve more than one single objective. The optimum solution is usually one that best satisfies several conflicting objectives. Therefore, in practice, the value of linear programming is to serve as a high-powered computational tool. With the help of an electronic computer, a highly complex problem involving hundreds of decision variables can be solved in a matter of seconds of computer time. Thus, by trying different formulations involving different objective functions and constraints, it is possible to identify the attractive as well as unattractive solutions and to find the optimum solution with respect to different objectives. In the final analysis, human judgment must be exercised to compare the most attractive alternatives and to arrive at a decision that is the optimum solution for the problem at hand.

Chapter 10 describes some models and solution procedures for those problems that cannot be formulated into all linear functions.

PROBLEMS

P9.1. Listed below are the objective function and constraints. Determine the maximum value of Z subject to the constraints. Work this problem graphically and by the simplex method.

1. $2x + 3y \leq 48$
2. $3x + 2y \leq 42$
3. $5x + y \leq 50$
4. $x - 3y \leq 0$
5. $x \geq 0; y \geq 0$

$$Z = 6x + 2y$$

P9.2. A contractor is planning a job that will require a large amount of gravel and sand. It is estimated that the job will need

Coarse gravel	20,000 yd^3
Fine gravel	29,000 yd^3
Sand	20,000 yd^3

There are two pits from which the contractor can obtain material. The contractor plans to haul from these two pits and separate (screen) on the job the material needed. Analysis shows that the material at each pit has the following composition:

Material	Pit A (%)	Pit B (%)
Coarse gravel	20	30
Fine gravel	14	50
Sand	25	20
Waste	41	0

It costs the contractor $8 per cubic yard for material and hauling from pit A and $16 per cubic yard from pit B. How much material should be hauled from each pit to minimize the hauling cost? What will the total hauling cost be? Which constraint is slack? What does this slack constraint mean in terms of material of the job? Solve this problem graphically and by the simplex method.

P9.3. Develop a mathematical programming model for the scheduling problem shown in Figure 7.1.

P9.4. A manufacturing plant requires 300,000 gallons of water per day. The water is filtered before it is delivered to the plant. The manufacturing process requires that the water be chlorinated and softened before it can be used. Two different brands of water additive contain some amount of a chlorination chemical, SMILO, and a softening agent, LATHERO. One package of the product put out by the Hooker Drug Co. contains 8 lb of LATHERO and 3 lb of SMILO. One package of the Amerikan Kemical Korp. product contains 4 lb of LATHERO and 9 lb of SMILO.

One-hundred and fifty pounds of LATHERO and 100 lb of SMILO are needed daily to maintain the water at an acceptable level of softness and to meet recommended levels of chlorination. A package of the Hooker Drug Co. product costs $8 and a package of the American Kemical Korp. product costs $10. What should be the plant manager's policy to meet the requirements for the water additives at least cost?

P9.5. Assume that the type of occupancy for a high-rise building is to be decided and you are asked to determine the number of units of each type that will maximize the profit. You are to provide the objective function and the constraints for a linear programming problem. The occupancy that you are to consider is the following.

1. First floor will be used for entrance, coffee shop, etc.
2. One floor for offices for the company (second floor)
3. Hotel rooms
4. One-room efficiency apartments
5. Two-bedroom apartments
6. General office space
7. Restaurant
8. Basement (recreational space for occupants plus a meeting room)

How would you approach the problem, what kind of information would you require, and how would you establish the objective function and constraints?

P9.6. A reservoir is designed to provide hydroelectric power and water for irrigation. The outflow or release from the reservoir is used to generate hydroelectric power and after passing through the turbines it can be diverted toward an irrigation project or it can continue to flow downstream. At least 1 unit of water must be left to flow down the stream each month. The reservoir has a capacity of 10 units of water. It currently contains 5 units of water and must contain 5 units of water at the end of the year. The maximum amount that can be released in any month is 7 units. The return for each unit of water released for hydroelectric power and irrigation and the estimated inflow for each month of a year are shown in Table P9.6.

TABLE P9.6

	Jan.	Feb.	Mar.	Apr.	May	June	July	Aug.	Sept.	Oct.	Nov.	Dec.
Inflow (units of water)	2	2	3	4	3	2	2	1	2	3	3	2
Hydroelectric power (10^6/unit of water released)	1.6	1.7	1.8	1.9	2.0	2.0	2.0	1.9	1.8	1.7	1.6	1.5
Irrigation (10^6/unit of water released)	1.0	1.2	1.8	2.0	2.2	2.2	2.5	2.2	1.8	1.4	1.1	1.0

Formulate the linear programming model that would be solved to determine how much water to release each month of the year so that the returns from hydroelectric power and irrigation will be maximized.

P9.7. A concrete transit-mix company owns three plants with capacities and production costs as follows:

Plant Number	Daily Capacity (yd)	Production Cost ($/yd)
I	160	10
II	160	9
III	80	13

The company is under contract to supply concrete for a bridge construction and is scheduled to deliver to the various job sites the following quantities of concrete:

Job Site	Amount (yd)
1. South bank pier	100
2. South bank abutment	40
3. North bank abutment	80
4. North bank pier	140

Based on distance, traffic, and site delays, the following haul costs are estimated:

Haul Cost ($/yd)

Job Site Number	Plant Number		
	I	II	III
1	2	1	3
2	1	2	3
3	2	1	2
4	3	2	1

Schedule tomorrow's production to minimize total cost to the company.

P9.8. Give reasons why you agree or disagree with the following.

    (a) In mathematical programming problem formulation, the decision sequence in model building is first to define the objective function, then the decision variables, and finally find the constraints.

    (b) In general, the number of constraints in a mathematical programming problem formulation is unknown.

    (c) In Section 9.4 the problem solution is not affected by the addition of the new constraint $x_1 + 0.5x_2 \geq 100$.

    (d) In Section 9.4 the problem solution is not affected by the addition of the new constraint $x_1 + 0.5x_2 \leq 100$.

    (e) None of the preceding problems can be solved more efficiently using the dual formulation.

Nonlinear
Programming
10

10.1 DEVELOPING NONLINEAR PROGRAMMING MODELS

Nonlinear programming models are similar to linear programming models except that the objective function and constraining equations are not required to be linear functions of the decision variables.

10.1.1 The Product-Mix Problem Again

In the product-mix problem of Section 9.3, which was formulated as a linear programming model as shown in Equation 9.4, it was assumed that the selling price per unit of product was constant regardless of how many units of a product were produced. Sometimes, however, the price per unit depends on the number of units produced.

Assume that the price per unit of product I decreases as each additional unit is produced and that the price per unit is given by $\$(40. - 0.1x_1)$, where x_1 is the total number of units of product I. If all other problem information remains the same, the total income for the company is now

$$Z = (40. - 0.1x_1)x_1 + 11x_2$$

or

$$Z = 40x_1 - 0.1x_1^2 + 11x_2 \tag{10.1}$$

This income must be maximized subject to the constraints of Equation 9.4. Thus the problem is formulated as a nonlinear programming problem that has a nonlinear objective function and linear constraints.

Other types of nonlinear models that can be formulated include models with a linear objective function and one or more nonlinear constraints and models with a nonlinear objective function and one or more nonlinear constraints.

10.1.2 The Network Improvement Problem Again

The network improvement problem that was formulated and solved using linear graph analysis in Section 7.6 can also be formulated as a nonlinear programming problem.

If the solution for the network capacity model (Section 9.3.5) indicates that the total flow capacity of the network is less than the demand capacity of 6000 vph, the next step is to construct a model that will indicate which improvements should be made to the network and the cost of these improvements to satisfy the demand.

The objective is to find the minimum cost of increasing the network capacity to 6000 vph. In order to incorporate the network flow capacity as a function of new lane and road construction, it is necessary to introduce decision variables for the new construction effects on the capacity constraints.

Let $d_{i,1}$ be a decision variable assigned to an existing road i and let it take a value of either 0 or 1 with the following meaning:

$$d_{i,1} = 0 \quad \text{do not add any new lanes}$$

$$d_{i,1} = 1 \quad \text{add two lanes to road } i$$

Furthermore, let $d_{j,2}$ be the decision variable assigned to a proposed road j. Similarly, $d_{j,2}$ can take on either 0 or 1 with the following meaning:

$$d_{j,2} = 0 \quad \text{do not build proposed road } j$$

$$d_{j,2} = 1 \quad \text{build a new two-lane road } j$$

Thus the additional decision variables for this problem can be symbolically represented by $d_{5,1}, d_{6,1}, d_{7,1}, d_{8,1}, d_{9,1}, d_{10,1}, d_{11,1}$, and $d_{11,2}$. For the proposed road 11, the decision variables $d_{11,1}$ and $d_{11,2}$ have the following combined meaning:

$$d_{11,1} = 0 \quad \text{and} \quad d_{11,2} = 0 \quad \text{no construction}$$

$$d_{11,1} = 0 \quad \text{and} \quad d_{11,2} = 1 \quad \text{build two lanes}$$

$$d_{11,1} = 1 \quad \text{and} \quad d_{11,2} = 1 \quad \text{build four lanes}$$

According to the definition above, $d_{11,1}$ cannot be equal to 1 when $d_{11,2} = 0$.

The objective, then, is to minimize the sum of the costs for new road construction. The new construction costs for existing road i become \$(mileage road i) $Rd_{i,1}$ and the costs for road 11 become \$(mileage road 11) $(1.5Rd_{11,2} + 0.8Rd_{11,1})$. Thus the objective is

Minimize

$$(20d_{5,1} + 20d_{6,1} + 20d_{7,1} + 30d_{8,1} + 30d_{9,1} + 20d_{10,1}$$
$$+ 60d_{11,2} + 32d_{11,1})R \qquad (10.2)$$

subject to the capacity constraints:

$$\text{for existing roads}\begin{cases}\text{four-lane}\begin{cases}0 \le x_1 \le 2000 \\ 0 \le x_2 \le 2000 \\ 0 \le x_3 \le 2000 \\ 0 \le x_4 \le 2000\end{cases} \\ \text{two-lane}\begin{cases}0 \le x_5 \le 500 + 1500d_{5,1} \\ d_{5,1} = 0 \text{ or } 1 \\ 0 \le x_6 \le 500 + 1500d_{6,1} \\ d_{6,1} = 0 \text{ or } 1 \\ 0 \le x_7 \le 500 + 1500d_{7,1} \\ d_{7,1} = 0 \text{ or } 1 \\ 0 \le x_8 \le 500 + 1500d_{8,1} \\ d_{8,1} = 0 \text{ or } 1 \\ 0 \le x_9 \le 500 + 1500d_{9,1} \\ d_{9,1} = 0 \text{ or } 1 \\ 0 \le x_{10} \le 500 + 1500d_{10,1} \\ d_{10,1} = 0 \text{ or } 1\end{cases}\end{cases} \tag{10.3}$$

$$\text{for planned road 11}\begin{cases}0 \le x_{11} \le 500d_{11,2} + 1500d_{11,1} \\ d_{11,1} = 0 \text{ or } 1 \\ d_{12,2} = 0 \text{ or } 1 \\ d_{11,1} \le d_{11,2}\end{cases} \tag{10.4}$$

and subject to continuity conditions that are the same as in Equation 9.13 except for nodes 2 and 6, which now become

Node 2: $\qquad\qquad -x_2 + x_5 - x_{11} = 0$

Node 6: $\qquad\qquad +x_3 - x_4 + x_7 + x_{11} = 0$ $\qquad\qquad$ (10.5)

and subject to the constraint that the total network flow is equal to 6000; that is,

$$x_1 + x_5 + x_6 = 6000 \tag{10.6}$$

This results in an integer linear programing model with 11 flow variables, $x_1, \ldots, x_{11}$ and eight integer decision variables.

The solution of this model for the flow variables and the decision variables will provide the cost of increasing the network capacity to 6000 vph and will indicate which roads should be improved. This information will then provide the basis on which to make a request for the money to construct the needed roads.

This model appears to be one that can be solved by linear programming solution procedures. However, the model requires a special solution procedure because some of the variables can only have integer values.

Budget constraint model.   If the total amount of money requested for new construction by the highway department is not appropriated, then, the department must work within the amount appropriated. Therefore, the objective now becomes maximizing the flow capacity of the network by increasing the capacity of some of the roads subject to the budget constraint. Hence the objective function is

Maximize

$$y_1 = x_1 + x_5 + x_6 \tag{10.7}$$

subject to the constraints of Equations 10.3, 10.4, and 9.13, as modified by Equation 10.5. It is slso subject to the constraint that

$$(20d_{5,1} + 20d_{6,1} + 20d_{7,1} + 30d_{8,1} + 30d_{9,1} + 20d_{10,1} + 60d_{11,2}$$
$$+ 32d_{11,1})R \le S \tag{10.8}$$

where S is the amount appropriated for the network improvement. This model is very similar to the network capacity model of Section 9.3.5. The only difference is that the capacity and continuity constraints now include the provision for the proposed roads, and there is an additional constraint, Equation 10.8, due to the budget limitation.

Determining optimum direction of flows.   Mathematical modeling may also be used to determine the optimum direction of flow through the branches in a network. Suppose that branch i in a flow network has a capacity of c_i, as shown in Figure 10.1(a). Furthermore, suppose that the flow can be either from node A to node B or from B to A; but simultaneous flow in both directions is not possible. Part of the system, then, is to determine the direction of flow that results in maximum flow.

Since flow in either direction is possible, the undirected branch in Figure 10.1a can be replaced by two branches directed in opposite directions, as shown in Figure 10.1b. Let d_j be a decision variable that has a value either of 1 or 0; 1 means that there is a flow along directed branch j, and 0 means that there is

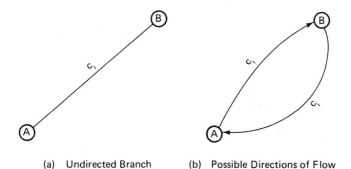

(a)   Undirected Branch   (b)   Possible Directions of Flow

Figure 10.1   Linear graph model when flow direction is unknown.

no flow through the branch. The following equations can thus be written to describe the capacity constraints:

$$0 \leq d_j x_j \leq c_i$$
$$d_j = 1 \text{ or } 0$$
$$0 \leq d_k x_k \leq c_i \tag{10.9}$$
$$d_k = 1 \text{ or } 0$$

Since simultaneous flow in both directions is not possible, d_j and d_k cannot both be equal to 1. This condition can be expressed as follows:

$$d_j + d_k \leq 1 \tag{10.10}$$

Thus Equations 10.9 and 10.10 together describe the flow condition in the original branch i. Note, however, that the flow equations in Equation 10.9 are nonlinear equations since the terms $d_j x_j$ and $d_k x_k$ are products of two unknown variables. The resultant mathematical model for the network will be a nonlinear programming model.

10.2 SOLUTION OF NONLINEAR PROGRAMMING MODELS

Many nonlinear problems can be solved by techniques that are slight variations of those applied to the linear model. However, some problems may require much more sophisticated techniques. At present, there exists no efficient solution procedure for the general nonlinear problem. Substantial progress has been made, however, for some special cases.

10.2.1 Graphical Solution

A problem with only two variables can be solved graphically. The graphical solution process for nonlinear problems is analogous to that for the linear problem. The graphical solution for the nonlinear product-mix problem formulated above is summarized in Figure 10.2. The only difference between this solution procedure and that for the linear problem is that the line representing income is now nonlinear instead of linear. The maximum income is $5285.60, with $x_1 = 172$ and $x_2 = 124$.

10.2.2 Quasilinearization

One of the most useful nonlinear programming techniques is quasilinearization. The application of this technique requires that the nonlinear objective function be approximated by linear segments and that each nonlinear constraint be approximated by a set of linear constraints.

Often times the objective function is separable so that it can be written as the sum of individual functions. Thus a separable function of n variables

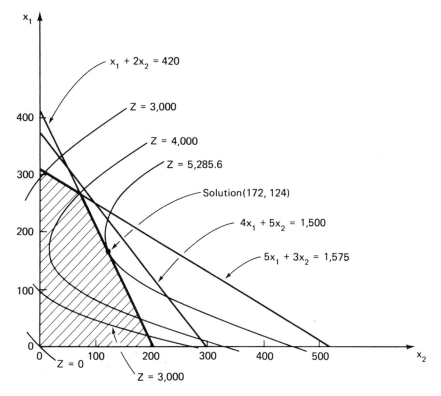

Figure 10.2 Graphical solution for nonlinear programming problem.

$f(x_1, x_2, \ldots, x_n)$ can be written as

$$f_1(x_1) + f_2(x_2) + f_3(x_3) + \cdots + f_n(x_n)$$

where each function depends on only one of the decision variables. Each individual nonlinear function can be approximated as shown in Figure 10.3, where for concave functions $f_i(a_i)$ is the value of $f_i(x_i)$ for $x_i = a_i$; and s_i is the slope of the linear segment. For convex functions the variables are indicated with a prime. When linearizing the objective function only concave functions can be maximized and only convex functions can be minimized. Two of the most popular methods for piecewise linearization are presented here. For Figure 10.3a one method is to approximate maximize $f_i(x_i)$ as

Maximize

$$f_i(x_i) = s_{i1}x_{i1} + s_{i2}x_{i2} + s_{i3}x_{i3} + \cdots = \sum_j s_{ij}x_{ij} \qquad (10.11)$$

This must be constrained such that

$$a_{i1} + x_{i1} + x_{i2} + x_{i3} + \cdots = x_i \qquad (10.12)$$

and

$$x_{ij} < a_{ij+1} - a_{ij} \qquad \text{for all segments } j \qquad (10.13)$$

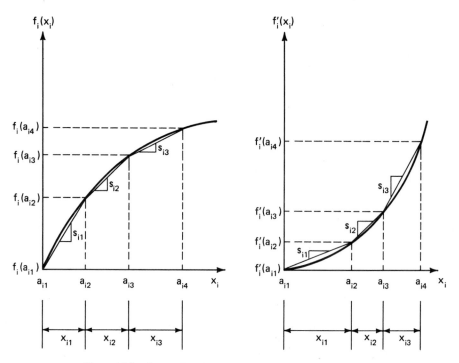

Figure 10.3 Approximation of nonlinear objective functions.

The overall approximation for $f(x_1, x_2, \ldots, x_n)$ is then the sum of the approximation for each $f_i(x_i)$. Equations 10.11 and 10.13 would then be added to the set of constraints for the original problem for which Equation 10.11 is the approximation of the ith part of the objective function.

A second method for piecewise linearization is to approximate maximize $f_i(x_i)$ as

Maximize

$$f_i(x_i) = f_{i1}(a_{i1})w_{i1} + f_{i2}(a_{i2})w_{i2} + \cdots = \sum_j f_{ij}(a_{ij})w_{ij} \qquad (10.14)$$

This must be constrained subject to

$$a_{i1}w_{i1} + a_{i2}w_{i2} + a_{i3}w_{i3} + \cdots = x_i \qquad (10.15)$$

$$w_{i1} + w_{i2} + w_{i3} + \cdots = 1 \qquad (10.16)$$

Equations 10.15 and 10.16 would then be added to the original set of constraints.

The procedure for Figure 10.3b would be similar. The only difference would be that the objective is to minimize rather than maximize.

For each individual function approximated these methods increase the number of constraints by two. The increase in the number of variables depends on how close the approximation must be to the original function. The more segments (i.e., the more variables added), the closer the approximation will be to the original function. The engineer must make an evaluation as to the relative value of a closer approximation for an increased cost of computation.

Figure 10.4 illustrates how a nonlinear constraint can be approximated by a set of linear constraints. Again the more linear constraints used the closer will be the approximation to the original constraint. The linear constraints are then substituted into the original constraint set for the nonlinear constraint.

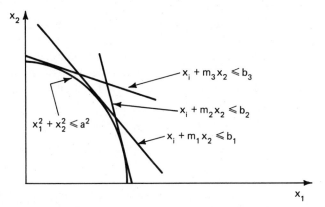

Figure 10.4   Approximation of nonlinear constraints.

10.2.3 Sequential Search Techniques

The nonlinear models in Chapter 8 were solved by calculation methods for the optimum point on the objective function. Thus they only examined one point: the optimum. Many problems involve a large number of variables or have nonlinear constraints and cannot be solved by calculus methods directly. Sequential search techniques are procedures for systematically examining the objective function. The simplex method in linear programming is a sequential search technique that examines the objective function at potentially optimal points. All search techniques start with a feasible solution (i.e., a solution that satisfies the set of constraints but may not be optimal) and then examines the objective function in the vicinity of the initial solution to determine if the value of the objective function can be improved by changing the value of one or more decision variables. If the objective function can be improved the value of the decision variables are changed, but not enough that they would violate any of the constraints, and the objective function is again examined in the vicinity of the new solution. This procedure is repeated until there is no improvement of the objective function for any change in the values of one or more variables and then the last solution is taken as the optimal. Some search techniques allow only one variable to change value after each examination of the objective function. Other search techniques allow two or more variables to change value after each examination of the objective function.

For some search techniques the objective function is examined by computing the partial derivatives of the objective function with respect to each decision variable and comparing them to determine the variable to change. If the objective function is being maximized, a positive derivative would indicate the variable value should be increased and a negative derivative would indicate that the variable value should be decreased. The variable that causes the largest rate of improvement in objective function is chosen to be changed. For other search techniques the objective function is examined by computing and comparing the values of the objective function for values of the decision variables that are greater than and less than the current values of the variables.

The nonlinear product-mix problem as summarized by Equations 10.1 and 9.4 will be used to demonstrate the general procedure of numerical search techniques. The partial derivative of the objective function with respect to each decision variable is

$$\frac{\partial Z}{\partial x_1} = 40 - 0.2x_1 \tag{10.17}$$

$$\frac{\partial Z}{\partial x_2} = 11 \tag{10.18}$$

If $x_1 = 0$ and $x_2 = 0$ are chosen as the initial solution, then $Z = 0$ and $\partial Z/\partial x_1 > \partial Z/\partial x_2$ and x_1 should be increased to increase the value of Z. If x_1 were increased to 200 and x_2 is unchanged, the constraints are still satisfied and $Z = 4000$. New $\partial Z/\partial x_2 > \partial Z/\partial x_1$, and x_2 should be increased to increase the value of Z. If x_2 were now increased to 100 and x_1 is unchanged, then $Z = 5100$ and the constraints are still satisfied. The values of x_1 and x_2 are increasing toward the optimum values of 172 and 124, respectively.

However, by varying only one variable at a time the procedure may terminate at a point before it arrives at the optimum. For example, the procedure above would arrive at the point $x_1 = 200$ and $x_2 = 110$; then $Z = 5210$, $\partial Z/\partial x_1 = 0$, and $\partial Z/\partial x_2 = 11$, which would indicate that x_2 should be increased. However, x_2 cannot be increased without decreasing x_1. Thus the procedure would terminate and indicate that the optimum had been obtained, although it can be seen in Figure 10.2 that $x_1 = 200$ and $x_2 = 110$ is not the optimum.

This example illustrates one of the major disadvantages of a numerical search technique. The technique may stop at a location other than the optimum. Therefore, most numerical search techniques would be used with several different initial solutions to determine if each starting solution would result in the same optimal solution. If all final solutions are the same, the engineer may have some confidence in the solution, although there is no complete assurance that an optimum has been obtained as there is with calculus methods or linear programming.

Some search techniques have a procedure for computing the amount that a variable is to be changed, step size, after each examination of the objective function and others require that the step size be specified before the start of the search procedure.

The problems analyzed in this chapter were completely quantified; that is, it was assumed that all the pertinent information could be expressed in the form of mathematical statements. Sometimes this complete quantification is impossible, for example, when choosing the location for a fire station or police station. A sequential search technique can be used in evaluating the best location for such a facility. A site is chosen and evaluated using quantified data such as cost, and subjective data such as how well the public will accept this location. The site may then be moved to what is felt to be a better location and the quantified and subjective data relative to this location are evaluated. This procedure would continue until no better site could be found.

10.3 SOLUTION ALGORITHMS

There are many solution algorithms available for nonlinear programming problems (Kuester and Mize, 1973) and most computing centers will have algorithms available. The nonlinear model of the product mix problem was solved using the SYMQUAD algorithm, which is available at several computing centers as a part of the Multi Purpose Optimization System, MPOS (Cohen and Stein, 1976), This algorithm is restricted to models with quadratic objective functions and linear constraints. In this algorithm the original quadratic model is transformed into the Kuhn–Tucker conditions (Cohen and Stein, 1976; Beightler, Phillips, and Wilde, 1979) for an optimal solution to this model. These conditions constitute a set of linear equations which are then solved by a modified simplex procedure similar to the linear programming simplex procedure.

The MPOS has language features that simplify input and output for the optimization algorithm. The nonlinear product-mix problem data was input as shown in Table 10.1. The data are almost obvious.

The first line indicates that the second line will be the title of the problem. The third line indicates the algorithm to use. The fourth line indicates that the

TABLE 10.1   Input Data for Nonlinear Product-Mix Problem

TITLE
NONLINEAR PRODUCT MIX PROBLEM
SYMQUAD
VARIABLES
X1   X2
MAXIMIZE
$40X1 - 0.1X1 * X1 + 11X2$
CONSTRAINTS
$4X1 + 5X2$ .LE. 1500
$5X1 + 3X2$ .LE. 1575
$X1 + 2X2$ .LE.   420
OPTIMIZE

variable names will appear in the next line. The sixth line indicates that the objective function is to be maximized. The seventh line is the objective function. The eighth line indicates that the constraints will appear on the following lines. The last line indicates that all constraints have been input and that the model is now to be optimized. The results from the algorithm are shown in Table 10.2.

TABLE 10.2   Output Results for Nonlinear Product-Mix Problem

```
***************************
* PROBLEM NUMBER 1 *
***************************
```

USING SYMQUAD
NONLINEAR PRODUCT MIX PROBLEM

SUMMARY OF RESULTS

MATH. VAR.	VAR. NAME	BASIC NON-BASIC	VALUE	OPPORTUNITY COST
X1	X1	B	172.50000	—
X2	X2	B	123.75000	—
Y1	SLACK 1	B	191.25000	—
Y2	SLACK 2	B	341.25000	—
Y3	SLACK 3	NB	—	5.50000

MAXIMUM VALUE OF THE OBJECTIVE FUNCTION= 5285.625000

There is no single procedure for solving all nonlinear programming problems and it is difficult to provide a systematic listing of available algorithms. The success of a particular algorithm depends on the number of variables and constraints and the type of computation facility available. All procedures are iterative and may involve calculating first and, in some cases, second derivatives of the objective function.

10.4 SUMMARY

There is no solution algorithm available for nonlinear programming which has the applicability that the simplex method has in linear programming. However, there are specialized solution techniques for certain kinds of nonlinear problems, and those problems with only two variables can be solved graphically.

The highway problem illustrates that the exact form of the equations in the model will depend on how the problem is stated. If the objective is to determine the maximum network capacity, one model will be appropriate, but if the objective is to minimize cost of network improvement, a model with slightly different equations becomes appropriate. In reality, the problem could be expanded to involve other considerations, such as construction schedule and roadway maintenance. Thus, in this case, all the problem components have not been fully defined at this point. The consideration of these other aspects involves the management of the system and are discussed in Chapter 13.

Because of the many types of search techniques and algorithms, it is preferable to develop the mathematical model for the problem and then search for a solution technique rather than try to model so as to use a particular solution algorithm.

A detailed discussion of the techniques of nonlinear programming is beyond the scope of this book; however, a unified approach to nonlinear programming has been presented by Beightler, Phillips, and Wilde (1979).

PROBLEMS

P10.1. Listed below are the objective functions and constraints:

Maximize
$$Z = 2x + 4y - 0.1y^2$$
subject to:
$$2x + 3y \leq 48$$
$$3x + 2y \leq 42$$
$$10x + y \leq 50$$
$$x - 3y \leq 0$$
$$x \geq 0, \qquad y \geq 0$$

(a) Solve this problem graphically.
(b) Use piecewise linearization and solve the problem as a linear programming problem.
(c) Compare the two solutions.

P10.2. Solve Problem P10.1 using the sequential search procedure demonstrated in the text with a step size of 2 units; that is, x or y can be increased or decreased 2 units each time the partial derivatives are evaluated.

P10.3. Develop a network capacity model, an improvement cost model, and a budget constraint model for the road network described in Section 7.6 if the following additional features must be considered.

1. Because of surface deterioration the capacities of roads 4, 7, and 8 are reduced to 1700, 300, and 300 vph, respectively. The capacity of road 4 can be increased at the rate of 300 vph for each \$0.5R per mile spent for road repair. The capacities of roads 7 and 8 can each be increased at the rate of 200 vph for each \$0.1R per mile spent for road repair. The improved capacity of each road cannot be greater than its original capacity.
2. The total budget for road repair and new road construction is S.

P10.4. Assume that for the product-mix problem of Section 9.3 that the cost of production per unit of each product increases as each additional unit is produced of each product. The production cost per unit is given by $\$(3 + 0.05x_1)$ for product I and $\$(2 + 0.06x_2)$ for product II.

(a) Formulate the nonlinear model that can be used to determine the number of units of each product to produce such that profit would be maximized.

(b) Solve the problem graphically.

(c) Use the sequential search technique demonstrated in the text to solve the problem with a step size of 20 units.

P10.5. The daily labor allocation problem is faced by project managers whenever planned labor resources are not met on the construction site. Some activities may be worked with reduced crew profiles between the trades involved if minimum requirements are met; others may be postponed. In some cases, available labor can be used to accelerate progress on "favored" activities.

Develop an integer linear programming model for the optimal allocation of available labor (from n trades) to work the scheduled m activities. What would be suitable objective functions for your model?

Use the following notation:

a_{ij} = number of trade j workers planned and scheduled to work activity i to make 1 day's progress in 1 day

x_{ij} = number of trade j workers available and allocated to work activity i

c_j = number of trade j workers actually reported for work this project day

d_{ij} = decision variable to work activity i with trade j; $d_{ij} = (0, 1)$ if activity i (is not, is) worked by trade j

l_{ij}, h_{ij} = management defined range limits (low, high) for the number of trade j workers for activity i if it is worked

In addition, to ensure reasonably compatible crew profiles among the trades working an activity, management defines a range of ratios (r_{low}, r_{high}) between the various trades (see Figure P10.5). Usually, this can be done by referring to a specific trade in each activity that basically determines the productivity output for that activity.

P10.6. Two cities are exploring the possibility of cooperating for a joint wastewater management plan. It may be more economical to pump part or all of the sewage from one city to the other for treatment and disposal than to treat it in the city of origin. Each city has a treatment site available. The plant sites are 3 miles apart. City 1 produces 10 million gallons per day (mgd) of wastewater and city 2 produces 15 mgd. The environmental standards require that all wastewater be treated.

The annual cost (dollars per year) of wastewater treatment at city i is

$$0.5Q_i^{0.75} + 0.3Q_i^{0.65}$$

where Q_i is the wastewater treated at site i in gallons per day (gpd). The pipeline cost between the two cities is $12Q^{0.7}$ in dollars per mile per year and the pumping cost is $35Q^{0.6}$ in dollars per year, where Q is the quantity of wastewater pumped from one city to the other in gpd.

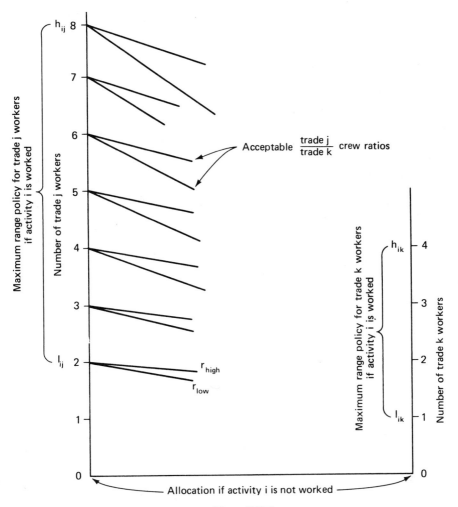

Figure P10.5

The cities must also dispose of the sludge, which consists of solids and other materials removed from the wastewater by treatment. These sludges amount to 0.001 pound per gallon of wastewater treated. The sludge can be incinerated or deposited in a sanitary landfill. Incinerators can be constructed at either or both wastewater treatment sites. At city 1 an incinerator costs $1800S_1^{0.9}$ dollars per year, where S_1 is the number of tons of sludge incinerated per day. At city 2 the incineration cost is $1600S_2^{0.9}$.

The sanitary landfill is 7 miles from city 1 and 4 miles from city 2. It costs $15 per ton to dispose of sludge in the landfill and it costs $0.50 per ton per mile to transport the sludge.

(a) Construct an optimization model that can be used to develop a wastewater and sludge management plan for the cities.

(b) Use quasilinearization to transform the nonlinear model into a linear model and show why quasilinearization cannot be used for this problem.

(c) Use a sequential search technique as suggested in the text to solve the model of part (a).

P10.7. Consider Problem P8.7.

(a) Solve the problem graphically.

(b) Solve the problem using quasilinearization.

(c) Solve the problem using the sequential search technique demonstrated in the text. Use a starting point of $x = 14$ and a step size of 2 months. Then use a starting point of 20 and a step size of 2 months. Compare the solutions.

P10.8. The sequential search technique demonstrated in the text used the partial derivatives of the objective function to determine which variable to increase or decrease in order to move toward the optimal value of the objective function. Solve Problem P10.1 using a procedure that does not use partial derivatives; that is, use the procedure that evaluates the objective function for values of the decision variables that are greater than and less than the current values. Use a step size of 2 units.

Decision
Analysis
11

11.1 ELEMENTS OF A DECISION PROBLEM

The primary objective in a decision problem is to choose the optimum plan or policy from a specific set of possible alternatives. However, the decision must be made in a logical and subjective manner so that all possible alternatives and their consequences are considered and the final decision can be explained and justified to the interested parties. In addition, contingency plans must also be prepared for all potential consequences of a decision.

For example, consider the decision problem of the contractor discussed in Example 4.3. The contractor must choose among three alternatives:

1. To move the equipment away from the river flat area to avoid loss due to potential flooding
2. To leave the equipment at the location and build a protective platform
3. To leave the equipment at the location and not build a protective platform

Each alternative results in a certain cost to the contractor, and the cost of the last two alternatives will depend on the extent of flooding during the spring. Therefore, in making his decision, the contractor must consider the chances of flooding and the potential loss that may result. Furthermore, should it be decided to leave the equipment on location, contingency plans must be developed for replacing the equipment in case of serious loss from a flood.

Two common elements characterize all engineering decision problems:

1. Probabilistic events
2. Insufficient data

Probabilistic events are occurrences that are beyond the control of the decision maker, although the probability or percentage chance of these occurrences can usually be predicted from either historical data or engineering foresight. For example, inclement weather is beyond the control of the engineer who is planning a construction schedule, yet it is a factor that must be taken into consideration. Natural events such as flooding, earthquakes, and hurricanes are vital factors that must be considered in many engineering decision problems, including the structural design for high-rise buildings, location of dams and reservoirs, and location of transportation systems.

The amount of technical data that is available to a decision-maker is usually limited by many factors. Among the more common factors are: technical skill of the decision maker, state of the art in science and technology, limited financial resources, limited research capability, and limited time available for research and experimentation. Although technical data can usually be obtained at a certain cost to the decision maker, the value of the additional information must be weighed against the cost of obtaining such information. For example, traffic surveys may be conducted to provide data on the volume of traffic flow, degree of congestion, commuter traffic behavior, percentage of truck and through traffic, and so on. Such surveys require time and money but provide in return information that traffic engineers may use to make sound decisions on the design of city traffic systems.

A decision problem involves a set of alternative actions that are connected in some way to a set of possible outcomes. If the decision maker knew which outcomes would occur with each act, a choice could be made that resulted in the most valued outcome. Actions may lead to outcomes where other decisions must be made and many actions lead to outcomes that depend on chance; therefore, the outcome is known only in terms of probability of occurrence if certain actions are taken. The necessary first step in the decision-making process, therefore, is to identify the alternative actions and outcomes and their interrelationships. The value of each outcome and the probabilities of its occurrence for the alternative actions should be quantified. These parameters, together with their interrelationships, constitute the problem model, which may then be analyzed to determine the set of actions the decision maker should take to achieve his most valued outcome.

As has been emphasized in earlier chapters, the engineer must formulate the problem for each situation that is encountered. Although all decision problems have common features, the specific formulation must reflect specific alternatives and decisions for that particular problem.

11.2 THE DECISION MODEL

Example 4.3 has already illustrated the usefulness of linear graphs in modeling a decision problem. The graphical model, called the *decision tree*, delineates all the alternatives and their possible consequences. It helps to facilitate the task of decision analysis. To further illustrate the purpose and technique of constructing the decision tree, consider the following problem.

11.2.1 Problem Statement

A contractor working on an outdoor construction project in a coastal area is reviewing progress on August 1. If normal progress is maintained and no time is last due to hurricanes, the job will be completed on August 31. However, due to the poor weather conditions in the area after August 16, there will be only a 40 percent chance of finishing on time. It is estaimated that there is a 50 percent chance of a minor hurricane, which will cause a delay of 5 days, and a 10 percent chance of a major hurricane, which will cause a delay of 10 days. It must be decided now whether to start a crash program on August 2 at an additional cost of $75 per day and finish the project on August 16. As an alternative, the normal schedule can be maintained and progress reviewed on August 31. At that time if a hurricane has occurred and the project is delayed, there will be a choice of accepting the delay at a certain penalty cost or trying to crash the program then. The penalty cost for delay of completion will be $400 per day for the first 5 days and $600 per day for the second 5 days. The additional cost of a crash program after the hurricane will be $200 per day. The total additional cost is computed as the sum of delay penalty cost and crash cost.

It is also estimated that the possible results (outcomes) of a crash program after a minor hurricane causes a 5-day delay will be as follows:

Crash Program Result	Probability	Total Additional Cost
Save 1 day	0.5	$1600 + 800 = $2400
Save 2 days	0.3	$1200 + 600 = $1800
Save 3 days	0.2	$800 + 400 = $1200

The possible results of a crash program after a major hurricane causes a 10-day delay is estimated as follows:

Crash Program Result	Probability	Total Additional Cost
Save 2 days	0.7	$(2000 + 1800) + 1600 = $5400
Save 3 days	0.2	$(2000 + 1200) + 1400 = $4600
Save 4 days	0.1	$(2000 + 600) + 1200 = $3800

The problem description above adequately outlines all the alternatives, the consequences, as well as all the relevant data needed to make the decision. Therefore, this description constitutes a descriptive model of the decision problem. It is obvious, however, that even for a problem as simple as this, it is difficult to obtain an overall view of the problem from such a descriptive model. As an interesting experiment, before reading any further, assume that you are the contractor and try to solve the problem.

11.2.2 Model Formulation

Figure 11.1 illustrates the different steps used to construct the decision tree in the problem above. These steps are described below.

1. A *decision node* (□) is drawn to represent the most immediate decision that the contractor must make. A branch is drawn from the node to represent each alternative that is available to him at this decision point, as shown in Figure 11.1a.

2. The potential results of each of the decision alternatives are then modeled. The outcome of immediately launching a crash program is that the project will be completed at an added cost of $75 per day for 15 days, or a total of $1125. This is represented by assigning a *terminal node* (△) with a value of $-$1125$, as shown in Figure 11.1b. The second alternative has three possible outcomes: no delay, 5-day delay, and 10-day delay with a probability of 0.4, 0.5, and 0.1, respectively. This is represented as a *chance node* (○), where each possible result is represented by a branch and the probability of occurrence is entered along the branch.

3. The only possible result of no delay on August 31 is that the project is completed on time with no delay cost. This is represented by a terminal node with a value of $-$0.0$. For the other two outcomes, the contractor must decide next whether to launch a crash program or to maintain a normal pace. Thus each is modeled by a decision node with branches issuing from the node to represent the possible alternatives, as shown in Figure 11.1c.

4. The decision tree is finally completed by again using a branch to represent each possible outcome, and the total cost resulting from that outcome is written next to the terminal nodes, as shown in Figure 11.1d.

Constructing the decision tree, however, is only the last and perhaps the easiest stage in modeling a decision problem. The greatest challenge lies in identifying and stating the sequential decisions and their possible consequences that constitute the problem model. Estimating costs and benefits and assigning probabilities to the chance occurrences are tasks that test the skills of the decision maker, and the accuracy of these estimates ultimately controls the validity of the final decision.

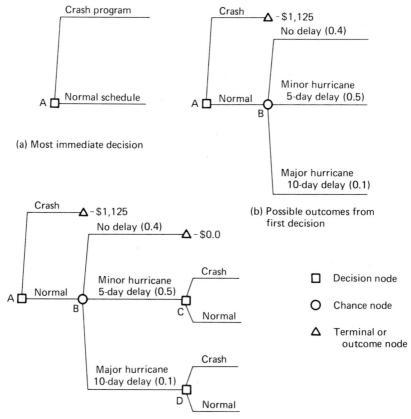

(a) Most immediate decision

(b) Possible outcomes from first decision

(c) Outcomes require more decisions

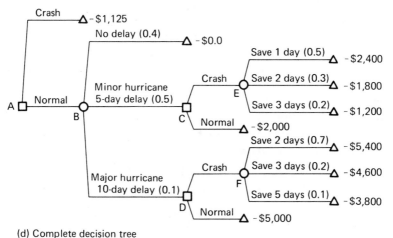

(d) Complete decision tree

Figure 11.1   Development of decision tree model of contractor's problem.

The concept that any problem model must of necessity be an abstraction of reality is also well illustrated by this example. Although the exact delay caused by a hurricane may range anywhere from 1 day to 10 days or more, the contractor has simplified the problem by considering only the two most likely cases—a minor hurricane resulting in a 5-day delay, or a major hurricane causing a 10-day delay. This abstraction is easily justifiable if the nature of the construction job is such that delays must be counted in intervals of several days; jobs such as pouring and drying concrete and exterior painting usually need several days of continuous good weather. In addition, considerations must be given to the repair of hurricane damages and to the loss in momentum due to the delay.

On the other hand, delays of any discrete number of days can be built into the decision model. The amount of time that can be saved by the crash programs may also be counted in half-day intervals. In reality, the contractor may also have the option of keeping his program under continuous review depending on the progress between August 1 and 16. Indeed, the model can be expanded to include some or all of these alternatives, with a resulting increase in the complexity of the model. However, in order to keep the problem on a manageable scale for the purpose of analysis, some degree of abstraction must be exercised using engineering judgment.

11.3 FUNDAMENTALS OF PROBABILITY

It is beyond the scope of this book to present a detailed discussion of the theories of probability. However, since probability is such an integral part of the method of decision analysis, it is useful to review here some of the basic fundamentals of the subject.

The probability that a chance event will occur is measured in a continuous scale ranging between 0 and 1. An event that has absolutely no chance of occurring has a probability value of zero. On the other hand, an event that is sure to occur has a probability value of 1. Thus any event that has a chance of occurring will have a probability value between 0 and 1. The more likely it is to occur, the higher will be its probability value. In symbolic language, the probability that an event A will occur is denoted as $P(A)$.

In the example discussed in Section 11.2, the following probabilities were cited for the three possible results in case the contractor decided to maintain construction at his normal pace.

$$P(\text{no delay}) = 0.4$$

$$P(\text{minor hurricane}) = 0.5$$

$$P(\text{major hurricane}) = 0.1$$

These probability values provide a measure of the relative chance of these three events occurring. They can be determined in many ways. They may simply be based on an educated guess by the contractor, whose judgment is based on present weather conditions and on recollection of weather conditions at the same period in previous years. They may also be the result of a statistical analysis of the historical records of local weather conditions. Suppose that the past 50 years of weather records for the area are available to the contractor. The number of years in which there was either a minor or major hurricane during the period August 16–31 can be counted. The above probabilities can then be computed as follows:

$$P(\text{no delay}) = \frac{\text{number of years with no hurricane}}{\text{total number of years counted}} = \frac{20}{50} = 0.4$$

$$P(\text{minor hurricane}) = \frac{\text{number of years with a minor hurricane}}{\text{total number of years counted}} = \frac{25}{50}$$
$$= 0.5$$

$$P(\text{major hurricane}) = \frac{\text{number of years with a major hurricane}}{\text{total number of years counted}} = \frac{5}{50}$$
$$= 0.1$$

The greater the number of years of weather records used in the analysis, the more reliable will be the computed probability value. However, it is important to realize that these probability values indicate the chance occurrence of these events during the *past* years. In projecting these figures for the present, the contractor is assuming that past weather conditions are a good indicator of what is going to occur this year. A more realistic set of probability values for the events above may possibly be determined from a meteorological study of the existing weather conditions and from long-range weather forecasts.

The methods above apply equally well to the occurrence of other uncertain events such as earthquakes, floods, flood damages, stream flows, traffic accidents, bearing strength of the soil, and the exact tensile strength of a concrete or steel beam. Figure 11.2 is a typical frequency histogram showing, for example, the number of rainy days in August for the past 50 years. Thus the probability that there will be x number of rainy days in this August can be estimated as follows:

$$P(x \text{ rainy days}) = \frac{\text{number of years with } x \text{ rainy days}}{\text{total number of years counted}}$$

In many instances, the chance variable may have a value within a continuous range rather than a discrete set of possible values. For example, the exact tensile strength of a given steel beam may have a value between 1.00 and 1.15 kips/in.^{2}. In such cases, the pattern of occurrences may be modeled by a mathematical function, $f(x)$, which is called the *probability density function* of that variable. Then the probability that the variable x may take on a value between, say, x_1 and x_2 can be computed by integration as follows:

$$P(x_1 \leq X \leq x_2) = \int_{x_1}^{x_2} f(x)\, dx$$

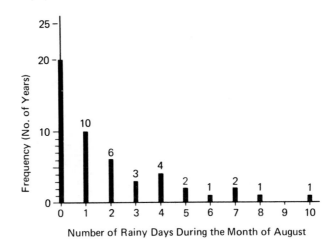

Figure 11.2 Frequency histogram.

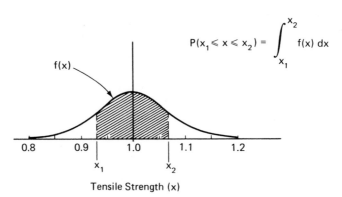

Figure 11.3 Probability density function.

Figure 11.3 illustrates such a probability model for the tensile strength of a steel beam.

Another basic property of probability may be learned from the problem above concerning the contractor. It has been assumed that there is no chance that more than 10 days can be lost due to inclement weather. That is,

$$P(\text{delay longer than 10 days}) = 0$$

In fact, it has been definitely assumed that there is to be either no delay, a delay of 5 days, or a delay of 10 days. That is,

$$P(\text{either no delay, a delay of 5 days, or a delay of 10 days}) = 1$$

Therefore, the summation of the probability values of the above three mutually exclusive (no two events can occur simultaneously) events must be equal to 1. In general, let $x_1, x_2, x_3, \ldots, x_n$ be n mutually exclusive events that could possibly occur at a chance node in the decision tree, and let $P(x_i)$ denote the probability that event x_i occurs; then

$$\sum_{i=1}^{n} P(x_i) = 1 \tag{11.1}$$

11.4 DECISION ANALYSIS BASED ON EXPECTED MONETARY VALUE

Consider the problem of the contractor evaluating the course of action that would be taken at decision node C. This portion of the tree is redrawn in Figure 11.4 for easy reference.

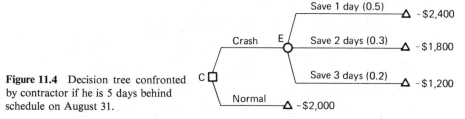

Figure 11.4 Decision tree confronted by contractor if he is 5 days behind schedule on August 31.

The contractor is now already 5 days behind schedule, and a decision must be made whether to launch a crash program immediately or to keep the normal pace. If the latter is chosen, there would be a loss of $2000 for sure. If a crash schedule is followed there is a probability of 0.3 of a loss of only $1800 and a probability of 0.2 of a loss of only $1200. However, there is a probability of 0.5 of a loss of as much as $2400, the alternative chosen will, of course, depend on the value placed on money and the willingness to take chances. But a decision must be made based on the comparative value of the two choices before the final outcome that will actually occur is known.

11.4.1 Expected Monetary Value

One way of comparing the values of alternatives is to compare their expected monetary values. In the present example, the expected monetary value of the choice to crash is computed as follows:

Expected monetary value (EMV) at node E

$$= 0.5 \times (-\$2400) + 0.3 \times (-\$1800) + 0.2 \times (-\$1200)$$
$$= -\$1980$$

It means that if the contractor has the opportunity to make the same decision in many other jobs under similar circumstances, there would be an average loss of $1980 of choosing to crash. However, if the normal pace is selected every time, there would be a $2000 loss each time. Thus if the decision maker is a believer in EMV, a logical choice is to start a crash program.

In general, let $x_1, x_2, x_3, \ldots, x_n$ be the monetary value associated with each of the n branches at a chance node and $P(x_1), P(x_2), P(x_3), \ldots, P(x_n)$ be the corresponding probabilities. Then the expected monetary value (EMV) of that chance node is defined as follows:

$$\text{Expected monetary value} = \sum_{i=1}^{n} x_i P(x_i) \tag{11.2}$$

11.4.2 Comparative Analysis of Alternatives

The decision problem modeled by the decision tree in Figure 11.1 can be solved by successively comparing the EMV of the alternatives at each decision node. Starting from the tips of the decision tree, the EMV of the chance nodes E and F are first computed. That is,

EMV at node $E = -\$1980$ (from above)

EMV at node $F = 0.7 \times (-\$5400) + 0.2 \times (-\$4600)$

$+ 0.1 \times (-\$3800) = -\$5080$

These EMVs are then entered into the decision tree at the respective node, as shown in Figure 11.5. By comparing the EMV of the two alternatives at node C, the best decision is to crash the project. A hatch mark can be drawn on the rejected alternative to denote this choice. Thus, should the contractor have to make the decision at node C, the choice would be to crash the project with an EMV of $-\$1980$. This is equivalent to saying that the decision node C has an EMV of $-\$1980$. Similarly, should the contractor have to make the decision at node D, the choice would be to keep the normal pace with an EMV of $-\$5000$. Again, a hatch mark is used to cross out the rejected alternative and an EMV of $-\$5000$ is assigned to node D.

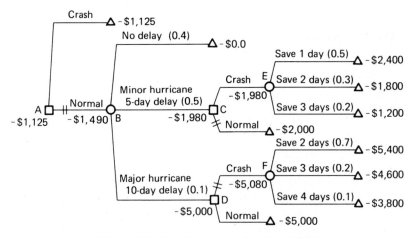

Figure 11.5 Decision analysis based on EMV.

Proceeding one step further, the EMV at chance node B can next be computed as follows:

$$\text{EMV at node } B = 0.4 \times (-\$0) + 0.5$$
$$\times (-\$1980) + 0.1 \times (-\$5000) = -\$1490$$

Finally, at decision node A, the alternative to crash imposes a certain loss of $1125, whereas to keep the normal pace would lead to an EMV of $-\$1490$. Thus the better choice is to crash the program immediately.

In general, the analysis procedure consists of the following steps:

1. Starting from the tips of the tree, compute the EMV at the chance nodes closest to the tips.
2. At each decision node, compare the EMVs of the alternatives and choose the alternative with highest profit or minimum loss. Assign the EMV of the chose alternative to that decision node.
3. Proceed node by node toward the root of the tree. The EMV for the decision problem is obtained at the root of the tree.
4. Trace back through the tree to determine the optimum decision as indicated by the branches that do not have hatch marks. This set of decisions constitutes the optimum strategy. In this case, there is only one decision: crash the project.

11.4.3 Implications of the EMV Criterion

In using the EMV as a decision criterion, a decision maker must always keep in mind that this approach carries the following two implications:

1. The decision maker is betting on the law of averages, since the EMV of an alternative means that if he or she chooses this alternative many times under similar conditions he or she would receive this much in return on the average. This may be quite different from the amount that is actually obtained if the choice is only made once because only one final outcome results.
2. The EMV is a completely objective measure of the value of money and implies that every dollar within a sum of money provides the same amount of satisfaction. It does not consider personal differences about the value of money.

For example, consider the decision problem illustrated in Figure 11.6. Alternative A has a 30 percent chance of making $100,000 profit and a 70 percent chance of losing $20,000. Alternative B has a 50 percent chance of making a $5000 profit and an equal chance of gaining $6000. According to a strict EMV analysis, alternative A has an EMV of $16,000 and alternative B has an EMV of only $5500; hence, alternative A is the better choice. On the other hand, it

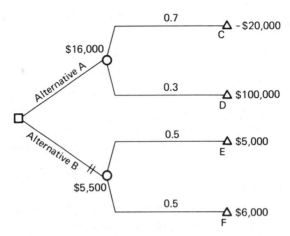

Figure 11.6   Decision problem for which EMV may not be a suitable criterion.

is obvious that in practice many people (including all the authors) would be quite happy to accept alternative *B* but absolutely refuse to accept alternative *A*. However, alternative *A* may be the preferred choice of one who draws little satisfaction from the meager sum of $20,000.

In reality, a decision maker often bases the financial decisions on the available capital and on the willingness to take risk. An attempt is made to prevent a total loss of his capital, but there must be a willingness to risk some loss for the opportunity to profit. For example, consider a young, aggressive investor who has a capital of $50,000. Confronted with the decision problem above, the decision would probably be strongly influenced by the opportunity of gaining $100,000 and thus tripling the capital. The prospect of losing $20,000 would not seriously deter the selection of alternative *A*, because even there is a loss of $20,000, there is still $30,000 remaining as capital. Therefore, the computed EMV of $16,000 for alternative *A* and $5,500 for alternative *B* may appropriately indicate the relative values of the two alternatives to this aggressive investor.

Suppose next that the decision maker is a young, conservative investor who has a capital of only $10,000. Although there would undoubtedly be interest in gaining $100,000, there would also be a strong deterrence by the prospect of losing $20,000. It would not only wipe out the capital, but would result in a $10,000 debt. Since alternative *B* means a certain gain of either $5000 or $6000, which is more than 50 percent of present capital, there would be a strong inclination to choose alternative *B*. To this investor, the computed EMVs fail to measure truly the relative values of the two alternatives.

The EMV criterion failed in this example because the monetary values assigned to the various outcomes at the terminal nodes *C*, *D*, *E*, and *F* do not truly reflect the values of these outcomes to the decision maker. A loss of $20,000 provides different degrees of satisfaction (or dissatisfaction) to different persons depending on their capital and risk behavior.

11.5 DECISION ANALYSIS BASED ON UTILITY VALUE

To provide a personal measure of the relative worth of the decision results, the monetary values may be transformed to their equivalent utility values by a utility function. Each decision maker has his or her own utility function, and it must reflect the person's risk behavior and outlook toward money. For example, let the utility function in Figure 11.7a represent the risk behavior of the conservative investor in the preceding example. A loss of $20,000 has an equivalent utility value of -30 utiles, and a gain of $100,000, $5000, or $6000 has an equivalent utility value of 75, 5, or 6 utiles, respectively. Utiles are defined as units that express personal values. By substituting these utility values for their corresponding monetary values in the decision tree, the expected utility values for alternatives A and B can now be computed as follows:

$$\text{Expected utility value of alternative } A = (0.7)(-30) + (0.3)(75)$$
$$= 1.5 \text{ utiles}$$
$$\text{Expected utility value of alternative } B = (0.5)(5) + (0.5)(6)$$
$$= 5.5 \text{ utiles}$$

The results are illustrated in Figure 11.7b. Thus, based on the expected utility values, alternative B is the better choice for this conservative investor.

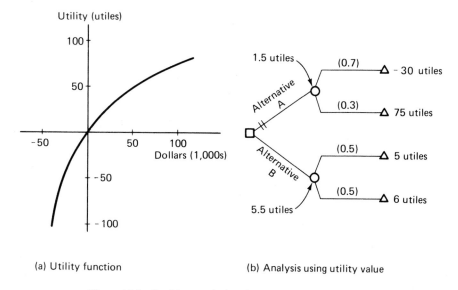

(a) Utility function   (b) Analysis using utility value

Figure 11.7   Decision analysis using expected utility value.

It is no easy matter to develop an accurate utility function for an individual. The risk behavior and the value of money must be thoroughly understood. One method is to test individual preferences under a wide range of risk situations involving different amounts of monetary values. This method is best illustrated by considering an example. The following is a step-by-step description of how a utility curve can be established for the contractor's decision problem in Figure 11.5:

1. Identify the upper and lower limits of the monetary values involved in the decision problem and arbitrarily assign utility values to two different monetary values within these limits; for example,

	Monetary Values	Arbitrarily Assigned Utility Value
Upper limit	0	0
Lower limit	$-\$6000$	-100

These figures establish two points on the utility curve. Since utility values provide relative measures, these two points merely establish a reference datum for the utility values.

2. Create a hypothetical decision problem that has two possible alternatives and three possible outcomes. Two of the outcomes should be assigned the two monetary values to which utility values have already been given. Figure 11.8a presents such a hypothetical problem. In alternative A, the contractor has a 70 percent chance of losing $6000 ($-100$ utiles) and a 30 percent chance of losing nothing (0 utiles). In alternative B, the contractor can purchase insurance to cover any potential loss. What is the maximum amount that the contractor would be willing to pay to cover the 70 percent chance that he might lose $6000? Suppose that the answer is $5400. The contractor's answer means, in essence, that the two alternatives have equal worth if alternative B is to lose $-\$5400$; that is, the utility value of $-\$5400$ is equal to the expected utility value of alternative A. Hence

$$\text{Utility value of } -\$5400 = 0.7(-100) + 0.3(0) = -70$$

Thus a third point is established on the utility curve.

3. Create another hypothetical problem that also has two alternatives and three possible outcomes, and two of these outcomes have monetary values of which the utility values are already known. By using the same procedure as above, a new point can be established. This procedure is repeated until

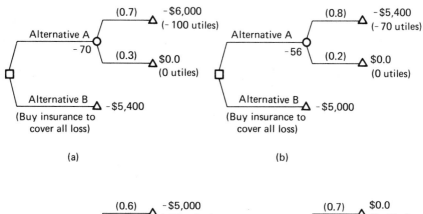

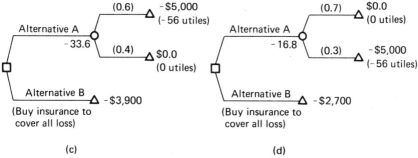

Figure 11.8 Hypothetical decision problems.

enough points are available to establish the utility curve (see Figure 11.8b to d.)

4. Finally, a best-fitting curve is drawn to represent the utility function, as shown in Figure 11.9.

Having established the utility curve, the equivalent utility for all the possible outcomes in the decision problem can be obtained directly from the curve. The alternatives are then evaluated according to their expected utility values. Figure 11.10 shows an analysis of the contractor's problem using utility values from the curve in Figure 11.9. If this utility curve truly represents the contractor's risk behavior, the proper choice would be to crash the project. This is the same conclusion reached by EMV in Figure 11.5. However, the decision at node *C* would be changed. Hence the results obtained using utility values may be, but are not necessarily, different from those obtained using EMV. The shape of the utility curve has a pronounced effect on the results when using utility values.

The foregoing procedure for establishing the utility function can be easily modified for decision problems involving profits instead of losses. For example,

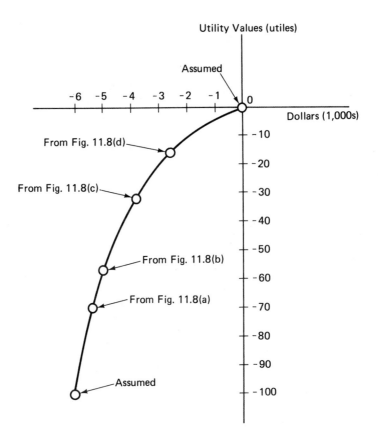

Figure 11.9   Utility curve for the contractor.

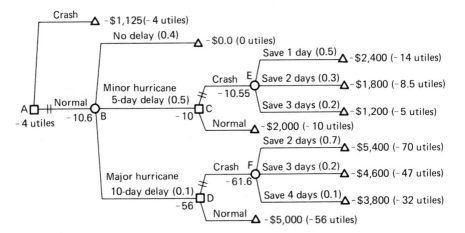

Figure 11.10   Analysis of construction problem using expected utility value.

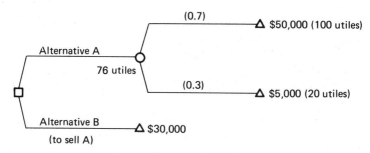

Figure 11.11 Hypothetical decision problem to aid in establishing a utility function involving profits.

the hypothetical problem in Figure 11.11 may be confronted by an individual. In alternative A, there is a 70 percent chance of making a $50,000 profit and a 30 percent chance of making a $5000 profit. Suppose that the utility values for $50,000 and $5000 profits have already been known to be 100 and 20, respectively. In alternative B, the individual has the chance to sell alternative A to another person. What is the minimum amount for which there is a willingness to sell A? Suppose that the answer is $30,000. The utility value of $30,000 is thus computed as $(0.7)(100) + (0.3)(20) = 76$.

11.6 BAYES' THEOREM

Let A and B be two events such that the occurrence of one event influences directly the occurrence of the other event. Then the conditional probability that event A would occur after knowing that event B has already occurred is denoted as $P(A|B)$, which reads the probability of A given B. *Bayes' theorem* states that

$$P(A|B) = \frac{P(B|A)P(A)}{P(B)} \tag{11.3}$$

This theorem can easily be proved using the diagram in Figure 11.12. Let area c bounded by the rectangle represent the sample space, area a the number of occurrences of event A, area b the number of occurrences of event B, and the common area d represent the event that both A and B occur together. Thus

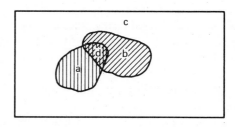

Figure 11.12 Venn diagram for events A and B.

$$P(A) = \frac{a}{c}$$

$$P(B) = \frac{b}{c}$$

$$P(A|B) = \frac{d}{b}$$

$$P(B|A) = \frac{d}{a}$$

Also,

$$\frac{P(B|A)P(A)}{P(B)} = \frac{(d/a)(a/c)}{b/c} = \frac{d}{b}$$

Therefore,

$$P(A|B) = \frac{P(B|A)P(A)}{P(B)}$$

This theorem is used extensively in decision analysis for computing the conditional probability of chance events. It is especially useful for problems in which engineering tests are conducted to predict the state of uncertain variables; for example, such as problems in which sounding is used to determine the depth of water, drilling to determine the depth of bedrock, or field testing to determine the strength of concrete. These tests usually cannot reveal the true state of the uncertain variables. The results can, at best, provide more reliable estimates of the probability of occurrence of the various possible states of the variables.

As an example, suppose that the contractor in the earlier problem decides to engage the consulting service of an expert meteorologist whose ability to predict the true state of events is given in Table 11.1. If the true state is T_1,

TABLE 11.1   Conditional Probabilities $P(S_i|T_i)$

Predicted State	True State		
	T_1: No Hurricane	T_2: Minor Hurricane	T_3: Major Hurricane
S_1: No Hurricane	0.8	0.3	0.1
S_2: Minor Hurricane	0.2	0.6	0.2
S_3: Major Hurricane	0.0	0.1	0.7

there is an 80 percent chance that the expert would predict S_1 and a 20 percent chance that the prediction would be S_2. Similarly, if the true state is T_2, then there is a 30 percent, 60 percent, and 10 percent chance of the prediction of the states S_1, S_2, and S_3, respectively. If the true state is T_3, then the probabilities are 10 percent, 20 percent, and 70 percent for predicting states S_1, S_2, and S_3, respectively.

Figure 11.13 is a decision tree of the problem now facing the contractor.

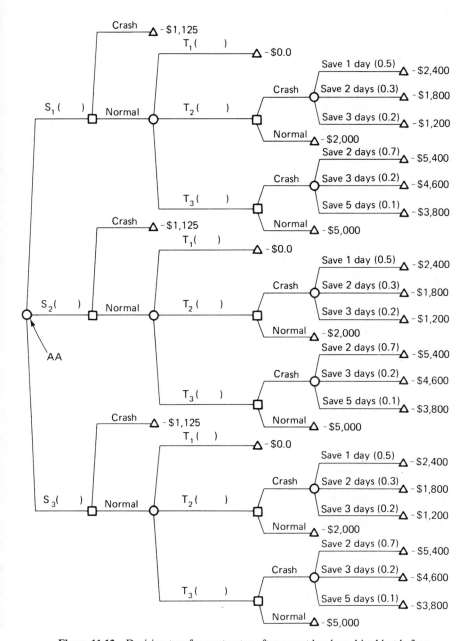

Figure 11.13 Decision tree for contractor after expert has been hired but before he has made his prediction.

The decision is to be based on the meteorologist's prediction. Before analysis can begin, the probability must be determined for each of the chance events. Let the conditional probability that T_j is the event that actually occurs after S_i has been predicted be denoted as $P(T_j | S_i)$. It is intuitively obvious that $P(T_1 | S_1) \neq P(T_1 | S_2)$, and so on. The immediate problem therefore, is to determine these conditional probabilities from the available information in Table 11.1.

The probability $P(S_i)$ for $i = 1$ to 3, can first be computed by enumerating all the different conditions under which the meteorologist would predict state S_i and summing the probabilities of these conditions. Figure 11.14 presents the graph model of all the different combinations of events. The probability that a true state T_j occurs remains the same as in the preceding problem. The probability that the true state is T_j and that the meteorologist predicts S_i, $P(S_i$ and $T_j)$, is then computed as $P(T_j) \times P(S_i | T_j)$. Thus, the probability of each possible combination of S_i and T_j is computed and given in Figure 11.14 in the column $P(S_i$ and $T_j)$. As a check, the sum of the probabilities under this column must be 1. The probability that the meteorologist would predict S_i is then simply computed as follows:

$$P(S_i) = P(S_i \text{ and } T_1) + P(S_i \text{ and } T_2) + P(S_i \text{ and } T_3)$$

Hence

$$P(S_1) = 0.32 + 0.15 + 0.01 = 0.48$$
$$P(S_2) = 0.08 + 0.30 + 0.02 = 0.40$$
$$P(S_3) = 0.0 \ + 0.05 + 0.07 = 0.12$$

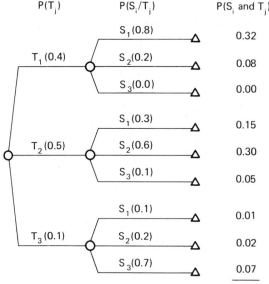

P(T$_j$)	P(S$_i$/T$_j$)	P(S$_i$ and T$_j$)

Sum = 1.00

Figure 11.14 Graph model of possible combinations of S_i and T_i.

Next, the conditional probabilities $P(T_j|S_i)$ for $i = 1$ to 3 and $j = 1$ to 3 can be computed directly using Bayes' theorem.

$$P(T_1|S_1) = \frac{P(S_1|T_1)P(T_1)}{P(S_1)} = \frac{(0.8)(0.4)}{0.48} = 0.667$$

$$P(T_2|S_1) = \frac{(0.3)(0.5)}{0.48} = 0.312$$

$$P(T_3|S_1) = \frac{(0.1)(0.1)}{0.48} = 0.021$$

$$P(T_1|S_2) = \frac{(0.2)(0.4)}{0.40} = 0.20$$

$$P(T_2|S_2) = \frac{(0.6)(0.5)}{0.40} = 0.75$$

$$P(T_3|S_2) = \frac{(0.2)(0.1)}{0.40} = 0.05$$

$$P(T_1|S_3) = \frac{(0.0)(0.4)}{0.12} = 0.0$$

$$P(T_2|S_3) = \frac{(0.1)(0.5)}{0.12} = 0.417$$

$$P(T_3|S_3) = \frac{(0.7)(0.1)}{0.12} = 0.583$$

These conditional probabilities can now be entered into the decision tree and decision analysis can next proceed according to the procedures described previously to obtain an expected value of $-\$931.92$, as shown in Figure 11.15.

It is interesting to note that the previous calculation for conditional probabilities using Bayes' theorem can be formulated directly in algebraic form as follows: Let p_{ij} be the conditional probability for predicted state S_i and true state T_j. Then it is a simple matter to write algebraic equations for each probability statement made in the problem formulation. Thus Table 11.1 can be recast as shown in Table 11.2.

The history of hurricanes:

$$P(\text{no delay}): \qquad p_{11} + p_{21} + p_{31} = 0.40$$
$$P(\text{minor hurricane}): \quad p_{12} + p_{22} + p_{32} = 0.50 \qquad (11.4)$$
$$P(\text{major hurricane}): \quad p_{13} + p_{23} + p_{33} = 0.10$$

The history of the past performance of the expert gives

$$\begin{array}{lll} p_{11} = p_{21} & p_{12} = 3p_{32} & p_{23} = 3p_{13} \\ p_{31} = 0 & p_{22} = 6p_{32} & p_{33} = 7p_{13} \end{array} \qquad (11.5)$$

$$\sum p_{ij} = 1.00 \qquad (11.6)$$

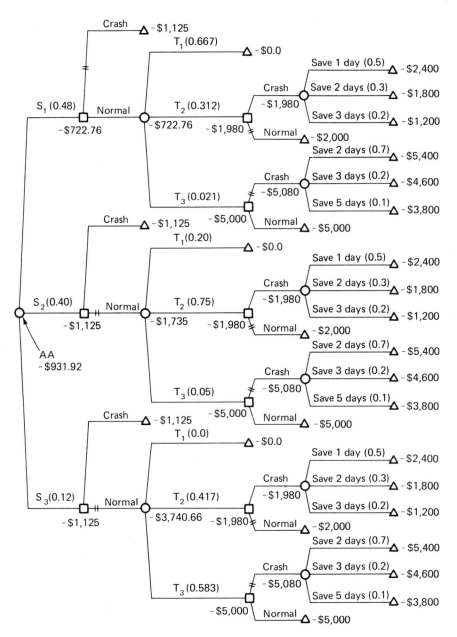

Figure 11.15   Decision analysis after expert has been hired but before he makes his prediction.

TABLE 11.2   Conditional Probabilities p_{ij}

	True State			
Predicted State	T_1: No Hurricane	T_2: Minor Hurricane	T_3: Major Hurricane	$\sum_{j=1}^{3} p_{ij}$
S_1	p_{11}	p_{12}	p_{13}	P(predicting no delay)
S_2	p_{21}	p_{22}	p_{23}	P(predicting minor hurricane)
S_3	p_{31}	p_{32}	p_{33}	P(predicting major hurricane)
	P(no delay)	P(minor hurricane)	P(major hurricane)	(certainty)

TABLE 11.3   Values of Conditional Probabilities p_{ij}

Predicted State	True State			$\sum_{j=1}^{j=3} p_{ij}$
	T_1	T_2	T_3	
S_1	0.32	0.15	0.01	0.48
S_2	0.08	0.30	0.02	0.40
S_3	0	0.05	0.07	0.12
	0.40	0.50	0.10	1.00

Solving these equations gives the results shown in Table 11.3 and

$$P(T_1 | S_1) = \frac{0.32}{0.48} = 0.667$$

$$P(T_2 | S_1) = \frac{0.15}{0.48} = 0.312 \tag{11.7}$$

$$P(T_3 | S_1) = \frac{0.01}{0.48} = 0.021 \qquad \text{etc. as before.}$$

11.7 VALUE OF INFORMATION

It has been assumed until now that the probabilities and values of the outcomes, either in monetary or utility terms, have been known for the decision problem. With regard to practical problems, it is often necessary to conduct studies or investigations to determine these probabilities and values. The question then arises as to how much should be spent to obtain this information.

Recall the contractor problem from the previous sections. The contractor can hire a meteorologist to obtain a more reliable prediction of weather conditions. How much would the contractor be willing to pay the meteorologist? One way of viewing the problem is to put Figures 11.5 and 11.15 together, as shown in Figure 11.16, where the heavy mark on branch *a* can be considered as a toll gate for which the contractor must pay a toll if this alternative is chosen.

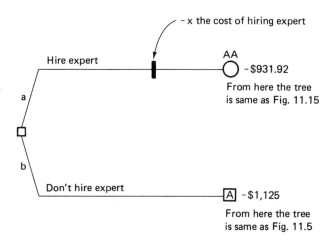

Figure 11.16   Decision tree to determine the value of new information.

The problem is then to determine the maximum toll the contractor would be willing to pay and still choose to hire the expert. Let this toll be represented by $x. If the contractor were indifferent between the two alternatives, his expected value of both would be equal. The expected value after hiring the expert is $(-x - 931.92)$ and the expected value before hiring the expert is $-$1125$. Hence

$$-x - \$931.92 = -\$1125$$
$$x = \$1125 - \$931.92 = \$193.08$$

which is the upper limit the contractor should pay for the services of the meteorologist. Thus the value of sampling or investigating to obtain new information is equal to the increase in the expected value that will be derived from this information. In this case, the cost of hiring the meteorologist was considered after the expected values had been computed. Another way to consider this cost would be to add this amount to the cost associated with each outcome that is a possible result of the decision to hire the meteorologist. That is, the cost $x would be added to each possible outcome in Figure 11.15 and then the expected value would be computed. The result is the same if the contractor is an EMV'er. However, if the contractor is a non-EMV'er, then all the computations must be repeated using utility values at the terminal nodes.

11.8 SUMMARY

This chapter presents a method for choosing the optimum plan or policy from a specific set of possible alternatives.

The expected monetary value is not always a good criterion because of the personal values of the decision maker which may be quite different from person to person, may vary with time, and is, in general, a nonlinear function of the monetary values. These personal values are indicated by the utility function which is a powerful concept, but one that is difficult to assess precisely.

The approach to decision making in probabilistic situations should include:

1. Determining the possible results of each of the alternatives
2. Determining the probability of each outcome
3. Determining the utility curve for the decision maker and assigning utility values to the results
4. Calculating the expected utility values of the alternatives

The selection criterion is the maximum expected utility value. A decision tree, a special type of linear graph, can be used to graphically represent the decision problem.

Bayes' theorem provides an easy means for computing conditional probabilities. The decision maker should continue to acquire new information as long as the increase in expected utility attributable to this new information is greater than the cost.

If it is not possible to estimate the probabilities of the outcomes, then expected values cannot be used as a criterion in decision making. However, if the problem is important enough to receive formal analysis, the decision maker must either make a subjective estimate of the probabilities or obtain data that will permit an estimate to be made.

PROBLEMS

P11.1. What is the contractor's optimum decision policy for the problem illustrated in Figure 7.2 and explained in Example 4.3?

P11.2. In Problem P11.1 the contractor hires an expert meteorologist whose ability to predict accurately normal water level, high water, and floods has the same conditional probabilities as given numerically in Table 11.1 for no, minor, and major hurricanes. For this situation determine the maximum fee the contractor would be willing to pay the meteorologist.

P11.3. A county highway engineer is allotted a fixed annual budget for maintenance of the county road network. The county road network contains different types of road surfaces and conditions.

The engineer is uncertain whether to plan the maintenance activities on a 1-, 2-, or 3-year time horizon. He can postpone any activity for any road and tolerate deterioration in quality until the road must be replaced, initiate temporary or semipermanent maintenance that either ensures the continuation of existing conditions or delays the time when the road must be replaced, or initiate a road replacement program with the same or better type roadways.

How would you go about formulating the problem as a n-year decision problem and what type of information would be required?

P11.4. A plant manager is about to commence an urgent above normal production run. To assure meeting the desired production rate, she has decided to transfer and incorporate into the production line a piece of equipment from another line at a cost of $5000. The manager is considering whether to overhaul this equipment before placing it in the new production line.

The piece of equipment costs $800 to overhaul, whereas if she incorporates the item into the process line and it then breaks down it will cost $1500 to cover the cost of repair and lost time. The manager estimates that there is a 66 percent chance that the equipment motor is reliable but is assured that it will be reliable if it is overhauled. A dynamometer test of the motor costs $100 but will only indicate whether the motor is in good or bad condition with a 10 percent chance that the test results prove invalid. The manager estimates that there is a 70 percent chance that the dynamometer test will indicate a reliable motor.

(a) Model the decision problem faced by the plant manager.

(b) If the plant manager wishes to base her decision on an EMV policy, what should be her optimum strategy?

P11.5. A group of investors is considering two alternative plans for a high-rise building complex. Plan I calls for a 70-story apartment building and a separate adjacent 40-story office building. Plan II calls for a single 100-story high-rise building with 43 stories for offices and 55 stories for offices, stores, and shops and 2 stories for mechanical and operational systems. The estimated costs and lifetime return of the two plans are as follows:

Plan I

Estimated Cost ($\times 10^6$)	Probability	Estimated Lifetime Return ($\times 10^6$)	Probability
$100	0.6	$300	0.5
95	0.3	250	0.4
90	0.1	200	0.1

Plan II

Estimated Cost (× 10⁶)	Probability	Estimated Lifetime Return (× 10⁶)	Probability
$150	0.7	$450	0.2
120	0.2	350	0.4
100	0.1	250	0.3
		200	0.1

If the investors prefer plan II, provided that the expected net profit differs by less than $10 million, which plan should they choose?

P11.6. The board of supervisors for a county is planning to build a dam costing $5 million on Mountain Creek. To protect the dam, a separate spillway is required and the board must decide whether to build a large spillway costing $3 million or a smaller spillway costing $2 million. Based on historical records, it is estimated that there is a 0.25 probability that one or more serious floods would occur during the life of the dam and a 0.10 probability that one or more major floods would occur. The probabilities that the two possible spillway types will fail during these two levels of floodings are estimated as follows.

	Serious Flood		Major Flood	
	Fail	Safe	Fail	Safe
Large spillway	0.05	0.95	0.1	0.9
Small spillway	0.10	0.90	0.25	0.75

If the spillway fails to function properly during a serious or major flood the dam will be destroyed. The replacement cost of the dam will be the same as the original cost. In addition to a total loss of the dam and its spillway, other property damages will be incurred. It is estimated that in case of failure during a serious flood, there is a 70 percent and 30 percent chance for other property losses amounting to $1 million and $3 million, respectively. In case of failure during a major flood, there is a 70 percent and 30 percent chance for other losses amounting to $3 million and $5 million, respectively.

(a) Model this decision problem with a decision tree.

(b) What is the optimum decision based on the EMV criterion?

(c) How would you caution the board about basing its decision on the EMV criteria?

Suppose now that the risk behavior of the board is as follows:

1. Against a 90 percent chance of losing $7 million and a 10 percent

chance of losing $20 million, the board is willing to pay insurance of $12 million.

2. Against a 70 percent chance of losing $12 million and a 30 percent chance of losing $20 million, the board is willing to pay insurance of $15 million.

3. Against a 50 percent chance of losing $12 million and a 50 percent chance of losing $20 million, the board is willing to pay $17.5 million in insurance.

(d) Draw a utility curve assuming a utility value of 50 for $-\$7$ million and a utility value of -100 for $-\$20$ million.

(e) What should be the board's decision in part (b) when this utility curve is used?

P11.7. A heavy-machinery manufacturing company is considering submitting a bid proposal for the supply and installation of the machinery system for the swing span of a bridge. The chief engineer of the company has estimated that a bid proposal can be prepared at a cost of $5000, but that such a bid will have only a 20 percent chance of being accepted. As an alternative, the company can invest $20,000 on an extensive research study of the project before preparing the bid proposal. Such a proposal will have a 30 percent chance of being accepted.

If the company does get the job, it can either expand its personnel and staff or subcontract work to other companies. If a major portion of the work is to be performed by subcontractors, it is estimated that there is a probability of 70 percent, 20 percent, and 10 percent for making a profit of $1 million, $1.5 million, and $2.0 million, respectively. If the work is to be accomplished mostly by expanding staff, there is a probability of 60 percent, 35 percent, and 5 percent for profits of $0.5 million, $2 million, and $3 million, respectively.

(a) Suppose that you were the company president and that your company has total assets of $10 million. Prepare a utility curve to reflect *your* risk characteristics.

(b) Based on the utility curve above, what is your optimum policy?

(c) What is the maximum dollar investment that you would be willing to provide for the preliminary proposed study?

P11.8. Discuss the subjective nature of the decision-making approach implicit in decision tree analysis? What alternative approaches exist for the decision maker?

System Simulation

12

12.1 SIMULATION CONCEPTS

The role of models in systems analysis is illustrated in Figure 12.1. Because engineering systems are so complex, a system model is first constructed to represent the real system and its environment. The model should include all the relevant system components and clearly define the component interrelationships, as well as indicate the constraints within and those imposed on the system. During the construction and operation of the model the designers have complete control over the components, structure, and constraints within their model. It is with this model that they test their designs and study the behavior of the system under various conditions.

Simulation is the process of conducting experiments with a model of the system that is being studied or designed. It is a powerful technique for both analyzing and synthesizing engineering systems. In an analysis problem, the system model is generally fixed and the objective is to determine the system response when a set of input variables is allowed to take on different values. The simulation process is then basically an iterative procedure and may be described as an input–output study with feedbacks provided to guide the changes in the input parameters, as illustrated in Figure 12.2. The inputs define the set of events and conditions to which the system can be subjected in the real world, and the outputs predict the system response. By studying the outputs at the end of each iteration, the designer learns more and more about the system and may then use the newly acquired knowledge to define new sets of inputs to be processed through the model.

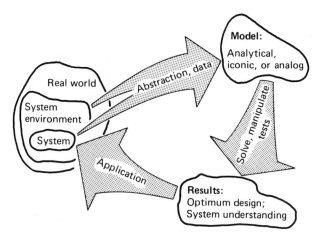

Figure 12.1 Role of models in systems analysis.

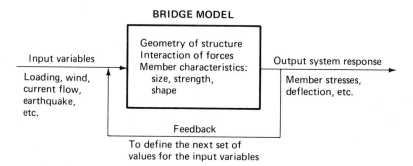

Figure 12.2 Analysis by simulation.

As an example, consider the problem of determining the failure conditions of a steel bridge. The simulation model may be in the form of a set of mathematical equations relating the interaction of the forces among the members, the geometric structure of the bridge, the size and tensile strength of the members, and so on; the input variables may include loading conditions, velocity and direction of wind, the flow velocity of water and debris in the channel, and the occurrence of earthquakes. For each specific set of combination of the input parameters, the model may be used to determine the stresses in the members and the deflection of the structure, which provide a direct measure of the performance of the system under the given set of conditions.

In system synthesis, the designer is interested in determining how the system components can best be put together so that the system can meet the performance standard. In this case, the system model itself is a variable; but a set of input–output characteristics has been specified as a design standard. The simulation process is again an iterative procedure, but the output of a simulation study is now used to decide which system parameter can best be changed in

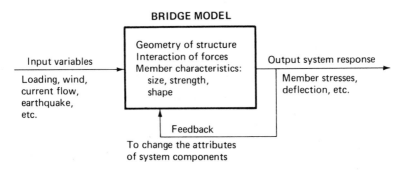

Figure 12.3 Synthesis by simulation.

order to improve the performance of the system. As shown in Figure 12.3, the feedback is now directed to changes within the model itself. For example, in the bridge example above, the problem may be to choose an optimum combination of sizes and strengths for the structural steel members. These design parameters directly affect the strength of the total structure, its weight, as well as its cost. If one design fails to meet the performance standard according to one simulation result, these system parameters may be changed and the simulation repeated until a set of feasible alternatives have been established and an optimum solution is identified.

The major advantage of the simulation approach lies in the fact that once a simulation model has been constructed, it may be used for both system analysis and synthesis and to test the design under a wide spectrum of environmental conditions. Moreover, after the design has been completed and the system implemented in the real world, the simulation model may be used to locate the sources of unpredicted system problems and to plan for system improvement.

12.2 SIMULATION MODELS

Simulation models can take many forms and can be of many different levels of complexity. A good model should represent the characteristics of the system so that the problem under consideration can be solved. Simulation models can be broadly grouped into three types:

1. Iconic
2. Analog
3. Analytical

Iconic models are physical replicas of the real systems on a reduced scale. This type of model is common in engineering. In aircraft design, wind tunnels are used to simulate the environment around an aircraft in flight. By subjecting a model of the aircraft to a well-chosen range of aerodynamic conditions, the designer can gain insight into the performance characteristics of the design. In the design of large engineering structures, such as skyscrapers, dams, bridges, and airports, three-dimensional architectural models are often prepared to provide a realistic view of the design. Such models are extremely useful both as a design tool and as a visual aid in presenting the project to the interested public.

Iconic models are particularly important for studying systems in which the interrelationship of the components are not well understood or in which components are too complex to be modeled mathematically. Thus, by constructing a replica of the system and experimenting with it under a set of controlled conditions, an insight can be gained into the system behavior. The end results may well be the discovery of some natural law that governs the components of that system.

In many engineering problems it is impossible to build a physical replica of the real system. For example, in studying the response of engineering structures to various intensities of earthquakes, it is impossible to build a small model of the earthquake zone using rocks and soils and to generate earthquakes at the command of the experimenter. However, if the dynamic property of quake waves is known, an instrument may be constructed to generate a similar type of force motion. At the University of Illinois at Urbana-Champaign, an electro-hydraulic system is used to simulate earthquakes (Sozen et al., 1969). It consists of a 12-ft^2 table driven by a hydraulic ram and is designed to activate a 10,000-lb mass in one horizontal direction with a sinusoidal or random vibration. The simulation is used to determine the response of reinforced concrete structures to earthquake shocks. Such a simulation model, in which the real system is modeled through a completely different physical media, is called an *analog model*.

In problems in which the characteristics of the system components and system structure can be mathematically defined, an analytical model constitutes a powerful simulation tool. It may be composed of systems of equations, boundary constraints, and heuristic rules, as well as numerical data. Basically, the model consists of a set of design variables and a set of system constants. Some of the design variables are independent parameters, the values of which are specified as input to the simulation process. The remaining variables are dependent variables that are used to measure system performance and response and thus they constitute the outputs. However, to construct an analytical model requires that fundamental properties of the system components and their interactions be understood.

With the availability of high-speed, large-memory electronic computers, analytical models are becoming versatile design aids in all disciplines of engineering. For example, analytical simulation models are playing an increasingly important role in air pollution control (Middleton, 1971). Although many electronic sensors have been developed to provide measurement of pollution levels, a sample network within a city is usually widely scattered and is not sufficiently dense to identify all heavily polluted areas. Moreover, such a system cannot predict the effect of new sources of pollution, such as proposed industrial plants or highways, on the overall community. Analytical simulation models have been a useful tool in both monitoring and predicting pollution levels. The model may consist of mathematical functions relating such system factors as the rates of pollutant decay, weather condition, topography of the locality, and the distribution of commercial, industrial, and residential districts. Once the major sources of pollutants are identified, such a model may be used to determine the pollution level anywhere within the community during different parts of the day and under different weather conditions. The simulation results may even include a contour map of the pollution level throughout the municipality. Moreover, once the model is constructed, it may also be used to study the effectiveness of the various methods of pollution control as well as the adverse effects of new pollution sources.

12.3 SIMULATING FLOOD CONDITIONS

To illustrate the general principle of simulation, consider again the decision problem confronting the contractor in Example 4.3. The contractor has the following options available:

Option 1.   Move the construction equipment away from the river flat area, store it, and then move it back four months later at a total cost of $1800.

Option 2.   Leave the equipment at the river flat area but elevate it from the ground with a platform, thus protecting it from damage by high water. If the actual water level in that spring is either normal or high, the total cost to the contractor would be the cost of building the platform (i.e., $500). However, if a flood occurs, the total cost would be $60,000.

Option 3.   Leave the equipment at the river flat area, and do not build a platform to protect it. If the actual water level is normal through the period, the total cost to the contractor would be $0. However, if high water occurs, the total cost would amount to $10,000, and if a flood occurs, the cost would amount to $60,000.

The probability of normal, high-water, and flood conditions is estimated to be 0.73, 0.25, and 0.02, respectively.

This decision problem can be analyzed by the methods of decision analysis discussed in Chapter 11. It is given as a practice exercise in Problem P11.1. The problem can also be easily analyzed by the method of simulation.

A simple simulation model of the river condition can be constructed by using 100 pieces of small paper chips and a paper bag. A letter N is marked on 73 paper chips to represent the probability of normal water level. A letter H is marked on 25 paper chips to represent the probability of high water, and a letter F is marked on the remaining 2 paper chips to represent the probability of flood. The 100 paper chips are then thoroughly mixed in the paper bag. The bag and its contents constitute a model of the river. With this simple model, the water level for as many years as needed can be generated. For example, to generate water level for one year, simply shake the contents of the bag thoroughly and then draw one chip from the bag. The letter on the chip then indicates the water level. To simulate the condition for a second year, simply return the chip to the bag, shake it thoroughly, and then pick another chip from it. The process can be repeated as often as necessary.

Table 12.1 shows the result of simulation experiment conducted in this manner. The simulated period is 100 years. The total cost to the contractor for each of options 2 and 3 above for each simulated river condition is also listed in the table. Over the 100-year period, if the contractor consistently decided to

TABLE 12.1   Simulating 100 Years of Flood Conditions*

Year	Simulated† River Condition	Total Cost to Contractor		Year	Simulated Water Level	Total Cost to Contractor	
		Platform Is Built	No Platform			Platform Is Built	No Platform
1	N	500	0	51	N	500	0
2	N	500	0	52	H	500	10,000
3	H	500	10,000	53	N	500	0
4	N	500	0	54	N	500	0
5	H	500	10,000	55	H	500	10,000
6	H	500	10,000	56	N	500	0
7	N	500	0	57	N	500	0
8	N	500	0	58	H	500	10,000
9	H	500	10,000	59	N	500	0
10	N	500	0	60	H	500	10,000
11	N	500	0	61	N	500	0
12	H	500	10,000	62	F	60,500	60,000
13	N	500	0	63	N	500	0
14	H	500	10,000	64	H	500	10,000
15	N	500	0	65	N	500	0
16	N	500	0	66	N	500	0
17	N	500	0	67	H	500	10,000
18	N	500	0	68	H	500	10,000
19	N	500	0	69	N	500	0
20	N	500	0	70	H	500	10,000
21	N	500	0	71	N	500	0
22	H	500	10,000	72	H	500	10,000
23	N	500	0	73	H	500	10,000
24	N	500	0	74	H	500	10,000
25	N	500	0	75	N	500	0
26	N	500	0	76	H	500	10,000
27	H	500	10,000	77	H	500	10,000
28	H	500	10,000	78	N	500	0
29	N	500	0	79	N	500	0
30	H	500	0,000	80	N	500	0

TABLE 12.1   (Continued)

Year	Simulated† River Condition	Total Cost to Contractor Platform Is Built	Total Cost to Contractor No Platform	Year	Simulated Water Level	Total Cost to Contractor Platform Is Built	Total Cost to Contractor No Platform
31	H	500	10,000	81	H	500	10,000
32	H	500	10,000	82	N	500	0
33	N	500	0	83	N	500	0
34	N	500	0	84	N	500	0
35	N	500	0	85	N	500	0
36	H	500	10,000	86	N	500	0
37	N	500	0	87	N	500	0
38	N	500	0	88	N	500	0
39	N	500	0	89	N	500	0
40	N	500	0	90	N	500	0
41	N	500	0	91	N	500	0
42	N	500	0	92	N	500	0
43	N	500	0	93	N	500	0
44	N	500	0	94	N	500	0
45	N	500	0	95	F	60,500	60,000
46	N	500	0	96	N	500	0
47	H	500	10,000	97	N	500	0
48	H	500	10,000	98	N	500	0
49	N	500	0	99	H	500	10,000
50	N	500	0	100	N	500	0

*Average annual cost if platform is built $1700
 Average annual cost if no platform is built $4200
 Average annual cost if equipment is moved $1800
†N (normal), H (high water), F (flood).

protect the equipment with a platform, the average annual cost would be $1700. If he consistently decided to leave the equipment at the river flat area without platform protection, the average annual cost would be $4200. Of course, the average annual cost of option 1 (moving the equipment away from the area) would be $1800.

Thus, should the constructor decide to leave the equipment in the river flat area, it is more economical, on the average, to spend $500 to build a platform to protect it against high water. Basically, the constructor is paying $500 for insurance against a 25 percent chance of losing $10,000. On the other hand, the average cost of moving the equipment ($1800) is only slightly higher than the average cost ($1700) of a platform protection scheme. At an additional average cost of $100, the contractor is protected against 20 percent chance of losing $60,000 due to flood damage. A prudent contractor, who has a relatively small capital, should therefore spend $1800 to move the equipment. However, ultimately, the decision will depend to a great extent on the contractor's willingness to take risks and estimates on the probability of high water or flood occurring.

The average annual costs shown in Table 12.1 were computed from a single simulation experiment for a 100-year period. Due to the probabilistic nature of the events, simulation of another 100-year period will probably yield different annual costs. The shorter the simulation period, the more variable will be the results of the different simulation periods. On the other hand, the longer the simulation period, the smaller will be the differences in the results. Thus, in this type of simulation experiments, a long simulation period is usually needed to yield realistic results. In this particular example, if the simulation period is extended to a much longer period, such as 1000 or 10,000 years, the simulation result should always yield an annual cost of $1800, $1700, and $3700 for options 1, 2, and 3, respectively. These figures, in fact, represent the expected monetary value (EMV) of the three options and can be easily computed by the method explained in Chapter 11.

Therefore, although the method of simulation is useful in studying problems involving probabilistic events, a direct solution based on a mathematical formulation of the problem is generally preferable. However, there are many problems that cannot be formulated as a mathematical solution, and simulation becomes the only means of solving such problems. An example of such a problem will be given in the next section.

12.4 A WAITING-LINE SIMULATION EXAMPLE

Problems that involve waiting lines or queues are frequently found in a variety of engineering design and planning situations. These problems are characterized by the analysis of arrivals at a point in a system and the processing of those arrivals. The concern in these cases is associated with the length of waiting line or queue that may develop. This type of problem may be commonly found in traffic flow and construction operations.

As an example, assume that a drive-in banking facility is being planned. For such a problem one of the planning and design concerns is related to the amount of space that is required to accommodate persons in vehicles who are waiting to undertake banking business. To keep the problem simple, assume that there is only one service window and customers are serviced in the order of arrival. For this problem, assume that the average arrivals will be one every 2 minutes or 0.5 per minute. Based on experience, it has been estimated that a teller can process 1.0 customer per minute. The design and planning question is the length of line that may develop.

In addressing a problem of this type, the analysis must consider not only the average arrivals and servicing but also the range of values that may occur

given those averages. For the purpose of this example, assume that the arrivals
are distributed randomly in time; therefore, the probability of a specific number
of arrivals can be estimated by using the Poisson equation, which is

$$P_{(x)} = \frac{e^{-M}M^x}{x!}$$

where $P_{(x)}$ = probability of x arrivals
     M = mean arrival rate
     e = base of the natural system of logarithms (2.71828)

Given the mean arrival rate of 0.5 vehicle per minute, the probability for
each number of arrivals is

x	$P_{(x)}$
0	0.6065
1	0.3033
2	0.0758
3	0.0126
4	0.0016
5	0.0002

It must be recognized that if the computations are carried out to greater
accuracy, there is a remote possibility of greater arrivals. For this example,
however, the probabilities will be used as shown. In reality, the probability of
5 arrivals is the value for 5 or more arrivals.

For the processing of customers, a distribution of probabilities could also
be developed based on a study of bank operations. Assume for this problem
that the following distribution of processing of customers has been ascertained:

Number of Customers Processed per Minute	Probability of Occurrence
0	0.20
1	0.60
2	0.20

To illustrate the simulation of the waiting line or queue, random numbers
will be utilized in the generation of arrival and processing input. The use of
random numbers will be more fully explained in Section 12.6. Based on the
probability of occurrence, blocks of numbers are assigned to each number of
arrivals or customers processed. These assignments are as follows:

x	$P_{(x)}$	Numbers Assigned
	Arrivals	
0	0.6065	0000 to 6064
1	0.3033	6065 to 9097
2	0.0758	9098 to 9855
3	0.0126	0956 to 9981
4	0.0016	9982 to 9997
5	0.0002	9998 to 9999
	Processing	
0	0.20	00 to 19
1	0.60	20 to 79
2	0.20	80 to 99

For each of the time periods, a random number is selected from the table of random numbers in Appendix A. Each of the columns in Appendix A consist of six digits; thus the four digits on the left will be used to indicate arrival data and the two digits on the right will be used to indicate processing input information. The random number in the table is matched with the assigned blocks of numbers to determine the arrivals and numbers of customers processed in each time period. Assume that the simulation is initiated using the six digits

TABLE 12.2   Simulation of Drive-In Bank

	Arrivals		Processing		Number Waiting
Time Period	Random Number	Number of Arrivals	Random Number	Number Processed	at End of Time Period
1	7911		23	1	0
2	1319	0	99	2	0
3	8563	1	03	0	1
4	6434	1	45	1	1
5	4410	0	71	1	0
6	7410	1	17	0	1
7	1236	0	38	1	0
8	5211	0	95	2	0
9	4206	0	73	1	0
10	7376	1	14	0	1
11	1230	0	71	1	0
12	1873	0	87	2	0
13	5318	0	32	1	0
14	2555	0	91	2	0
15	3763	0	55	1	0
16	4965	0	06	0	0
17	5667	0	25	1	0
18	7619	1	84	2	0
19	4608	0	50	1	0
20	5774	0	61	1	0

at the lower right corner of the last page of random numbers. Subsequent random numbers will be selected by moving upward in the column. The simulation of the arrival and servicing of customers for 20 time periods is shown in Table 12.2. The length of waiting line is determined by subtracting the number processed from the number of arrivals for that period plus the number waiting from the previous period. In this case it would be necessary to run this example for several hundred time periods to obtain stabilized results.

12.5 SIMULATING A CONCRETE PLANT OPERATION

Consider a problem involving the operations of a company producing ready-mixed concrete. The company now owns a batching plant to mix the appropriate quantities of cement, sand, gravel, water, and special additives to produce concrete mixes that are then loaded into delivery trucks for delivery to the customer. The company now owns five delivery trucks. The company management has recognized that the full production capacity of the batching plant has not been utilized because of the limited number of trucks. Their problem is to determine the optimum number of delivery trucks that can be added to their fleet so that its batching plant can be fully utilized.

Figure 12.4 schematically illustrates the production process in the company's system. Company policy has dictated that orders be filled on a first-come-first-served basis, and in order to have all trucks back in time for clean up, no order is to be processed after 3:30 P.M. When an order is received at the central office, it is immediately placed in the waiting list of orders. Eventually, when its turn arrives for processing, the order is transmitted to the batching plant, which then initiates the following procedure. The batching plant operator calls the parking lot and requests the required number of trucks to fulfill the order. If

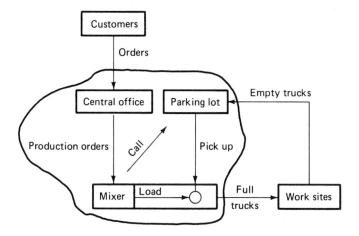

Figure 12.4   Ready-mix concrete plant production process.

delivery trucks are available, they are dispatched to the batching plant where the order is then loaded for delivery. However, if the required number of trucks are not available, those trucks that are available are loaded and dispatched to the delivery site. Then the batching plant has to stop operations and wait for trucks to return before completing the order. As a result of the existing shortage of trucks, there is a continuous backlog of orders and the batching plant is idle a large part of the day.

The general procedure of a simulation study of this problem is illustrated in Figure 12.5. A simulation model is constructed to represent the operational procedure within the concrete plant. A set of fictitious customer orders is generated to approximate the inflow of orders in the real system during each day.

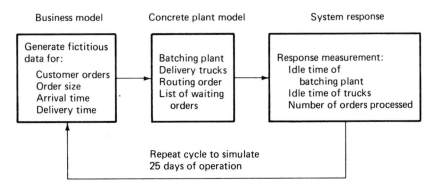

Figure 12.5   Simulation of a concrete plant operation.

This set of fictitious orders is processed through the system, and the system response is measured according to the following set of parameters: idle time for the batching plant and the delivery trucks, number of orders processed, and number of truckloads delivered. Thus, by repeating the experiments for an extended simulation period of, say, 6 months, an accurate measure of the average idle time for the batching plant in one day may be obtained. Moreover, the same set of experiments can be repeated with a different number of trucks in the model. The outcome of the simulation process then provides a measure of the system response depending on the number of delivery trucks available.

12.5.1 Modeling Customer Orders

The purpose of modeling customer orders is to provide a means whereby a fictitious flow of customer orders can be generated for each day in the simulation process. The model should be representative of the real situation so that the flow of orders generated by it would closely resemble that normally encountered in the concrete plant. Within the scope of this problem, three elements of the customer orders are of major importance: time between orders, number of truckloads in each order, and the round-trip delivery time required for the orders.

TABLE 12.3 Records from the Real System

Day	Order Number	Arrival Time	Interval from Last Order	Number of Truck-loads	Round-Trip Delivery Time (minutes)						Deviation from Average
					Truck-load 1	Truck-load 2	Truck-load 3	Truck-load 4	Truck-load 5	Average	
Monday	1	8:15	15	2	64	69				67	-3, 2
	2	8:32	17	1	75					75	0
	3	9:03	31	3	73	77	70			73	0, 4, -3
	4	9:47	44	3	123	116	123			121	2, -5, 2
	5	10:00	13	2	105	104				105	0, -1
	6	11:16	76	4	27	34	34	30		31	-4, 3, 3, -1
	7	1:03	107	1	48					48	0
	8	2:06	63	2	55	59				57	-2, 2
	9	2:18	12	1	83					83	0
	10	2:49	31	3	50	46	54			49	1, -3, 5
	11	4:06	77	2	61	65				66	-5, -1
Tuesday	12	9:40	100	3	73	78	77			76	-3, 2, 1
	13	10:12	32	4	66	71	70	69		69	-3, 2, 1, 0
	14	10:45	33	4	41	43	45	48		44	-3, -1, 1, 4
	15	11:55	70	3	60	65	65			63	-3, 2, 2
	16	1:24	89	2	91	103				97	-6, 6
	17	2:22	58	1	7					7	0
	18	4:27	125	3	99	105	94			99	0, 6, -5

TABLE 12.3 (Continued)

Day	Order Number	Arrival Time	Interval from Last Order	Number of Truck-loads	Round-Trip Delivery Time (minutes)						Deviation from Average
					Truck-load 1	Truck-load 2	Truck-load 3	Truck-load 4	Truck-load 5	Average	
Wednesday	19	8:25	25	4	47	41	40	45		43	4, −2, −3, 2
	20	10:35	130	4	68	66	62	66		66	2, 0, −4, 0
	21	12:03	88	3	88	88	85			87	1, 1, −2
	22	3:02	179	5	53	59	52	53	53	54	−1, 5, −2, −1, −1
	23	3:50	48	2	39	38				39	0, −1
	24	4:15	25	5	30	39	32	37	32	34	−4, 5, −2, 3, −2
Thursday	25	8:05	5	4	74	71	70	72		72	2, −1, −2, 0
	26	10:33	148	3	63	62	63			63	0, −1, 0
	27	1:04	151	2	94	88				91	3, −3
	28	1:28	14	4	81	87	88	82		85	−4, 2, 3, −3
	29	1:40	12	3	38	40	41			40	−2, 0, 1
	30	2:23	43	1	21					21	0
	31	4:10	107	4	71	69	69	70		70	1, −1, −1, 0
Friday	32	8:30	30	5	72	76	75	73	69	73	−1, 3, 2, 0, −4
	33	9:28	38	4	107	94	100	94		99	8, −5, 1, −5
	34	12:12	164	4	60	69	66	65		65	−5, 4, 1, 0
	35	12:33	21	3	55	57	54			55	0, 2, −1
	36	2:05	92	1	82					82	0
	37	2:15	10	4	87	91	85	82		86	1, 5, −1, −4
	38	4:13	118	2	55	59				57	2, −2

Therefore, these elements should be included in each order generated by the model.

The data used in model construction can be obtained by direct sampling from the real system. The arrival times of all orders actually received during five successive days at the concrete plant are listed in column 3 of Table 12.3. In column 5, the sizes of the orders are listed in terms of the number of truck loads ordered. The time intervals between the arrival of two successive orders are computed and recorded in column 4. In columns 6 to 10, the actual time required for the delivery trucks to make a round trip are also recorded. The delivery time depends on the distance to the customer's construction sites, as well as on traffic conditions along the way. The latter factor accounts for the differences in time among trucks delivering to the same customer. Thus the delivery time can be conveniently divided into two components. The average delivery time for an order depends primarily on the delivery distance. The deviation of each delivery from the average is due to unpredictable traffic conditions.

The data in Table 12.3 can now be used to develop models that can be used to generate a fictitious flow of customer orders. The first step is to develop a frequency distribution for each attribute to be included in a customer order. The data in the fourth column can be used to develop the frequency histogram of Figure 12.6 for the arrival intervals. This is accomplished by grouping the data into classes. Thus the data in the fourth column have been grouped into six classes. The first class contains the 12 arrival intervals that are less than 30 minutes; the second class contains the arrival intervals that are between 30 and 60 minutes; the third class contains the arrival intervals between 60 and 90 minutes, and so on. Each class can be represented by its midpoint, for example, each time interval in the first class can be represented by a time interval of 15

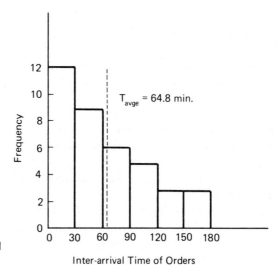

Figure 12.6   Histogram of arrival intervals.

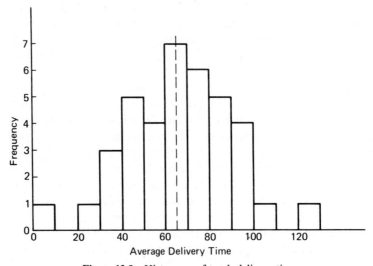

Number of truck loads (x)	Number of orders	Probability p(x)
1	6	0.158
2	8	0.211
3	10	0.263
4	11	0.289
5	3	0.079
	Total = 38	

p(x) = Number of orders for x truck loads/38

Figure 12.7 Model of order sizes.

Figure 12.8 Histogram of truck delivery times.

minutes. In Figure 12.7 the order size frequencies are converted to probability densities and shown on a pie chart. Figures 12.8 and 12.9 present the frequency histograms for average delivery time for an order and the deviation of each delivery from the average, respectively.

The models can be developed directly from these frequency distributions. For example, using Figure 12.6, 38 pieces of paper of identical size can be prepared to represent the 38 orders. On each piece the arrival interval before the next order is written. Thus, there should be twelve with the number 15, nine with the number 45, and so on. These paper pieces are then placed in a cup. This cup is now essentially a simulation model for generating arrival intervals for customer orders.

A similar model can be constructed for the distribution of order sizes. According to Figure 12.8, the cup should contain six paper pieces marked with one truckload, eight marked with two truckloads, ten marked with three truckloads, and so on.

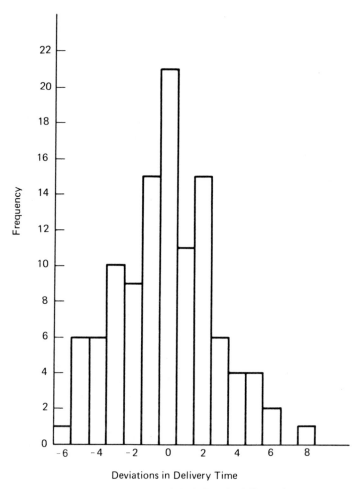

Figure 12.9   Histogram of deviations in delivery time.

A third cup should contain 38 paper pieces representing the 38 average delivery times according to the frequency distribution in Figure 12.8. This cup of paper pieces is then the simulation model for generating the average delivery time of an order. Similarly, a fourth cup should contain the 111 pieces of paper representing the deviations from average according to Figure 12.9.

To generate one fictitious customer order, the cups are covered and shaken thoroughly so that the paper pieces are well mixed. One piece is then drawn from each of the first three cups. The piece from the first cup gives the arrival interval of the order; the second gives the number of truckloads for the order; the third gives the average delivery time. These paper pieces are then returned to their respective cups. To generate the deviation in delivery time for each truckload, a paper piece is drawn from the fourth cup. The time marked on it is recorded and then returned to the cup before a second paper piece is drawn for the next truckload. Figure 12.10 gives a listing of jobs generated for one day of simulated operation.

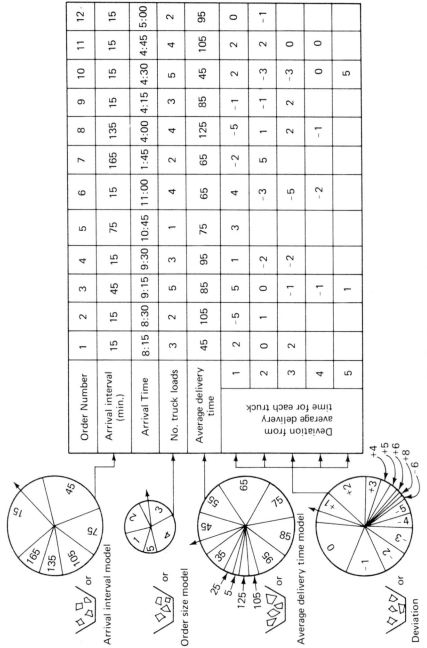

Order Number	1	2	3	4	5	6	7	8	9	10	11	12
Arrival interval (min.)	15	15	45	15	75	15	165	135	15	15	15	15
Arrival Time	8:15	8:30	9:15	9:30	10:45	11:00	1:45	4:00	4:15	4:30	4:45	5:00
No. truck loads	3	2	5	3	1	4	2	4	3	5	4	2
Average delivery time	45	105	85	95	75	65	65	125	85	45	105	95
Deviation from average delivery time for each truck — 1	2	-5	5	1	3	4	-2	-5	-1	2	2	0
2	0	1	0	-2		-3	5	1	-1	-3	2	-1
3	2		-1	-2		-5		2	2	-3	0	
4			-1			-2		-1		0	0	
5			1							5		

Figure 12.10 Simulated customer orders for one day.

The procedure above can be repeated as many times as necessary. The reliability of the simulated data depends on how representative the original sample data are of the real situation. To improve the reliability of the models, it is necessary to collect actual sample data from at least several days of normal operation.

12.5.2 Modeling the Plant Operations

The simulation model of the operations within the concrete plant will consist of a series of logistic rules that describe the routing and processing of customer orders through the plant. Figure 12.11 is a general schematic simula-

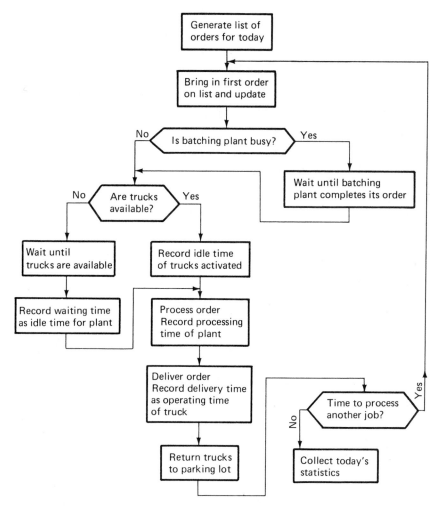

Figure 12.11   Flowchart for simulation model of concrete plant operation.

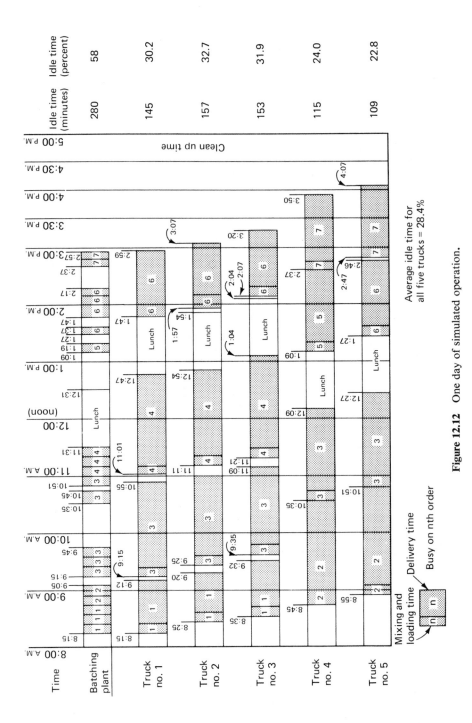

Figure 12.12 One day of simulated operation.

tion model of the plant operations. In addition, the model should provide a mechanism whereby the operational status of the batching plant and delivery trucks can be continuously recorded. For example, the time diagram in Figure 12.12 may be used to supplement the schematic model in Figure 12.11. Also shown in Figure 12.12 are the results for one day of simulated operation using the list of orders generated in Figure 12.10. During the day, the batching plant is idle 58 percent of the working time and the trucks idle on the average of 28.4 percent of the working time.

The results above represent one simulated day. In order to increase the reliability of the results, many days must be simulated. The mean of all the results is then used to indicate the operational characteristics of the system when only five delivery trucks are available.

Simulation can be extremely tedious when computations are performed manually, as in the example above. However, the schematic model in Figure 12.11 can be easily programed for computer use, as described in Section 12.6. Using a modern, high-speed electronic computer, one day of simulated operation can be performed in much less than 1 second of computer time. Therefore, the power of analytical simulation relies almost exclusively on the availability of an electronic computer.

Another important lesson should be learned from the example in Figure 12.12. The first order of the day arrived at 8:15 A.M., and hence all equipment was idle during the first 15 minutes of working time. This is normally not the case, since there should always be some orders left over from the preceding day. To overcome this problem, a so-called *run-in* period of several simulated days should be allowed before statistics are recorded on the system response characteristics.

12.5.3 The Simulation and Analysis Process

Once the simulation model has been constructed, it is a simple matter to use as many delivery trucks in the model as desired. Figure 12.13 shows a possible result of this simulation problem. Each point in the graph is the average computed from a large number (say 100 days) of simulated days. The percentage of idle time of the batching plant decreases with the increase in the number of delivery trucks. However, beyond a certain point, the rate of decrease becomes very small. Furthermore, because of operational constraints, such as the rule that no orders will be processed after 3:30 in the afternoon unless the delivery truck can return to the plant by 4:30 P.M., the percentage of idle time never approaches zero.

Contrary to the response of the batching plant, the idle time of the delivery trucks increases with the increase in the number of trucks. Therefore, the efficiency of one component of the system is sacrificed for the efficiency of another system component. Undoubtedly, the management will want to consider also the economics of the problem, since the ultimate objective is to maximize the

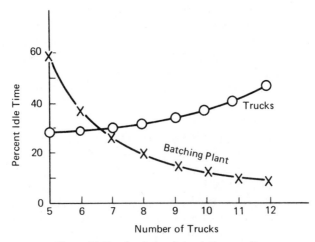

Figure 12.13   Analysis of simulation results.

profit for each dollar invested. The computation of capital and operational cost as well as daily sales volume for different numbers of delivery trucks can be easily incorporated into the above simulation model. Having determined all the relevant system response characteristics from simulation, the company management must use their results to find an optimum trade-off for the problem. The optimization techniques, such as mathematical programming and decision analysis, can be used to perform this function.

12.6 GENERATING INPUT DATA

A simulation study requires a sequence of generated input data with the same frequency distribution as the sequence of sample input data. In the concrete plant example, this was accomplished by drawing pieces of paper from a cup. Each piece of paper had a number written on it and the frequency distribution of the numbers in each cup was the same as the frequency histogram for the input attribute that was being simulated. Roulette wheels, random number tables, or numerical methods can also be used to generate input data for simulation.

12.6.1 Roulette Wheels

Consider, for example, the frequencies of order sizes in Figure 12.7. The frequencies can be transformed into probability densities by determining the relative probability of the occurrence of each class; for example, class 1 has a probability of occurrence 6/38, or 0.158. A roulette wheel can now be constructed such that the divisions on the wheel are proportional to the probability of each class, as shown in the pie chart in Figure 12.7. Now, in order to generate an order size, we just spin the wheel and record the number on which it stops. The same can be done for each of the other frequency histograms.

12.6.2 Random Number Tables

Random number tables are groups of 10 numbers (0 to 9), or hundreds of numbers (00 to 99), or any other larger similar group. The order in which the numbers are placed is absolutely random. Appendix A is a table of random numbers between 000000 and 999999. These numbers may be thought of as the result of spinning a roulette wheel that is marked with 1,000,000 equal divisions and labeled 000000 to 999999. Each number is equally likely and appears at random.

In order to use a random number table to generate input data, the random numbers must be assigned to each class of the input data in proportion to the probability of occurrence of that class. Consider again the probability model of Figure 12.7 for order sizes. The first class has a probability of occurrence of 0.158 and, therefore, 15.8 percent of the random numbers should be assigned to the class representing one truckload. This can be accomplished by assigning the numbers 000000 through 157999 to this class. The next class (two truckloads) has a probability of occurrence of 0.210, and therefore the numbers 158000 through 367999 should be assigned to this class. Similarly, the third class (three truckloads) is assigned 368000 through 630999, the fourth class (four truckloads) is assigned 631000 through 920999, and the fifth class (five truckloads) is assigned 921000 through 999999.

We can now draw sampling numbers from the random number table (Appendix A). From the first column, we find the sequence 258164, 547250, 279794, and so on. Each of these numbers represents an order size. Thus 258164 corresponds to an order size of two truckloads. Similarly, 547250 and 279794 represent order sizes of three and two truckloads, respectively. We began at the first number in the table and began to read down the column, but we could have begun anywhere in the table. Furthermore, it is not necessary always to read down the column. As long as the numbers are picked from the table in some regular fashion, such as down a column or from a horizontal row, they will appear at random with no member in the group enjoying any preference over any other.

The probability distribution of the order sizes will be the same in the generated sequence as in the sample sequence because each order size has random variables assigned to it in proportion to its probability of occurrence in the sample.

This same procedure can be used to generate the input data for arrival interval times, average truck delivery times, and deviations from average delivery times.

Numerous tables of random numbers are available, of which the most extensive are those published by the Rand Corporation (1955).

12.6.3 Numerical Methods

When simulation is performed with the help of electronic computers, it is convenient to describe the frequency histograms by appropriate mathematical functions. For example, the frequency histogram for the interarrival time of orders in Figure 12.6 may be adequately described by the exponential density function

$$F(t) = \begin{cases} \dfrac{1}{t_a} e^{-t/t_a} & \text{for } t > 0 \\ 0 & \text{elsewhere} \end{cases} \tag{12.1}$$

where t is the time interval between successive orders, and t_a is the average time interval, which is equal to 64.8 minutes in this example. Similarly, the histogram in Figure 12.8 can be described by a normal density function as follows:

$$f(D) = \frac{1}{\sqrt{2\pi}\,\sigma_D} e^{-[(D-u)/\sigma_D)]^2} \tag{12.2}$$

where D is the average delivery time, u is the mean value of D and is equal to 66.1 minutes in this example, and σ_D is a constant that defines the lower and upper limit within which 68 percent of the values of D lie. Similarly, the following normal density function may be used to describe the histogram in Figure 12.9:

$$f(d) = \frac{1}{\sqrt{2\pi}\,\sigma_d} e^{-(d/\sigma_d)^2}$$

where d is the deviation in delivery time.

Since we can seldom afford to fill the computer's memory with a large number of random digits, computer programs are available to generate numbers that conform to these and other distribution functions (Balintfy et al., 1966).

12.7 THE SIMULATION PROCESS

The example of a concrete plant operation demonstrated the general procedure of system stimulation. The major procedure of system simulation study may be summarized as follows:

12.7.1 Problem Definition

A simulation study is usually conducted to serve two primary objectives:

1. To measure the system response under a wide range of system inputs
2. To measure the system response when the system components and their interrelationship themselves are altered

Therefore, in defining the problem it is vital to clearly identify the input parameters, the system components, their interrelationship, and the feasible alternatives of system design to be investigated. In addition, a set of quantifiable parameters, which can realistically characterize the system response, must be specified.

In the example above, the input parameters are the customer orders and their elements, which include arrival interval, number of truckloads in each order, and the round-trip delivery time. The system components include the batching plant and the delivery trucks, and the interrelationship of the components is established by the routing procedure. Finally, the parameters selected to measure the system response are idle time of the batching plant and the delivery trucks.

12.7.2 System Input Model Construction

Models must be constructed to simulate the behavior and pattern of the input parameters in the real system. The purpose of these models is to generate fictitious input data in the simulation process. The models may be composed of mathematical functions, logical steps, or numerical data. They are usually constructed from a set of sample data collected from the real system.

12.7.3 System Model Construction

A model is then constructed to realistically represent the operation of the real system.

12.7.4 Solution Process

The solution process comprises two major cycling steps, as shown in Figure 12.14. In the inner cycle, the system response is measured with respect to a specific set of values for the design parameters within the system model. To test the design rigorously under the complete feasible range of the input parameters, a large number of independent solutions must be conducted within this cycle. In the concrete plant example, each inner cycle represents only one simulated day, which may or may not be representative of the system response. In

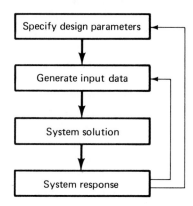

Figure 12.14 Solution process in simulation.

order to provide a reliable measure of the response trend over an extended period of operation, many days must be simulated. Alternatively, management may wish to determine the system response to an increase in the flow of customer orders. This constitutes an inner loop cycling process and the results will indicate the sensitivity of the system to changes in input data.

In the outer cycle, the system response is measured with respect to changes in the design parameters. If the simulation model is properly constructed, the values of the design parameters can be easily changed within the model. In the concrete plant example, the only variable design parameter is the number of delivery trucks. Thus, many days are first simulated using five trucks. The entire procedure is then repeated with six trucks, seven trucks, and so on. Studies of this nature indicate the sensitivity of the system to changes in the values for system parameters.

In some simulation problems, it is necessary to study the system response using several different system structures. It may then be necessary to build separate system models to represent the different designs. For example, if the management of the concrete plant wishes to study the relative benefits of several different job routing procedures in addition to the first-come-first-served method, it will be necessary to build a separate system model for each routing method. In these cases, the structure and logic of the system will vary. Studies of this nature indicate the sensitivity of the system to changes in system structure.

12.7.5 Analyzing Simulation Results

Simulation is neither an optimization nor a decision-making process. It merely provides an understanding of the system response under a set of a specified conditions. The results must then be carefully analyzed using system techniques to establish an optimum design or policy.

12.8 LIMITATIONS OF THE SIMULATION APPROACH

The simulation approach provides a powerful tool for analyzing complex systems that cannot be easily studied by any other means. However, the accuracy and reliability of the simulation results depend largely on how well the simulation models can truly represent the response characteristics of the system in the real world. On the other hand, the validity of the models and the reliability of the results are extremely difficult to evaluate because of the very complexity of the system. It is advisable, therefore, that the simulation results be always analyzed with a certain degree of reservation, and that the response of the system in the real world be continuously monitored so that the feedback from the real world may be used to update and validate the simulation models.

Because of the limitations noted above, the simulation approach should, wherever possible, be supported concurrently by a theoretical study of the system; the latter serves to provide a hypothesis or simplified mathematical model of the system response characteristics. The results from the two approaches then serve as mutual checks. Moreover, even a simple theoretical analysis can help to narrow the values of the input and design parameters to be tested in the simulation process and thus effectively reduce the simulation calculations.

PROBLEMS

P12.1. Simulate two days of operations for the concrete plant example described in this chapter. Generate the necessary data using the random number tables of Appendix A as explained in Section 12.6.

P12.2. By direct observation determine the probability of interarrival times for vehicles at an intersection. For what period of time is your model valid, and what factors might affect the probability distribution you have obtained?

P12.3. Figure P12.3 illustrates an entrance ramp that leads into a four-lane interstate highway. The frequencies of arrivals at the ramp-highway intersection during peak hours are as follows:

Eastbound Traffic Along Highway		Ramp Traffic	
Arrival Interval (seconds)	Frequency (cars)	Arrival Interval (seconds)	Frequency (cars)
0	0	0	0
2	50	4	2
4	40	8	2
6	30	12	4
8	30	16	3
10	25	20	5
12	25	24	6
14	15	28	4
16	10	32	6
18	10	36	8
20	5	40	10

There is a 0.75 probability that a car arriving in an east-bound lane will be in lane 1. To enter safely, a vehicle waiting to enter the highway requires 10 seconds before the next car in the east-bound lane 1 arrives.

(a) Simulate one 30-minute period of traffic flow to determine the average waiting time for the vehicles entering from the ramp. Also determine the maximum number of cars waiting at any one time.

(b) The traffic department is considering the possibility of installing a traffic light at point A that will direct cars into lane 2 such that the probability of a given car being in lane 1 becomes 0.25. Determine the ramp waiting time and maximum queue length with the light installed.

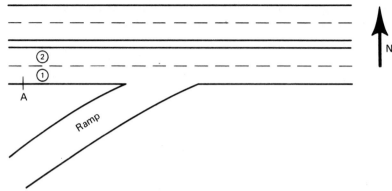

Figure P12.3

P12.4. The development of a new plant is under consideration and the probability distributions for order interarrival times and order sizes are as follows:

Order Interval (days)	Probability	Order Size	Probability
0	0	10	0.05
0–1	0.05	50	0.20
1–2	0.15	100	0.05
2–3	0.35	200	0.20
3–4	0.20	300	0.40
4–5	0.15	400	0.05
5+	0.10	500	0.05

Management is concerned with the problems associated with the selection of production rate and finished product inventory size.

(a) For a given production rate of 350 units per 5-day week and inventory size of 300 units, determine how frequently the orders cannot be immediately fulfilled and the average delay. Simulate the plant operation for a period of 2 months.

(b) For the same production rate, what would be the maximum inventory size to ensure that all orders can be immediately filled on receipt? Simulate the plant operation for a period of 2 months.

P12.5. A building owner has decided that a single elevator in his building is not adequate and causes unnecessary delay for people going to offices in the building. You are asked to study the problem and make recommendations. You plan to simulate the system behavior with an analytic model. The study can be divided into three steps:

1. Model the present system with one elevator.

2. Test this model of the present system to see if it is a reasonable representation of the real thing.

3. Study variations in the model to determine their effect.

In this problem you are to work out in detail step 1. Consider carefully what should be the input and output for the simulation model. Draw a flow diagram of your simulation model. The output should be in the form of numbers, and you are to use the simulation model to derive the output. Then discuss how you would proceed in step 2. For step 3 repeat the analysis with two elevators and the same input to see how the output from the simulation model is affected.

Tables P12.5a and b contain data that were collected at the entrance to the elevator on the first floor. You may assume that this data is typical of the traffic entering the building and for the performance of the elevator. Everyone entering the building would like to use the elevator. This is a very small elevator,

TABLE P12.5a   Number and Arrival Time of Passengers During Observation Period

Arrival Time* (minutes)	Number of Passengers	Arrival Time* (minutes)	Number of Passengers
0		22.25	2
2.00	1	24.25	3
3.25	3	24.75	2
4.75	2	26.00	4
6.50	3	27.00	2
7.50	5	28.50	2
9.00	2	29.25	5
10.00	3	31.00	3
11.25	1	32.00	3
13.00	3	33.50	4
14.00	2	34.00	2
14.75	2	34.75	4
16.00	1	36.50	1
17.50	3	37.50	1
18.75	4	38.75	3
19.50	2	39.00	2
20.75	4		

*Time is started at zero for convenience; you may convert to clock time with an arbitrary starting time if you wish.

TABLE P12.5b   Frequency of Trip Durations of the Elevator with 1 to 5 Passengers

Number of Passengers	Trip Duration of the Elevator (minutes)											
	$\frac{1}{4}$	$\frac{1}{2}$	$\frac{3}{4}$	1	$1\frac{1}{4}$	$1\frac{1}{2}$	$1\frac{3}{4}$	2	$2\frac{1}{4}$	$2\frac{1}{2}$	$2\frac{3}{4}$	3
1	15	35	44	48	45	40	7					
2		15	28	38	43	35	26	15	7	2		
3			10	20	28	35	38	32	23	14	8	
4				4	8	14	22	30	34	31	25	
5					4	7	11	15	21	27	30	

TABLE P12.5b (Continued)

Number of Passengers	Trip Duration of the Elevator (minutes)											
	$3\frac{1}{4}$	$3\frac{1}{2}$	$3\frac{3}{4}$	4	$4\frac{1}{4}$	$4\frac{1}{2}$	$4\frac{3}{4}$	5	$5\frac{1}{4}$	$5\frac{1}{2}$	$5\frac{3}{4}$	6
1												
2												
3		3										
4		21	16	13	9	6	4	2				
5		28	26	23	21	18	16	13	11	9	6	3	1

however, with a maximum capacity of five people and notice that it is quite slow. These constraints are imposed to reduce the amount of calculations required but do not detract from the general principles involved.

P12.6. Table P12.6 lists a 20-year record of flood damage for a river basin. In designing a flood control project for the area, the amount of capital investment to be spent on flood control is related to the anticipated reduction in flood damage as a result of the flood control project. Assume that the cost of flood damage can be related to the discharge rate by the following simple expression:

$$D = KQ$$

where D is the flood damage in dollars, Q is the river discharge rate in cubic feet per second (ft^3/sec), and K is a multiplication factor, the value of which depends on the time of the flood, concentration and distribution of residential and industrial areas, and so on. Assuming that there is only one flood each year, simulate a 50-year period and compute the total flood damage for the 50 years. (*Hint:* First develop probability distributions for K and Q from the data given.)

TABLE P12.6   20-Year Flood Damage Record

Year	Discharge Rate (ft^3/sec)	Damage ($\times 10^6$)	Year	Discharge Rate (ft^3/sec)	Damage ($\times 10^6$)
1951	1000	$ 10	1961	2300	$18.5
1952	2500	30	1962	3400	34
1953	4200	25	1963	4500	50
1954	5700	40	1964	1300	8
1955	1200	12	1965	2000	14
1956	3600	28.5	1966	6000	60
1957	5000	70	1967	1700	17
1958	2600	26	1968	3500	28
1959	7000	70	1969	4300	56
1960	1500	195	1970	3000	30

P12.7. Three stockpiles of crushed aggregate of fine, medium, and coarse grades receive rock from a crusher and deliver rock to customers' trucks. The rock arrives ordinarily in the stockpiles at a uniform total rate of 600 tons per day over 6 hours commencing at 7 A.M. The crushed rock is delivered from the crusher normally in the proportions of 3:2:1 for fine, coarse, and medium, respectively.

When any stockpile drops to 50 tons or less a special order is placed to the crusher to provide an additional 50 tons of the grade required using spare crusher capacity only. These additional orders flow in at the rate of 50 tons per hour, commencing as soon as a request is made. If such an order is incomplete at the end of the working day, it is completed first thing the following working day.

If any stockpile reaches or exceeds 500 tons, action is taken as follows.

1. For the fine-grade stockpile 50 tons is disposed of into five 10-ton trucks as soon as possible from the same loading point that supplies customers' trucks. These five trucks are immediately available when needed and each requires 10 minutes to be loaded. However, they must always give priority to customers' trucks.

2. For the medium-grade stockpile, supply from the crusher is completely stopped until the stockpile is reduced to 450 tons. This stoppage would not apply to the crushing and delivery of a special 50-ton order to replenish a depleted stockpile.

3. For the coarse-grade stockpile the proportion of the three grades from the crusher is immediately altered from the normal 3:2:1 for fine, coarse, and medium, respectively, so that all rock goes into fine and medium grades in the proportions 4:2, respectively. However, when this alteration is made the total crusher rate of supply is reduced to 50 tons per hour, excluding special orders of 50 tons to replenish depleted stockpiles. This altered output is allowed to continue until the coarse stockpile is reduced to 400 tons or less.

The arrival of a truck represents an order and orders are supplied in full truckloads of one grade only. Trucks are either of 5 tons or 7 tons capacity, arriving in equal proportions. The mean rate of truck arrival is 100 trucks per 8-hour day starting at 8 A.M. The mean rate is uniform throughout the day and trucks arrive according to the following probability distribution:

Interarrival Time (minutes)	Probability
0–1	0.02
1–2	0.03
2–3	0.10
3–4	0.15
4–5	0.25
5–6	0.15
6–7	0.10
7–8	0.10
8–9	0.05
9–10	0.05
10+	0.00

Only one truck can be loaded at a time at each stockpile, with a total service time of 10 minutes.

The demand for the grades of fine, medium, and coarse rock is in the proportions 2:3:1.

By means of a schematic diagram, show how this rock-crushing plant operation can be modeled for the purpose of simulation. The diagram should illustrate the following:

1. Major system components
2. Component characteristics
3. Component interactions

P12.8. Simulation can be used to study the response of an existing system for different levels of input or it can be used to test alternative designs of proposed systems.

Discuss the following subjects with respect to each of the simulation problems above.

(a) What level of detail is to be used, and what variables should be considered?
(b) How are the results to be used?
(c) What confidence can be placed in results obtained using short observation periods for the characteristics of the selected variables and of the simulation runs?

Network Planning
and Project Scheduling

13

13.1 THE PLANNING AND SCHEDULING PROBLEM

Engineers are often responsible for the planning and supervising of projects. Although most projects are associated with the construction of engineering, industrial and housing works, others are related to the implementation of commercial products, advertising campaigns, research programs, planning studies, and the like. In all cases the projects can be defined as a collection of work tasks or activities, each of which must be performed before the project can be completed. In addition, for most projects, the individual work tasks can be performed after certain stages, conditions, or project implementation states have been achieved. Many project work tasks are interrelated logically and/or sequentially to each other and their planning and supervision poses system problems.

Project definition requires the enumeration and description of each work task involved in the project. The decision process for each work task entails selecting, evaluating, and enumerating the resources (material, equipment, labor, finance, etc.) necessary to ensure its completion. In addition, to develop a system model, it is necessary to determine for each work task the conditions that must exist for the task to begin. Finally, the mobilizing for and implementation of the project requires procuring the necessary resources and their use at specific times. This requires a major scheduling effort. Thus engineers are interested in:

1. The definition, planning, and scheduling of a project as the means for specifying the overall effort required for the project.

2. The detailed planning of the specific means and sequences by which project objectives will be accomplished

3. The enumeration of the resources required and the determination of the periods of time over which major, or scarce resources, will be required or utilized

4. The policies under which these resources will be committed to the project work tasks

Planning a project involves selecting the technological methods to be used for each work task and determining a specific work order for the accomplishment of the project work tasks from all the various ways and sequences in which this could be done. Scheduling a project requires determining the timing of the work task activities that comprise the project and coordinating them so that the overall project duration is established.

Supervising a project involves conceiving, implementing, and monitoring an information system that will aid management in the formulation of work directives and permit the project status to be evaluated at any time. Such an information system model, coupled with a planning model, can be the base on which project control can be established or on which corrective action can be initiated, if necessary, to ensure the smooth progression of the project as planned. The planning and scheduling process is iterative in nature throughout the life of the project.

In all these organizational phases, the engineer is concerned with identifying project components (work tasks) and structuring them, through the portrayal of work task relationships, into a coherent system plan. The obvious organizational aspects of these efforts suggest that the project plan or system model as produced be called an organizational system model.

A modeling method for organizational systems associated with project planning, scheduling, and supervision was introduced to the construction industry about 1960 and is now commonly known as the *critical path method* (CPM). Critical path methods require the formulation of linear graph models to represent specifically the unique features and *plan for the project* under consideration. The various project activities are collected together and synthesized in a connected linear graph to show the project logic and the sequential nature of the various activities. Once the project plan is defined through the CPM structural graph model, a variety of attributes associated with the project work tasks can be added to the model. In this way, planning, scheduling, procurement, and other organizational problems can be modeled as simple graphical problems.

A simple organizational problem is that concerned with the sequencing of work tasks (or activities) when each task has physical or technological prerequisites that must be met before the task can begin (i.e., scheduled). Thus, as shown in Chapter 7, sequencing activities leads to directed path network models.

13.2 NETWORK MODELING OF A PROJECT

The network modeling of a project is really one application of linear graph theory and models. In this particular case, the network depicts the tasks necessary to complete the project as well as their necessary technological sequencing and planned ordering. The project work task activities become the system components, and the system structure reflects the technological and managerial ordering and sequencing of those activities.

In the modeling of projects, there are two modeling formats or practices that have evolved. One modeling form is *circle notation*, in which each work task is represented by a node, and the relationship or ordering of the project work tasks are portrayed by the branches that connect these nodes. The circle notation (activity on node) representation of projects is commonly used in CPM computer programs.

The other modeling format is the *arrow diagram* (activity on branch) representation of a project. In this approach, the work tasks are modeled as branches, and the nodes indicate the relationships between the tasks. This modeling format is very popular because of its close similarity to the Gantt or bar chart representation of a project, and consequentially is more readily understood by construction and trade personnel.

Figure 13.1 shows several typical representations of activity ordering and sequencing in circle and arrow notation. It should be noted that in the arrow notation modeling format in Figure 13.1e and f, it is necessary to add a logic branch to depict the proper sequencing of activities. The logic branch is shown here by the dashed line and is known as a dummy activity. The use of the dummy activity becomes essential in arrow notation modeling to maintain the logic in the ordering and sequencing of the various activities. In the example of Figure 13.1e, activities 1 and 2 must be accomplished before beginning activity 3. The dummy simply reflects this logic.

When undertaking a project, it is first necessary to compile a list of work tasks that collectively comprise the accomplishment of the project. The number and extent of the work tasks defined depends on the project, the level of detail required, and the intended management use of its model. As an example, assume that a supermarket is to be constructed. The building will be approximately 20,000 ft² and will have parking facilities for the patrons. The supermarket is to be of a concrete block construction; however, poured concrete columns will be used to support the roof. Table 13.1 shows the list of work tasks that have been defined for the project, as well as other pertinent planning data. Obviously, the number of tasks has been limited to simplify the problem.

It is possible to further subdivide each work task if more detailed analysis is desired. For example, work task 3 could be separated into the specific tasks related to plumbing and electrical rough-in work requirements. The same is true for other work activities. The degree of breakdown of work activities

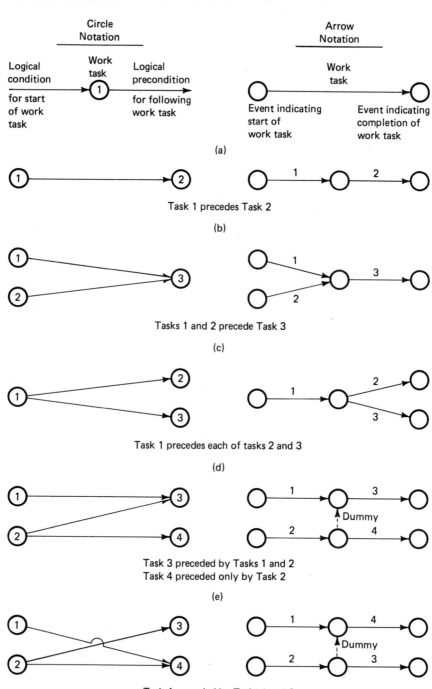

Figure 13.1 Representation of activities in circle and arrow notation.

TABLE 13.1   Planning Data for Supermarket Project

Work Task	Duration (days)	Direct Cost ($\times 10^3$)
1. Site clearance, grading, field office, and services hookup	14	\$ 8.8
2. Excavate, form, and install rebars for foundation	10	4.8
3. Utility rough-in	6	9.6
4. Pour and cure foundation	5	8
5. Form, pour, and cure columns	12	8
6. Construct block walls	20	25.6
7. Finish grade, lay mesh, place concrete, and cure floor slab	5	48
8. Erect roof system, including roofing, drainage, and insulation	15	64
9. Complete utilities	5	56
10. Glaze windows and install doors	5	16
11. Construct cold storage room	10	16
12. Construct interior partitions	6	1.6
13. Finish interior, including ceiling, painting, plastering, and flooring	20	56
14. Install heating, ventilating, and air conditioning equipment	10	32
15. Grade, curb, and pave parking area	10	80
16. Site cleanup and landscape	6	5.6
		\$440

depends on the level of accuracy of project management control that is required.

Given the list of project activities indicated in Table 13.1, the logic associated with their accomplishment must be determined in order to develop the model for the building project. In this case, it is necessary to first clear and grade the site as well as establish the field office. Once this is accomplished, it is possible to proceed with the excavation and forming for the foundation. Having completed that task, work may proceed on the rough-in of the utilities as well as the pouring and curing of the foundation. The columns may be constructed after the foundation is cured; however, work on the floor slab must await completion of the utility rough-in and the column construction. The construction of the walls may proceed after the columns are completed. The roof work will be undertaken after the walls and floor are finished. These work tasks are associated with the shell of the building, and further interior work may not proceed until the shell is completed.

Upon completion of the roof, work may begin on heating, ventilating, and air conditioning (HVAC) efforts, as well as the construction of interior partitions and the cold storage room. All three tasks must be finished prior to the installation of windows and doors or the work associated with completing the utilities. Once the windows and doors are installed, the interior finishing may be undertaken.

With regard to the parking facilities, this work task can be initiated upon the completion of the building shell, which is at the time of finishing the roof. Basically, the parking facility work task may proceed independent of the building activities. The final task of site cleanup and landscape may not begin until all other activities on the site have been completed.

It should be recognized that the project logic is a function of the design details and the general approach to the construction of the facilities. In developing a plan for the accomplishment of the work tasks, it is necessary to consider carefully the sequence of activities and the corresponding logic of the work task accomplishment. Reexamination of the plan can frequently result in savings in the total time for project completion.

Given the list of work tasks as identified in Table 13.1 and the preceding logic, the circle notation model of the project is shown in Figure 13.2a. The corresponding arrow notation model is depicted in Figure 13.2b.

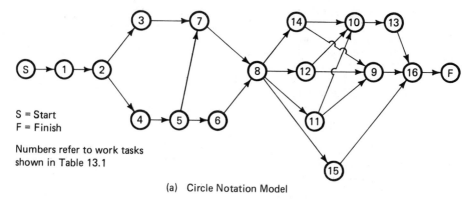

S = Start
F = Finish

Numbers refer to work tasks
shown in Table 13.1

(a) Circle Notation Model

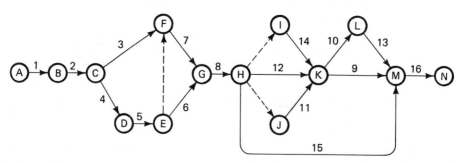

Numbers refer to work tasks
shown in Table 13.1

(b) Arrow Notation Model

Figure 13.2 Critical path models for supermarket project.

An examination of Figure 13.2a or b reveals that the completion of work task 8 reflects a significant state in terms of project completion. In this case, the completion of work task 8 represents the basic enclosure of the building. Upon the completion of that activity, work may begin on a variety of tasks associated with the construction of the interior of the building. A point in the schedule that represents such a significant state in the project is known as a *milestone*. It can also be noted that the milestones have a considerable impact on subsequent project activities.

13.3 CRITICAL PATH ANALYSIS

The addition of the estimated duration for each of the work tasks permits the circle and arrow notation models to be used for analyzing the scheduling of activities. This analysis requires that estimates of the time required to complete each work task be made. Such an estimate will require the engineer to consider the nature of the work task, the amount of resources that are required and the manner in which the task will be accomplished. For the supermarket project, the estimated duration of each work task is shown in Table 13.1.

In CPM, the duration estimate is a single value that reflects the probable time required for completion of the work task. It is recognized that actual time requirements for work task completion may vary due to such factors as work crew experience, equipment reliability, weather, and resource availability. It is also possible to recognize the variation in activity durations by estimating an expected range of time values. An analysis approach that utilizes this approach to work task duration, known as the *program evaluation and review technique* (PERT), is discussed in Section 13.6.

The project manager should monitor the conduct of a project so that changes can be made in the model to reflect actual time durations of specific project work tasks or the need to incorporate revised work sequences. In this manner the project manager can use the model to develop better work strategies for the use of project resources in meeting project milestone and completion dates.

Given the circle or arrow notation model and the estimated duration for each work task, it is possible to determine the following:

1. The earliest time at which an activity may be started (known as *earliest start time* or EST)

2. The minimum time in which the total project may be completed

3. The latest time at which an activity may be started if the project is to be completed in a minimum time (known as *latest start time* or LST)

4. The earliest time at which an activity may be finished (known as the *earliest finish time* or EFT)

5. The latest time at which an activity may be finished if the project is to be completed in a minimum time (known as the *latest finish time* or LFT)

Each of these pieces of information is important in scheduling an actual project, because they define when a particular activity may occur within the overall work schedule.

The earliest start time represents the earliest time that a given activity can begin after the initiation of a project. This time is a function of the other activities that must be completed prior to starting the particular activity under consideration. The earliest start time can be ascertained by determining the maximum necessary time required for preceding activities. This requires summing the time requirements along each linear graph path from the starting point to the activity involved.

The minimum time in which the project can be completed is given by the duration of the maximum earliest start time path. That earliest start time path from the start of the project which determines the project completion time is known as the *critical path*.

Determining the EST time requires the selective addition of activity durations along the various directed paths to each node of the network. The combinatorial calculation can be systematically developed so that only the set of preceding nodes need be considered at any time.

Let T_i^E represent the earliest start time at node i. For the arrow notation network of Figure 13.3b, the earliest start time at node B is determined by the path "from node A" and is the earliest start time at node A plus the duration of the activity between nodes A and B (activity 2). Therefore,

$$T_B^E = T_A^E + 14 = 14 \text{ from node } A$$
$$T_C^E = T_B^E + 10 = 24 \text{ from node } B$$
$$T_D^E = T_C^E + 5 = 29 \text{ from node } C$$
$$T_E^E = T_D^E + 12 = 41 \text{ from node } D$$

The early start time at node F (or the beginning of work task 7) is dependent on work tasks 3 and 5 having both been completed. This means that the early start time is the maximum of the preceding nodes plus the duration of the activities that follow these nodes. Therefore,

$$T_F^E = \text{maximum} \begin{cases} T_C^E + 6 = 30 \\ T_E^E + 0 = 41 \end{cases} \quad \therefore \quad T_F^E = 41 \text{ from node } E$$

A similar analysis would be true for node G, where

$$T_G^E = \text{maximum} \begin{cases} T_F^E + 5 = 46 \\ T_E^E + 20 = 61 \end{cases} \quad \therefore \quad T_G^E = 61 \text{ from node } E$$

Finally, $T_N^E = 117$ indicates that the time required for the completion of the project is 117 working days.

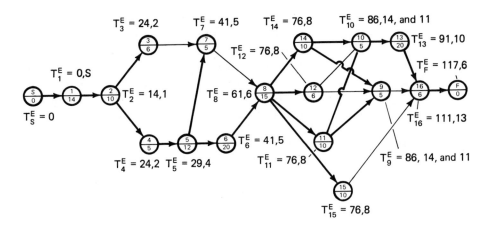

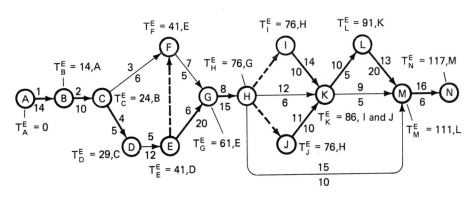

Figure 13.3   Early start trees of supermarket project.

It is convenient to include in the tabulation the label for the unique preceding node on the directed path that determines the earliest start time for the specific node under consideration. The calculations are left to the reader and are illustrated in Figure 13.3 for both circle and arrow notation graphs. In this way the preceding calculations and labels identify the earliest start time paths for each node. The collection of these paths produces a special graph structural property called a *directed tree*. In Figure 13.3 the directed trees are shown in both graphs by the heavy branches. For obvious reasons the trees are called the earliest start trees of the CPM network.

In linear graph theory a tree of a connected graph is a minimal collection of branches such that the graph is still connected and only one path exists between any two nodes. In CPM graphs the branches are directed in such a way the tree produced can be said to be rooted at a datum node.

The latest start time represents the latest time that an activity may be initiated in the schedule if the project is to be completed in a minimum time. Again the LST is specified in terms of the time from the start of the project; however, it is computed backward in time from the project finish node based on the minimum project time.

The actual LST is calculated by determining the minimum time path backward in time from the completion of the project. If T_j^L is the latest start time at node j, then from Figure 13.4 T_M^L is T_N^L minus the duration of activity 16 (i.e., $T_M^L = T_N^L - 6 = 117 - 6 = 111$). The latest start time at node N is the

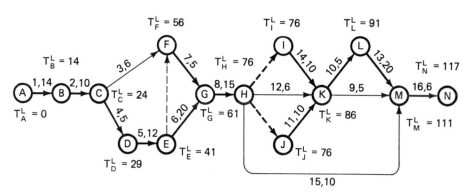

T_i^L = Latest start time at node i

a = Work task number

d = Work task duration

$\xrightarrow{a,d}$ Branch of latest start time

Figure 13.4 Late start tree for arrow notation model of supermarket project.

same as the earliest start time because node N represents the end of the project. Also, plotting minimum time paths for all nodes from the completion node of the project results in the latest start tree, as shown in Figure 13.4. Its derivation is left to the reader.

Figure 13.5 illustrates the calculations that might be made for each activity in the CPM network. The EFT for activity ij is computed as the sum of the EST(T_i^E) and the activity duration. The LST for activity ij is computed as the difference between LFT(T_j^L) and the activity duration. All activities that start at the same node have the earliest start time of that node. Also, all activities that end at the same node have the latest finish time of that node. The activities that take place on the critical path for the project will have an early start time that will be the same as the latest start time.

For this project, the critical path that determines minimum project duration passes through activities 1, 2, 4, 5, 6, and 8. Upon the completion of work task 8, there are two paths that are critical. One path passes through activity 11, and the other path passes through activity 14. At that point in the project, the critical path returns to a single path that passes through activities 10, 13, and 16.

For activities not on the critical path, however, the earliest start time will be less than the latest start time. The difference between the two is known as the *float time*. A variety of measures of float time have been developed and used. Thus if an arrow notation activity A_{IJ} of duration d_{IJ} is considered joining nodes I and J and T^E and T^L are earliest and latest start node time symbols, the following float definitions can be made:

$$\text{Total float TF}_{IJ} = T_J^L - (T_I^E + d_{IJ})$$

$$\text{Free float FF}_{IJ} = T_J^E - (T_I^E + d_{IJ})$$

$$\text{Interfering float} = \text{TF}_{IJ} - \text{FF}_{IJ}$$

$$= T_J^L - T_J^E$$

Total float for an activity corresponds to the concept of making available to the activity the greatest amount of available float time without jeopardizing the project duration. It therefore assumes event times T^L at node J and T^E at node I. It can be used up only once on the chain passing through the activity.

Free float for an activity corresponds to the concept of making available to the activity only that amount of available time that does not interfere with subsequent activities. It therefore assumes event times T^E times at both nodes I and J.

The use of interfering float by, and during, an activity indicates that subsequent activities are affected in that they can no longer utilize all their previously available total float.

The relationships between the different float measures are shown in Figure

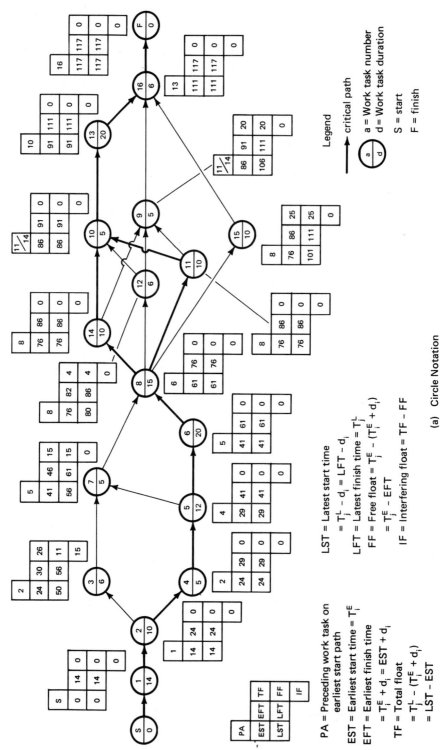

Figure 13.5 Network calculations for supermarket project.

(a) Circle Notation

Legend

→ critical path

(a/d circle) a = Work task number
d = Work task duration

S = start
F = finish

PA = Preceding work task on earliest start path

EST = Earliest start time = T_i^E

EFT = Earliest finish time
= $T_i^E + d_i$ = EST + d_i

TF = Total float
= $T_i^L - (T_i^E + d_i)$
= LST − EST

LST = Latest start time
= $T_j^L - d_i$ = LFT − d_i

LFT = Latest finish time = T_j^L

FF = Free float = $T_j^E - (T_i^E + d_i)$
= T_j^E − EFT

IF = Interfering float = TF − FF

PA		
EST	EFT	TF
LST	LFT	FF
IF		

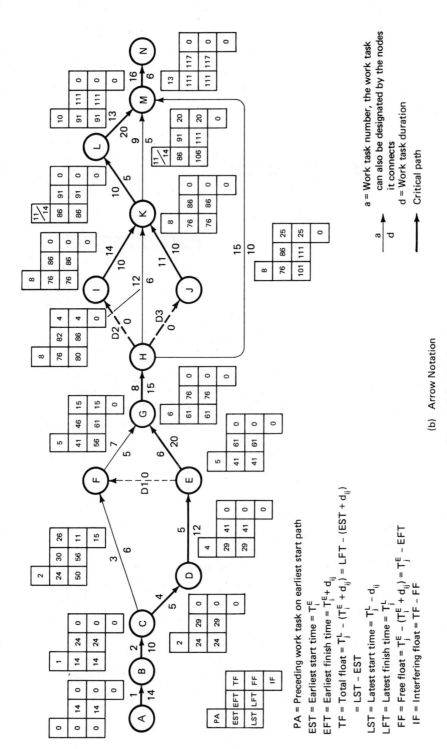

(b) Arrow Notation

Figure 13.5 Continued.

PA = Preceding work task on earliest start path

EST = Earliest start time = T_i^E

EFT = Earliest finish time = $T_i^E + d_{ij}$

TF = Total float = $T_j^L - (T_i^E + d_{ij}) = LFT - (EST + d_{ij})$

  = LST − EST

LST = Latest start time = $T_j^L - d_{ij}$

LFT = Latest finish time = T_j^L

FF = Free float = $T_j^E - (T_i^E + d_{ij}) = T_j^E - EFT$

IF = Interfering float = TF − FF

297

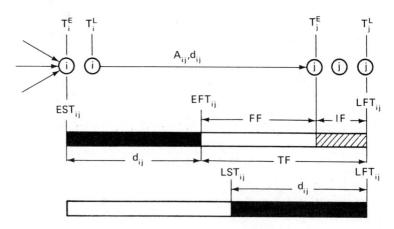

Figure 13.6   Relationships between floats.

13.6 and the various floats for the supermarket project are also shown in Figure 13.5.

13.4 PROJECT SCHEDULING

Given the information from the CPM network models, the project manager can develop a time schedule for undertaking and accomplishing each of the work tasks. When viewing a project on a time basis, it is frequently more convenient to portray the CPM model in a time-scaled context. This can be done by depicting the work tasks in a bar chart (or Gantt chart) form. Figure 13.7 shows a bar chart for the supermarket project, and indicates the scheduling of work tasks given an early start schedule.

The bar chart has the advantage that it indicates readily those work tasks that would be under way at any particular time in the project schedule. As an example, Figure 13.7 reveals that on day 78 of the project, work tasks 11, 12, 14, and 15 are being done simultaneously if the early start schedule is being followed. The various resources utilized by these activities are therefore simultaneously required at that time. This observation enables a cut-set concept to be applied to organizational networks in which capacity flows correspond to the resource use concepts. For this reason, the time-scaled bar chart format of the CPM model is more useful in examining resource requirements than is the circle notation network model. The bar chart model, however, must indicate the connectivity of the various activities necessary for the analysis of start and finish times. This is an excellent example of how one model can be transformed to provide information for another type of analysis. Further discussion of the question of resource scheduling is presented in the next section.

Another useful aspect of the bar chart format is the examination of the cash flow for a project. Once the schedule of activities has been decided by

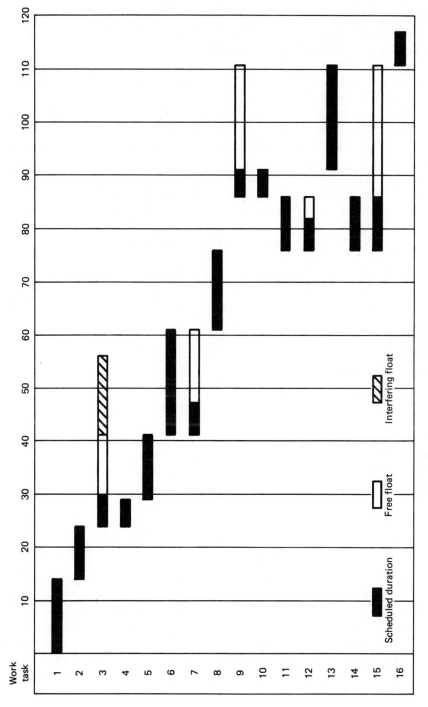

Figure 13.7 Bar chart for supermarket project for earliest start times.

locating each activity in calendar time, it is possible to determine the cash flow for the project. In many cases, the construction work is carried out under contract, with monthly payments based on the work that has been done in the field. The contractor performs work in the order specified by the CPM model under the supervision of the project engineer. In addition, the contractor must invest money to initiate construction before any payment is received; therefore, there is concern with cash flows and overdraft requirements.

Cost information for the supermarket project is shown in Table 13.1. Table 13.2 summarizes pertinent data that includes the cost of accomplishing the various activities in the project. As work is conducted on each activity, a cost is incurred to the contractor. However, the contractor is paid only on completion of certain phases of the project or on some schedule that is based on completed work. The contractor must anticipate the income versus the costs incurred on the project to determine the financing that may be required.

TABLE 13.2   Summary of Work Task Durations, Costs, and Start Times

Work Task	Duration (days)	Direct Cost	Value of Work	Start Time
1	14	$ 8,800	$ 10,000	0
2	10	4,800	5,455	14
3	6	9,600	10,909	24
4	5	8,000	9,091	24
5	12	8,000	9,091	29
6	20	25,600	29,091	41
7	5	48,000	54,545	41
8	15	64,000	72,727	61
9	5	56,000	63,636	86
10	5	16,000	18,182	86
11	10	16,000	18,182	76
12	6	1,600	1,818	76
13	20	56,000	63,636	91
14	10	32,000	36,364	76
15	10	80,000	90,909	76
16	6	5,600	6,364	111
		$440,000	$500,000	

For this project, assume that the costs are incurred uniformly over the period for each activity. Of course, if the level of effort is variable throughout the activity, it may be necessary to consider variations in monthly costs.

At any time during the project a vertical line may be drawn indicating the end of a particular work period. The summation of costs incurred along this cut-set line represents the expenditures for that period. Thus costs can be obtained for each work period, which in this case is in days. Also, the schedule of income can be anticipated.

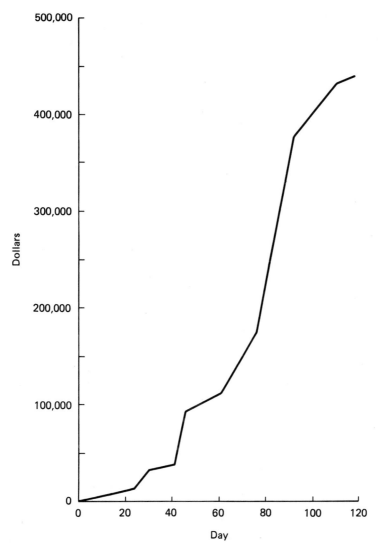

Figure 13.8 Accumulated project cost.

Figure 13.8 shows the schedule of accumulated direct cost plotted graphically over the duration of the project. Using this curve, the engineer can anticipate the project cash flow and the finance requirements. It must be recognized that the cost is a direct aspect of the manner in which the project is undertaken and accomplished. Also, the cash flow and finance requirements are a function of the arrangements for payment for the work accomplished.

The contractor, however, must provide financing to initiate and to continue the project. As work progresses claims are submitted (usually monthly) for payment that, if correct, are paid within the next month less a retainage (usually 10 percent). The retainage concept has been developed to ensure that the contractor has an inducement to complete the project, at which time all the accumulated retainages are paid. Assuming that the contractor is earning a net 12 percent profit, Figure 13.9 shows the calculation and graph of the financing

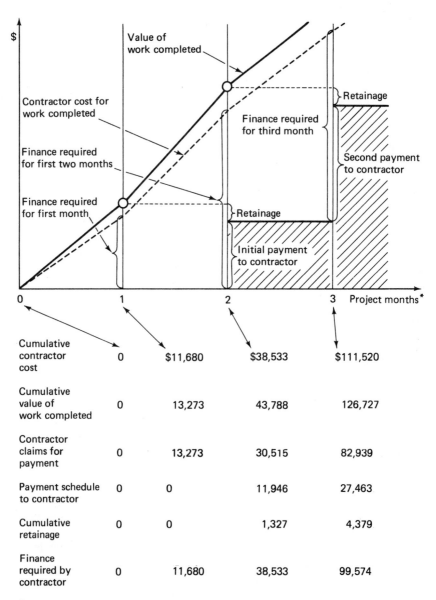

	1	2	3
Cumulative contractor cost	0	$11,680	$38,533	$111,520
Cumulative value of work completed	0	13,273	43,788	126,727
Contractor claims for payment	0	13,273	30,515	82,939
Payment schedule to contractor	0	0	11,946	27,463
Cumulative retainage	0	0	1,327	4,379
Finance required by contractor	0	11,680	38,533	99,574

*Assumes 20 workdays per month

Figure 13.9 Financing requirements of the contractor.

requirements of the contractor for the first three months of the project. Nowithstanding payments to the contractor, the overdraft requirements are still increasing and may not stabilize for some time.

In the event that the contractor must reduce the overdraft requirements, or the owner is tardy in making payments, it may become desirable to slow down the rate of working the project. The contractor may choose to shift activities within their total float as a means of reducing the cash requirements or the rate of progress in all current activities may be reduced and the project duration extended.

13.5 RESOURCE SCHEDULING

The project schedule, as shown in previous sections, inherently establishes the resources that will be required for the project. The project resource requirement follows directly from the initial resource allocation decisions to each work task, which are then integrated over time according to the project schedule. Thus for any given time segment within the project schedule, the total project resources required for that period is the sum of the resources required for each ongoing work task in that time segment. Thus the project schedule is based on the assumption that resources will be made available and the project total resource time profiles and the resource allocations to each work task are undertaken without regard to overall project resource allocation and management questions.

The resource requirements as given by the project schedule may be undesirable from the viewpoint of total resource requirements or the resource profiles may exhibit isolated peak resource demands and widely fluctuating requirements. Thus it may be desirable to modify the project schedule to reflect resource availabilities or resource management decisions.

Whenever resource availabilities are less than that required by a particular project schedule, a limited-resource problem exists. Limited resources actually are additional constraints on the conduct of the project. The constraints may be due to one of the following reasons:

1. It may be desirable to keep the level of resource usage nearly constant for the duration of its use on the project. Thus to avoid frequent hiring and firing, it is desirable to keep the site work force level stable for the duration of the project. Also, it may be desirable to schedule the use of certain equipment, such as a crane or concrete mixing plant, so that its commitment to the project is minimized and also, as far as possible, so that its on-site use is continuous and nearly constant for the time that it must be at the project site.

2. There may only be a limited amount of a particular resource available. The amount of equipment or the number of workers with certain skills are usually limited. This limitation may be due to a shortage of supply or

it may be imposed by the financial resources of the contractor. In any case, a project schedule must be developed that is compatible with available resources.

3. Although all required resources may be available, the cost of acquiring these or using them on site at the scheduled rate may cause cash flow and overdraft problems. In such financially constrained situations it may become necessary to shift work tasks within their floats or even to extend the project duration to lessen or eliminate such financial resource constraints.

In most practical situations there is no need to seek for optimal project resources allocations and profiles. Instead, the resource scheduling process may only involve the development of a more acceptable or feasible solution. Thus it may become economically feasible to accept a project overrun and potential penalties to increasing the level of special resources made available to the project.

Consider the problem of scheduling the work tasks and resources involved in constructing a reservoir that will be used to supply water and recreation facilities and to generate electricity. The CPM arrow network is presented in Figure 13.10 together with the durations of the work tasks. The critical path is composed of work tasks 7, 1, 3, 5, and 6 and the minimum project duration is 73 months. The critical path bar chart based on earliest start times is shown in Figure 13.11.

Assume that the number of workers required to complete each work task is as shown in Figure 13.10. Furthermore, assume that the workers can work on any work task (i.e., they are not specialists). The total number of workers required during each month can be determined by adding the number of workers required for each work task that is active that month. The values are shown at the bottom of Figure 13.11.

If there are 160 workers available to work on the project, the early start schedule is satisfactory. However, if only 120 workers are available, a new schedule must be determined that is consistent with this resource availability constraint.

One heuristic method for determining the new schedule is to examine the project schedule day by day and to solve each resource conflict by delaying those work tasks which pose requirements that exceed resource availability. This approach can be formulated as follows:

1. Start with the first day and schedule all jobs possible, then do the same for the second day, and so on.
2. If several jobs compete for workers, schedule first the one with the smallest float.
3. Then, if possible, reschedule work tasks not on the critical time and/or resource path to other time slots within the now revised floats or to

Figure 13.10 Arrow CPM network for reservoir construction.

Task Number	Work Task
1	Clear dam site
2	Clear rest of reservoir area
3	Construct earth dam
4	Construct concrete spillway
5	Powerhouse structure
6	Install generators
7	Construct access roads
8	Construct other roads
9	Construct recreation facilities
10	Administration building and utilities
11	Water treatment
12	Install plant equipment

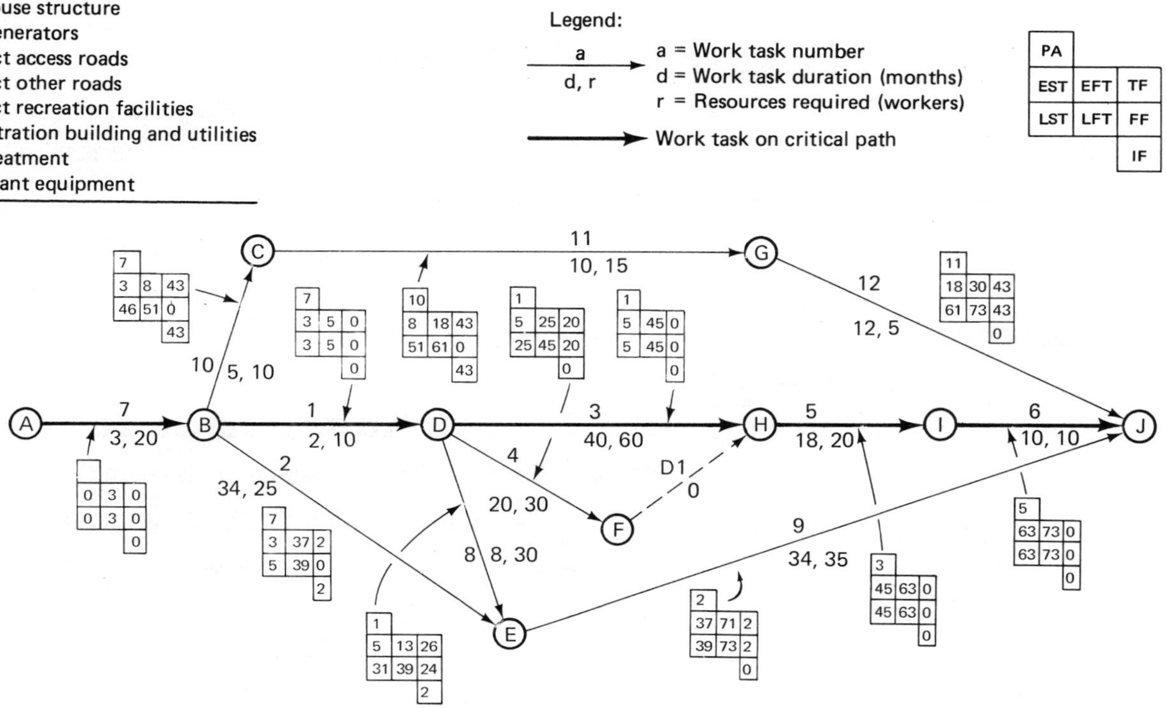

Legend:
a
d, r
a = Work task number
d = Work task duration (months)
r = Resources required (workers)
Work task on critical path
PA
EST EFT TF
LST LFT FF
IF
A
B
C
D
E
F
G
H
I
J
7
3, 20
10
5, 10
1
2, 10
2
34, 25
11
10, 15
3
40, 60
4
20, 30
8
8, 30
D1
0
12
12, 5
5
18, 20
6
10, 10
9
34, 35

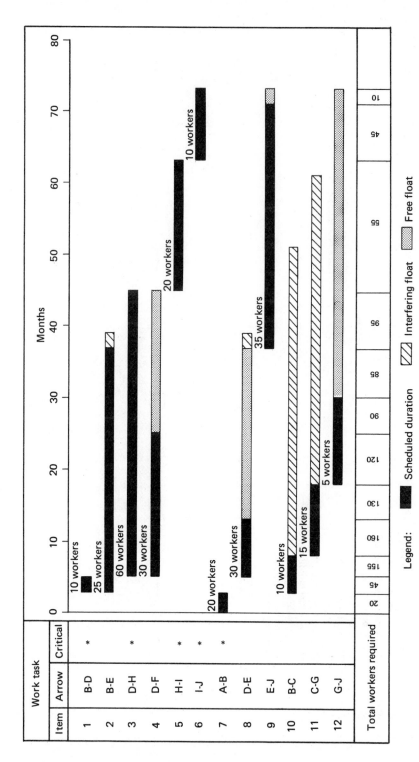

Figure 13.11 Bar chart for reservoir construction for earliest start times.

account for a now prolonged work task duration resulting from a changed work rate or to smooth out resource requirements.

Figure 13.12 is a revised bar chart that meets the requirement that not more than 120 workers can be employed at any time. It was developed according to the following steps:

1. One job can be started the first month, work task 7. It is scheduled and requires 20 workers. The bar for the work task is shaded for the duration of the work task and the number of workers is indicated at the upper-left corner of the bar.

2. No other work task can begin until the fourth month, work task 7 has been completed and work tasks 1, 2, and 10 can start. Because they require a total of only 45 workers, they are all started.

3. The next new work task cannot begin until the sixth month. Work task 1 has been completed and work tasks 3, 4, and 8 can begin. If 3, 4, and 8 are all started and 2 and 10 remain active, 155 workers would be required. Comparing the total floats for each work task:

Work Task	Workers	Float	
3	60	0	← smallest float
4	30	20	
8	30	26	

Therefore, schedule work task 3 with a total of 95 workers employed.

4. At the beginning of month 9, work task 10 is completed and this makes 10 more workers available. Work task 4 or 8 or 11 can start but not all three. The float comparison is:

Work Task	Workers	Float	
4	30	17	← smallest float
8	30	23	
11	15	43	

Therefore, schedule work task 4. Note that work tasks 4 and 8 have lost 3 months of float at the beginning of month 9.

5. Work task 4 is completed at the start of month 29, making 30 more workers available. Work tasks 8 and 11 can be started.

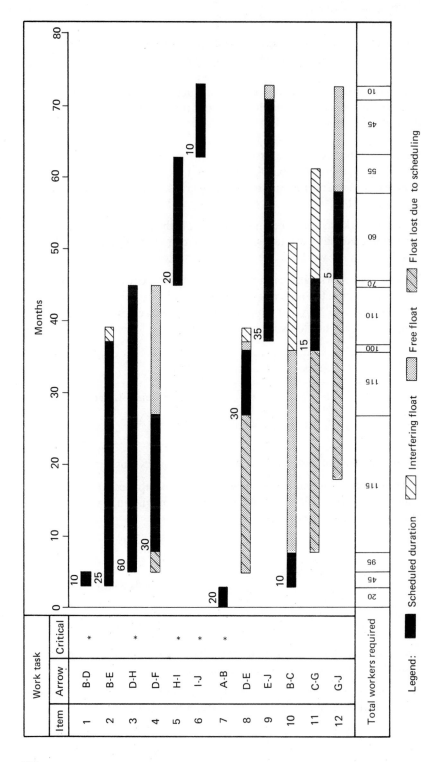

Figure 13.12 Bar chart for labor constrained schedule for reservoir construction.

Work Task	Workers	Float
8	30	3 ← smallest float
11	15	23

Therefore, schedule work task 8.

6. At the start of month 37, work task 8 is completed, releasing 35 workers. Work task 11 is started.

7. Work task 2 is completed at the start of month 38 and this makes 25 more workers available. Work task 9 is started, leaving 10 workers available.

8. At the beginning of month 46, work task 3 is completed and work task 5 is started.

9. Work task 11 is completed at the beginning of month 47 and work task 12 is started.

10. Work task 12 is completed at the beginning of month 59 and there are no new work tasks that can start.

11. At the beginning of month 64, work task 5 is completed and work task 6 is started.

12. Work task 9 is completed at the beginning of month 72, leaving a total of 10 workers employed for the last two months of the project.

In this case the work tasks can be scheduled so that the project can be completed within the minimum project duration.

The completion of each work task within a project requires a certain quantity of resources and a specific amount of time. If more resources are applied to work task, it may be possible to complete the work task in less time but at a greater cost. In actual projects a change in the environment may cause the construction work to fall behind schedule.

Assume that the project schedule is the one given in Figure 13.10 and that at the beginning of month 46 there is a labor strike that lasts for 3 months. The engineer must decide which activities should be completed at a faster pace and how much faster in order to complete the project in 73 months. At the beginning of month 46, activity 5 is just ready to start, activity 6 cannot be started until activity 5 is completed, activity 9 is 8/34 completed, activity 11 is 9/10 completed, and activity 12 cannot be started until activity 11 has been completed. The cut-set shown in Figure 13.13 indicates the portion of the project completed at time of delay.

All work tasks to the left of the cut-set have been completed and all work tasks to the right of the cut-set remain to be completed, including 1/10 of work task 11 and 26/34 of work task 9. The strike causes a construction delay of 3 months and the project must now be completed in 25 work months instead of 28 to be able to meet the 73 calendar-months deadline. In evaluating any

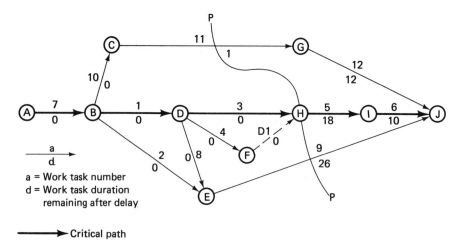

——————▶ Critical path

Figure 13.13 Cut-set indicating portion of project completed at time of delay.

changes in schedule, the engineer is interested only in that portion of the graph to the right of the cut-set.

Figure 13.13 indicates that work tasks 11 and 12 can be completed in 13 months and, therefore, can continue at their normal pace when work is resumed. However, work tasks 5 and 6 require 28 months, so their duration must be shortened by a total of 3 months. Also, the duration of the remaining portion of work task 9 must be shortened by 1 month. This reduction of the duration is called work task duration compression.

The duration of the work tasks can be compressed only by committing more resources to the work tasks than would normally be required. This increases the cost to the contractor and, of course, it is desired to find the minimum cost of decreasing the duration of the work tasks so that the 73 calendar-months deadline can be met.

A time–cost curve must be developed for each work task that has the potential of being compressed. A typical time–cost curve is shown in Figure 13.14, where C_n is the cost of completing the work task under normal conditions with a duration of t_n and C_c is the cost of completing the work task under crash conditions in the shortest duration possible of t_c. The slope of the curve is the cost for compressing the work task duration one unit of time. Therefore, the time–cost curve can be defined for each work task by specifying its normal and crash durations and its normal and crash costs.

Table 13.3 contains the time–cost data for work tasks 5, 6, and 9.

The project compression calculations are made as follows:

1. List the work tasks on the critical path; that is, identify the subgraph of the network model that must be considered. (Work tasks 5 and 6.)

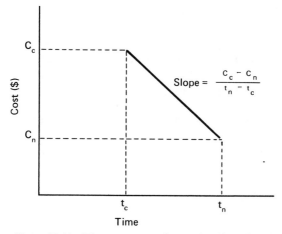

Figure 13.14 Time–cost curve for a selected work task.

TABLE 13.3 Time–Cost Data for Work Tasks 5, 6, and 9

Work Task	Duration (months)		Cost ($)		Cost Slope
	Normal	Crash	Normal	Crash	($/month)
5	18	16	2.5×10^6	2.7×10^6	0.1×10^6
6	10	7	6.0×10^6	7.5×10^6	0.5×10^6
9	26*	22	1.53×10^6	1.73×10^6	0.05×10^6

*Data for work task 9 is for 26 work months remaining at time of strike.

2. Remove from this list those work tasks that cannot be compressed because the normal and crash durations are identical or they have already been fully crashed in previous stages. (None.)

3. Select the work task with the smallest cost slope, since this will result in the cheapest compression. (Work task 5.)

4. Determine the amount by which this work task can be compressed. (2 months.)

5. Determine if this compression results in a new critical path. (Work task 9 becomes a critical path also; thus there are two critical paths.)

6. If a new critical path results, carry out the compression only to the point where the new critical path is the same as the compressed old critical path. (Compress work task 5 by 2 months.)

7. Compute the new project duration and cost. (The remaining project duration is now 26 work months and the remaining project costs are 10.23×10^6.)

8. Steps 1 through 7 are repeated until the desired project duration is achieved or until each critical path through the network model has been compressed

to its crash point. (Work tasks 6 and 9 must both be compressed by 1 month to achieve the 25-months duration because they are on different critical paths. A reduction of 1 month will not result in another new critical path, so the new duration is 25 months at a cost of 10.78×10^6.

The computations are illustrated graphically in Figure 13.15.

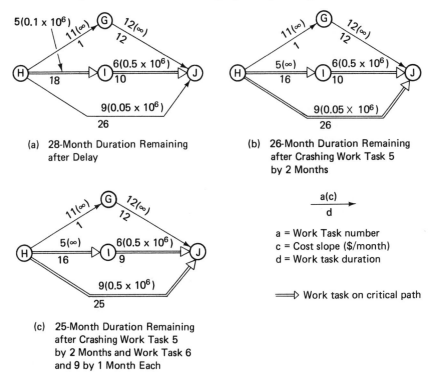

(a) 28-Month Duration Remaining after Delay

(b) 26-Month Duration Remaining after Crashing Work Task 5 by 2 Months

(c) 25-Month Duration Remaining after Crashing Work Task 5 by 2 Months and Work Task 6 and 9 by 1 Month Each

$$\frac{a(c)}{d}$$

a = Work Task number
c = Cost slope ($/month)
d = Work task duration

⟹ Work task on critical path

Figure 13.15   Network compression of reservoir construction.

13.6 PERT

In many cases, and especially on construction projects, work task durations vary from that initially estimated. This variation may be due to incorrectly estimated work content, variations in work productivities, weather delays, equipment breakdowns, and other factors. In CPM methods these variations are often handled by the addition of compensating periods to the work task durations before network calculations are performed and by the judicious scheduling of critical events with built-in floats. Thus in CPM methods a deterministic approach is used to cope with the essentially stochastic nature of the construction environment. In many cases this approach does not properly address the magnitude of risks involved in a planned schedule and management may require a more sophisticated analysis.

PERT (program evaluation and review technique) introduces uncertainty into the time estimates for work tasks and project durations. PERT uses a work task duration called the *expected mean time* (t_e), together with an associated measure of the uncertainty of this work task duration. This uncertainty may be expressed either as the standard deviation (σ_{t_e}) or the variance (v_{t_e}) of the duration. The expected mean time is intended to be a time estimate having approximately a 50 percent chance that the actual duration realized will be less, and a 50 percent chance that the actual duration will exceed it. Work tasks are determined using a probability distribution curve for the work task completion times. Three engineering time estimates are made and embedded within the assumed distribution curve for each individual work task. These three estimates of the duration of the work task enable the expected mean time, as well as the standard deviation and the variance, to be derived mathematically.

The *optimistic time* (t_a) is an estimate of the minimum time required for a work task if exceptionally good luck is experienced.

The most likely time (t_m) is based on experience and judgment, being the time required if the work task is repeated a number of times under essentially the same conditions.

The pessimistic time (t_b) is an estimate of the maximum time required if unusually bad luck is experienced; it may take account of an initial failure or delay, but should not be influenced by major hazards (such as floods) unless these are inherent in the work task.

The calculation of these three time estimates forces the planner to take an overall view of the particular difficulties involved in the work task. They tend to offset the effects on the planner's judgment of known target dates. They become the framework on which is erected the probability distribution curve for the work task duration.

A probability distribution curve that can represent this situation is called the *beta distribution*. In this curve mathematically simple (and slightly conservative) approximations can be made for the work task's expected mean time and its standard deviation.

The expected mean time is derived from the following equation:

$$t_e = \frac{t_a + 4t_m + t_b}{6} \tag{13.1}$$

The standard deviation (the statistical measure of uncertainty being the spread of the distribution curve about its mean value) is given by

$$\sigma_{t_e} = \frac{t_b - t_a}{6} \tag{13.2}$$

Finally, the variance is defined as the square of the standard deviation, so that

$$v_{t_e} = (\sigma_{t_e})^2 = \left(\frac{t_b - t_a}{6}\right)^2 \qquad (13.3)$$

From Equation 13.1, except for a symmetrical distribution (where $t_a + t_b = 2t_m$), the expected mean time (t_e) will be different from the most likely time (t_m), as indicated in Figure 13.16. Furthermore, since the optimistic (t_a) and pessimistic (t_b) times have small probabilities of attainment (1 percent chance is usually assigned to these two time extremes), they approximately define the range of distribution of feasible work task durations. The standard deviation, measuring the uncertainty, is one-sixth of this range; consequently, the smaller the range becomes, the more certain becomes the work task's duration. The work task durations used in CPM calculations may be considered as special cases of the more general time estimates used in PERT.

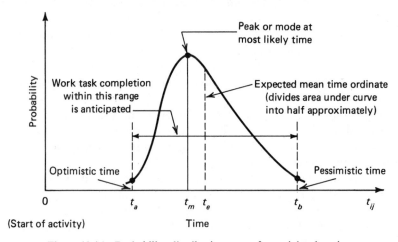

Figure 13.16   Probability distribution curve for activity durations.

By adopting work task expected mean times, the critical path calculations proceed as before. Associated with each duration in the PERT approach, however, is its standard deviation or its variance. The timing of events thus computed will be expected mean event times, and consequently subject to doubt; the measure of this doubt in the timing of the events requires the derivation of event standard deviations.

The project duration is determined by summing the work task expected mean times along the critical path and will thus be an expected mean duration. Since the critical path activities are independent of each other, statistical theory gives the variance of the project duration as the sum of the individual variances of these critical path work tasks. Therefore, for a project duration of expected mean time T_{X^P},

$$V_{T_{X^P}} = (\sigma_{T_{X^P}})^2 = \sum v_{t_e} = \sum (\sigma_{t_e})^2 \qquad (13.4)$$

from which the standard deviation of the project duration is easily determined. If more than one critical path exists, the project duration variance is taken as the maximum of those summed along the various independent critical paths. From this it is clear that the variance of the expected mean time of any event is the sum of the variances of those work tasks along the most time-consuming path (in terms of expected mean times) leading to that event.

Once the *expected mean time for an event* (T_x) and its standard deviation (σ_{T_x}) are determined, it is possible to calculate from probability theory the chances of meeting a specific *event scheduled time* (T_S). To do this the event completion time is considered to have a normal probability distribution with the mean value T_x and a standard deviation σ_{T_x}, determined as before from the series of individual work task beta-distribution curves. This hypothesis implies that the effect of adding a series of independent beta-distribution curves gives a curve of normal distribution; this is true only for an infinite series but is approximately true in practice with reasonable-sized networks.

Hence, to calculate the chances of meeting the time T_S, it is necessary to plot a normal distribution curve centered on time T_x, as illustrated in Figure 13.17. With this curve, the probability of meeting the desired scheduled time T_S is obtained by determining the percentage of the area cut off by this time from the total area beneath the normal distribution curve as shown.

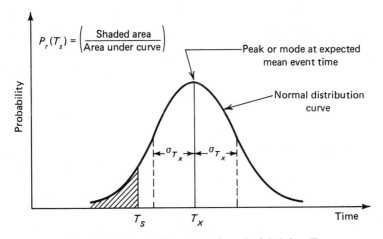

Figure 13.17 Probability of meeting scheduled time T_S.

Instead of plotting a normal distribution curve each time, the practical approach is to use standard probability tables prepared for normal distribution functions, of which a condensed version is given in Table 13.4 it is emphasized that determining probability to the nearest 1 percent is more accurate than is generally required in construction practice and hence condensed tables are adequate. To use this approach, the difference between the scheduled and

TABLE 13.4 **Approximate Values of the Standard
Normal Distribution Function**

Z	Probability	Probability	Z
−2.0	0.02	0.98	+2.0
−1.5	0.07	0.93	+1.5
−1.3	0.10	0.90	+1.3
−1.0	0.16	0.84	+1.0
−0.9	0.18	0.82	+0.9
−0.8	0.21	0.79	+0.8
−0.7	0.24	0.76	+0.7
−0.6	0.27	0.73	+0.6
−0.5	0.31	0.69	+0.5
−0.4	0.34	0.66	+0.4
−0.3	0.38	0.62	+0.3
−0.2	0.42	0.58	+0.2
−0.1	0.46	0.54	+0.1
0	0.50	0.50	0

expected mean times for the event is scaled down to the standard curve by computing a factor Z, where

$$Z = \frac{T_S - T_X}{\sigma_{T_X}} \tag{13.5}$$

Using this computed value of Z, a direct entry into Table 13.4 gives the probability of meeting the scheduled time T_S (interpolating in the table, if warranted).

An equivalent form of Equation 13.5 enables the scheduled time for an event to be determined, based on a given risk level; thus

$$T_S = T_X + Z\sigma_{T_X} \tag{13.6}$$

where the value of Z is obtained from Table 13.4 for a specific probability or risk level acceptable to management.

It will now be clearly understood that the scheduling of particular event times for a project requires an assessment of the uncertainty in the project and the acceptance by management of the risk levels in the desired schedule. As a preliminary to the actual scheduling of particular event times with specific risk levels, it is often an advantage to determine the latest expected finish time (T_{XL}) for each event, based on a neutral or 50 percent chance, and the calculations then proceed in exactly the same way as those for LFT in CPM calculations. The corresponding event variances $(\sigma T_{XL})^2$ are derived similarly to those for the T_{XE} variances, starting from the project completion event. An examination of these will show that, in some cases, the T_{XE} and T_{XL} values are different; this difference is called the *event slack*, and is expressed mathematically as

$$\text{Slack } T_{XL} - T_{XE} \tag{13.7}$$

Slack in PERT corresponds to the total float concept CPM, and is a measure of the flexibility available in a project schedule. An event with zero slack must therefore lie on the expected critical path.

The PERT approach is interesting in that it focuses attention on the uncertainty that is inherent in many projects. Thus decision makers, when using PERT, are reminded that the management of the project requires risk evaluation and continuous surveillance of progress with the resultant continuous dynamic application and reallocation of resources to each work task. The PERT approach thus stresses the fluid and dynamic use of resources to meet project event schedules and is an example of a stochastic decision process.

13.7 SUMMARY

CPM represents a network model of an organizational system that can be used for the management of a project. The development of this model requires the engineer to understand the components involved in the project and the manner in which these components interact. In this case, the components are the work tasks, and the interaction is defined by the sequence in which the activities may be undertaken.

The addition to the network model of an attribute such as time permits the engineer to evaluate the scheduling of activities. If constrains such as limits on the available resources are imposed, the model may be used to determine a suitable course of action. In this way, the engineer may use the model for the purpose of managing the project.

The circle or arrow notation model is effective when examining the sequencing of work tasks or computing the time schedule. The determination of work tasks which may be scheduled for the same time is not readily discernible from these models. The bar chart represents a transformation of information to another format whereby the work tasks are depicted according to some time frame. Again, this is a situation where the transformation concept has been applied, and the resulting model can be more readily evaluated.

Certainly, some variation in the duration of a work task can be expected. Although PERT is similar to CPM, it does recognize the time variation in the accomplishment of the work tasks, and it provides an analysis technique that is based on stochastic concepts.

Although the examples in this chapter are associated within construction projects, it is important to recognize that these concepts and analysis techniques can be applied to any situation that involves the organization of a number of work tasks. For example, the management of a planning or design project may require the definition of work tasks and the sequencing of the work effort. The application of CPM concepts and analyses would aid in the effective management of such a project.

PROBLEMS

P13.1. Figure P13.1 shows a CPM network for a project in arrow notation in which the durations are given in number of weeks.

   (a) Compute the following for each job in the schedule: EST, EFT, LST, LFT, TF, FF, and IF.

   (b) Suppose that all the work tasks have been scheduled to start as early as possible and that the tasks have been on schedule up to the end of week 5. There is a strike on week 6 causing a delay of 1 week. Draw a CPM diagram for the work tasks remaining to be done when work resumes on week 7.

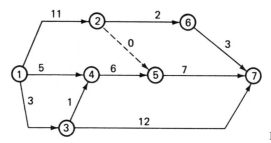

Figure P13.1

P13.2. A project is defined below in tabular form using arrow notation for the various project work tasks. The table also gives estimated work task times, costs, and crew sizes for the range of feasible rates for working the tasks.

	Duration (days)		Cash		
Work Task	Normal	Crash	Normal	Crash	Crew Size
1, 2	6	4	$700	$ 840	6
1, 3	12	10	300	420	7
1, 4	4	2	200	260	5
2, 3	8	6	900	1000	3
2, 4	4	2	600	760	4
2, 6	15	8	100	380	4
3, 5	6	3	660	960	6
5, 6	2	1	500	600	5
4, 6	6	4	400	500	7

The financial arrangements are such that the indirect cost for overhead will be $50 per day for the first 20 days and then will increase by $50 per day for each day thereafter. Thus the indirect cost will be $100 for day 21, $150 for day 22, and so on. The type of job allows any worker to perform any work task, but once a work task is started it must be completed. That is, the work tasks cannot be interrupted or segmented.

A complete financial and resource project study is required based on the following analyses.

(a) Perform the CPM analysis and develop the bar chart form showing all the TF, FF, and IF values.

(b) Draw graphs that show daily resource requirements (number of workers) and cumulative cost for both the EST and LST schedules.

(c) Determine the schedule of work tasks that will allow an orderly increase in workers to the peak requirement and then allow an orderly reduction in work force (i.e., try to minimize the fluctuation in the number of workers required). What is the schedule and number of workers required?

(d) Assume that there is a constraint of a maximum of 17 workers per day available for the job. Set up the schedule with the shortest completion time subject to this constraint.

(e) Obtain the direct, indirect, and total project costs, without regard to the resources available, for the normal and possible crash solutions.

(f) What is the minimum cost duration?

(g) What is the duration and schedule for each work task under part (f)?

P13.3. An engineer who has been assigned to supervise construction of a highway wishes to develop a CPM network model as a management aid.

In constructing a highway, the basic sequence of operation is: surveying, clearing and grubbing, moving dirt, placing subbase material and drainage, paving, curing pavement, forming shoulders, seeding embankments with grass, painting lane marks, installing signs, and clearing site. The construction plan calls for highway construction to proceed from one end of the route to the other.

The separate crews for each construction work task are to be organized to progress at the rate of 4 miles per week, the highway being 40 miles long.

Develop a circle notation network model in sufficient detail to indicate logically the construction planning and determine the minimum project duration required in the field. Then translate the model into one using arrow notation. Set out the network roughly to scale, with each horizontal chain representing the progress of one construction work task, such as surveying, clearing, and so on.

P13.4. A pump that supplies water for a production process must be replaced. The company would like to replace it during a shift when the plant is not working. The job requires at least two workers and a hoist to lift the pump, which weighs 1000 lb. The following work tasks have been identified.

1. Shut down pump
2. Close downstream valve
3. Close upstream valve
4. Drain downstream pipe connections
5. Drain upstream pipe connections
6. Remove upstream connections
7. Remove downstream connections
8. Remove nuts from anchor bolts
9. Lift pump and set it aside
10. Position replacement pump

11. Align the replacement pump
12. Tighten nuts on anchor bolts
13. Replace upstream connections
14. Replace downstream connections
15. Prime pump
16. Open upstream valve
17. Open downstream valve
18. Test pump

The pipe is connected to the pump with flange connections. The pipe is 6 in. in diameter. Develop a linear graph model of this process and discuss the selection of work tasks included in the process. Estimate the time required for each work task and determine the time required to replace the pump.

P13.5. In constructing a high-rise building there is a regular progression of work tasks and building trades from floor to floor. After a certain initializing period, a steady-state process and work force develops. The balanced progression of work tasks, trades, and crews up the building is desirable to project management and is the basis for scheduling crew sizes.

Several work tasks, trades, and crew sizes for the construction of a typical reinforced concrete brick wall building are as follows:

Work Task	Floor i Relational Logic	Work Task Crew				
		Labor-ers	Car-pen-ters	Iron Work-ers	Brick Ma-sons	Cement Fin-ishers
Erect formwork for floor and columns*	i	1	5	0	0	0
Make and place steel reinforcement cages*	i	0	0	5	0	0
Place concrete for floor and columns*	i	6	3	0	0	5
Dismantle formwork	$i - 2$	2	1	0	0	0
Wall bricklaying	$i - 3$	6	0	0	10	0
Set windows in walls	$i - 4$	1	2	0	0	0

*These work tasks are concurrent because this building has a large floor plan.

Assuming that each crew finishes its work tasks on a floor each week.

(a) Develop a network model portraying the construction logic for the first six floors of the building.
(b) Develop a bar chart model and show that a steady-state labor work force develops.
(c) Indicate those "line of balance" cut-sets that portray the relational logic among the work tasks.

P13.6. A contractor has been awarded a contract to build a large concrete dam to be used for both flood control and power generation on the Colorado River in southern Nevada. The river is located in a gorge with steep rock sides and bottom. A schematic outline of the dam is shown in Figure P13.6.

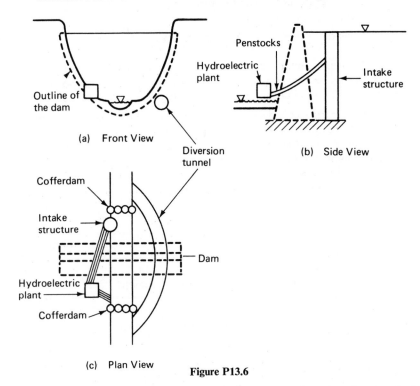

(a)   Front View

(b)   Side View

(c)   Plan View

Figure P13.6

The contractor wishes to develop a construction plan showing the logic and order of the work tasks that must be performed to complete the contract. An initial list of activities was prepared as follows.

1. Prepare foundation and canyon wall for dam
2. Place cofferdams to divert water into diversion structure
3. Place concrete for dam
4. Construct rock crushing and screening plant
5. Construct penstocks through the dam to the hydroelectric plant
6. Construct concrete mix plant
7. Construct housing for the work force
8. Construct intake structure at intake of penstocks
9. Transport labor to the site
10. Construct hydroelectric plant buildings
11. Transport equipment to the site

12. Test performance of dam and hydroelectric plant
13. Clean up site
14. Bidding process for turbines
15. Manufacture and delivery of turbines
16. Installation of turbines
17. Construct water diversion structure in wall of gorge

Develop a suitable construction plan in both circle and arrow notation form. It may be necessary to break down and redefine several work tasks into smaller work tasks in order to better portray the construction logic.

P13.7. A contractor has received a contract to construct an overpass over an interstate highway. The contract also includes the approach structures (embankments) for the bridge and the concrete slab for a distance of 0.500 mile on each side of the centerline of the bridge. The bridge is to consist of rolled steel beams of standard section with a concrete slab deck. The contractor has prepared an initial list of work tasks and their workday durations as follows:

Work Task	Time (workdays)
1. Clear site and construct temporary access road	6
2. Foundation, forms, and reinforcing steel for north abutment	16
3. Obtain bids for steel girders between abutments	15
4. Survey for locating improvements and grade stakes	4
5. Order steel girders	15
6. Foundation, forms, and reinforcing steel for south abutment	16
7. Place fill on north side	45
8. Foundation, forms, and reinforcing steel for center abutment	16
9. Construct forms for concrete deck over steel girders	10
10. Place steel girders between abutments	15
11. Pour concrete deck over steel girders	4
12. Place fill on south side	45
13. Place steel railings and other accessories on bridge	8
14. Pour concrete slab on south side of bridge	8
15. Prepare subgrade for concrete slab on north side of bridge	10
16. Pour concrete slab on north side of bridge	8
17. Finish grading for drainage for entire site	5
18. Prepare subgrade for concrete slab on south side of bridge	10
19. Pour concrete for center abutment	6
20. Place topsoil and sow grass seed for entire site	4
21. Pour concrete for south abutment	5
22. Pour concrete for north abutment	5

The times given for pouring concrete do not include curing times. In general, 2 days should be provided for curing before any work task is permitted on the concrete.

Develop a feasible construction plan and portray your plan in linear graph form using both arrow and circle notation. Then determine a bar chart model for the project indicating total, free, and interfering floats. Define a project schedule based on an earliest start policy.

P13.8. Give reasons why you agree or disagree with the following.

  (a) CPM networks use unidirectional concepts to model project logic (i.e., no cyclic paths can exist).

  (b) CPM networks are useful for situations involving many work tasks and heavily interfacing logic.

  (c) CPM networks are not useful for heavily repetitive projects such as pipe laying unless gross modeling concepts are used.

  (d) A trade-off must exist in any model between the accuracy of the modeling logic and representation and the usefulness of the model to the decision maker.

  (e) Projects with cyclic logic are best modeled with flow and/or simulation models.

  (f) Different CPM models are possible for the same project depending on the user and his purpose.

The Engineering Process: A Systems Approach

14

14.1 THE APPLICATION OF SYSTEMS CONCEPTS IN ENGINEERING

The general nature of the role of the engineer in addressing and solving societal problems is discussed in Chapter 1. In performing the broad engineering functions that are required in the solution of a problem, the engineer is involved in a range of planning and design activities. These activities should eventually lead to the implementation of solutions to the problems that have been identified. Subsequent chapters present concepts related to problem formulation, modeling, and analysis.

Chapter 2 develops a methodology for the portrayal of a problem in a representational form which reflects a system theory viewpoint. Once this has been established, the engineer can utilize the later chapters to develop analysis, optimization, and decision process approaches that may be relevant to the professional understanding and solution of a problem. This systematic approach is aimed at aiding the engineer in making decisions at any level of planning, design, or implementation activities.

Inherent in all of the material presented is the concept of the systems approach. As has been indicated previously, the concept suggests that the engineer examine the problem in a comprehensive context and determine the appropriate approach to be taken given the available resources. In all cases, the analysis methods or techniques which are selected should fit into a systematic framework that depicts a professional engineer's assessment and solution to a problem. The subsequent sections of this chapter serve to demonstrate the application of these systems concepts to the broad range of engineering functions.

14.2 THE IDENTIFICATION OF ENGINEERING FUNCTIONS

Engineers are involved in a variety of activities which require them to perform many different professional functions. The specific role that they perform will depend on their professional responsibilities in an organization. These responsibilities may be either of a technical or managerial nature and are related to professional activities associated with the identification and resolution of a problem.

One rather basic approach to identifying the range of engineering activities would be to categorize them in terms of the following functional areas:

1. Technical
2. Technical policy development
3. Engineering management

The technical function area is concerned with activities such as the performance of field experiments, data gathering, and engineering calculations. In addition, this area could include activities involving the use of prescribed models, techniques, and procedures in which fundamental engineering judgment may be required. The technical policy development functional area is critical in professional engineering because it requires the development of appropriate procedures, criteria, and models relevant to the task to be performed. Engineering management is the functional area concerned with the organization and staffing of a work group and in the management of the overall work effort.

It should be noted that this categorization of engineering functions is another application of the representation of an area in terms of functional components. In this case, the three components represent an organizational system involving professional effort. The use of three components to represent the organizational system is convenient here, but a system representation including a greater number of components could be developed if greater detail is desired.

These three general functional categories transcend all areas of engineering because of the nature of engineering work and its accomplishment. For example, consider the resolution of a problem in terms of the following three specific phases, which represent components in the solution process:

1. Problem definition
2. Problem resolution
3. Solution implementation

Certainly, a number of more specific steps or components might be considered as is suggested by the discussion in earlier chapters. For the purpose of presenting the concepts related to the application of the systems approach to the engineering process, the number of steps has been limited.

The interfacing of the engineering functional areas with the phases in the resolution of a problem can be represented in a two-dimensional engineering activities matrix format. This format enables engineering functions to be related to each of the phases in the problem-solving process. These relationships then define the specific engineering activities which form the statement of each combination of functional area and problem-solving phase.

Table 14.1 depicts the engineering activities matrix resulting from the interfacing of the two component groupings. An examination of this matrix reveals the general description of the engineering activities associated with each component pair.

For a small project, one engineer may be involved in all aspects of the effort. In the case of larger projects involving a number of engineers, the specific activities of any one engineer may be more restricted, so that the focus of the effort of that individual is more limited. In such situations, it is necessary for the engineer to work as part of a team and interface with others who are involved in the effort.

14.3 THE SYSTEMS APPROACH AND THE ENGINEERING PROCESS

The engineering activities matrix provides a venue for the full portrayal of the spectrum of engineering effort which is involved in the resolution of a problem from inception to solution implementation. An examination of the matrix reveals the diversity of professional activities which are necessary. This diversity of activities also presents the engineer with a wide variety of issues and questions that must be addressed or answered in the course of the total effort.

It is important to note that the systems concepts and analyses presented in earlier chapters have potential application in all of the areas of professional activity represented by the elements of the matrix. Although the particular application may differ depending on the professional activity, the concepts can serve to aid the engineer in decisions relative to planning and design questions that are inherent in those activities.

For example, at the engineering management level, it is likely that the engineer will be more concerned with the issues associated with setting up groups to handle the necessary phases of the engineering effort. Thus the engineer may be involved in the modeling and analysis of an organizational system and the evaluation of its potential capabilities in terms of the management problems that must be addressed. In this case, the organizational system is composed of the group or team and resources that have been made available for the accomplish-

TABLE 14.1 Engineering Activities Matrix

Engineering Functions	Phases of Problem Resolution		
	Problem Definition	Problem Solution	Solution Implementation
Management of engineering	Acquire, allocate, and manage resources relative to problem-definition activities	Manage resources necessary for the accomplishment of the problem-solution activities	Manage resources associated with the implementation of the solution
Technical policy development	Develop technical approach to definition of problem and direct all technical efforts in problem-definition phase	Identify technical analyses and procedures Implement and direct all technical activities	Develop methodology and approach to implementation. Be responsible for technical aspects of implementation
Technical	Carry out technical functions as required for problem definition	Perform technical analyses associated with the planning and design of alternatives and the selection of the preferred solution	Undertake and accomplish technical analyses relative to implementation efforts

ment of the necessary tasks. The pertinent analyses would be associated with questions involving the scheduling, allocation, and management of the available resources.

At the technical level, the application of the concepts would be used to ascertain answers to specific planning- and design-focused questions. Depending on the type of problem, the engineer could be involved with a physical or an organizational system. For this functional-level situation, any organizational system would have a very specific technical focus, such as the requirements for a transportation system.

In addition to demonstrating the broad application of systems concepts, the activities matrix can be a useful tool with respect to understanding the total effort that is involved in the solution of a problem. For this purpose, the matrix can be used to:

1. Define the activities that must be undertaken and the expertise required for a particular problem
2. Portray the role of specific engineers or groups relative to the overall effort
3. Define the scope of work to be accomplished if constraints or boundaries on the activities are established

For the first two uses, the activities matrix becomes a valuable planning aid, especially when a large number of engineers and other personnel may be utilized. As has been indicated in the preceding section, it may be necessary to increase the number of functional components or possibly consider the division of any element in the matrix into subelements if greater resolution detail is desired in terms of activities and expertise. Again, this subdivision of the components could reflect the concepts associated with defining the components in a problem formulation effort or in establishing the project team's ability to perform in terms of a scheduling of work tasks included in a CPM analysis.

In the last use of the activities matrix, an individual or group may be responsibile for some portion of the overall effort. When this situation occurs, it is necessary to define the scope of work that is to be accomplished and how it will interact with the remainder of the total effort. As discussed in Chapter 2, the manner in which the boundaries are drawn will affect the scope to be considered. In this case, the scope involves the activities that are to be accomplished.

14.4 DEFINING THE PROPOSED EFFORT

In the course of undertaking a planning or design project, an engineer may be required to develop a proposal that identifies the problem that is to be addressed and the attack that will be taken in the solution of the problem. The development of a proposal should require the engineer to outline a comprehensive approach to a given problem. Depending on the nature of the project, the proposal may range from being a rather informal statement of the problem and the proposed effort to a detailed document that may represent the basis for a contractual agreement for the work that is to be accomplished. Obviously, the more informal proposal would be appropriate when the engineer is dealing with a rather routine situation. In such cases, the engineer is thoroughly familar with the work that is to be done and the approach that is to be taken. In the latter case, however, the proposal serves to define the engineer's understanding of the problem and the scope of the effort required for the solution of the problem and to identify the resources that will be required in accomplishing the work. If several engineering firms are being considered for a particular project, the proposal may become the basis for making the decision on which firm will be selected for the job.

A formal proposal generally contains key information about the work that is involved in the proposed effort. It will normally require the engineer to apply many of the concepts that were put forth in earlier chapters. For example, a typical proposal might contain the following information:

1. Introduction
   a. Statement of the problem
   b. Definition of study goals
   c. Limitations on the scope of work or study
2. A summary of previous work that is pertinent
3. A statement of the approach
4. The definition of methodologies that will be employed
5. The identification of required resources
   a. Equipment
   b. Time schedule
   c. Budget

Certainly, a proposal with this information requires a rather extensive effort; and it means that the engineer must demonstrate a full understanding of the problem and carefully plan the manner in which the problem will be addressed.

The focus of the introduction is on the formulation of the problem that is to be the subject of the planning and design study. It is at this point in the effort that the engineer must demonstrate an understanding of the nature of the problem for which a solution is being sought. In addition, there should be a statement of why the problem exists and what benefits can be achieved by a solution. It may even be up to the engineer to justify the need for devoting resources to the study of the problem if the engineer is attempting to inititiate an effort in the area. In some cases, the engineer may be responding to a problem area that has been defined by others. The statement of the problem simply verifies the engineer's understanding of the problem.

Once a statement of the problem has been developed, it is important to specify the expected goals of the planning and design effort. The goals reflect the end product that is to be achieved. For example, the goals might be expressed in the form of the questions that are to be answered. If the study is to develop a particular plan or design, the goals might be related to determining the best plan or design for the given condition.

Obviously, it is likely that there will be constraints or limitations on the scope of the effort; thus such a statement should be included so that a reviewer of the proposal will understand the context in which the engineer will seek a resolution of the problem. As will be discussed later, the limitations are important in the determination of the resources that are required for the accomplishment of the work.

For some types of efforts, it may be necessary to summarize previous work that has been undertaken or accomplished. This is particularly important when the engineer is attempting to indicate the current state of the art with respect to a particular problem area or project. If there is some element of research associated with the effort, such a summary will show that the engineer is familar with the work of others. This is critical in that it establishes that the person developing the proposed work is aware of the previous work that may be applied to the solution of the problem. Such a section in the proposal may also give the engineer an opportunity to demonstrate how others have dealt with relevant problems of a similar nature.

The statement of the approach should reflect the conceptual manner in which the solution of the problem will be achieved. For example, one might indicate how the professional disciplines will interact in addressing the problem. If field studies are to be the basis of empirical results, this conceptual approach should be defined. The same would be true if the solution is to be achieved using a theoretical modeling approach or simulation.

Once the general approach is established, it is possible to indicate the specific methodologies that will be employed. Again, if field studies are to be used, the scope and nature of the work should be defined. The section should also contain a description of specific analysis techniques or programs that will probably be used in the work effort. In essence, this section should define the technical tools that will be used in the resolution of the problem and how they will be used. This would require the engineer to determine the appropriate methods or analyses that are applicable to the given problem. As was the case with the scope of the study, the resource requirements may impose constraints on the methodologies that are used. If a particular analysis method is very expensive or time consuming, it may be necessary to select another method that is more economical in terms of resource requirements.

The resource requirements are an estimate of the equipment, time, and money that will be required to do the work. Certainly, the resource requirements must be directly related to the scope of the problem that is to be addressed and the amount of work that is to be done. This part of the proposal would possibly necessitate that the engineer employ a CPM analysis to develop a time schedule for the effort. In addition, cost estimates for each portion of the work must be made. At this point, it may be necessary to evaluate the added cost of expanding the scope of the problem or utilizing a particular methodology or technique.

Finally, if the proposal is a formal document, it is usually necessary for the engineer to submit the proposal with a cover letter to the client, relevant public authority, or supervisor. The cover letter will often refer to some previous request for services or authorization to undertake the work involved in the preparation of the proposal.

14.5 THE CHALLENGE TO THE ENGINEER

Technical as well as societal problems are often complex and involve engineers in all phases of their identification and resolution via the many stages of decision processes, whether at the federal, state, city, or private levels. Thus engineers involved in the resolution of such problems are called upon to perform a variety of functions in the identification and resolution of societal problems. In many cases, engineers work in independent areas related to the many phases in a problem-solving process. In others, the efforts of an agency vested with authority to cope with the problem establishes the environment for an integrated effort by the engineer.

In any case, the individual engineer should address the problem in a manner that fully exploits responsibilities and obligations to society, organizational constraints, and the need to integrate efforts with others. The engineer, however, is restricted by the functional performance avenues available. Within these limits of the engineer's involvement a problem definition and solution-implementation process must be formulated that optimizes the understanding of the problem, the use of available resources, and which is compatible with the team efforts of other engineers and professionals.

Consider, for example, the changing focus and variety of engineering functions performed by engineers charged with addressing a societal problem such as waste management. Thus at the engineering management level, the following statement of issues may be pertinent to undertaking the problem-definition activities:

Development of a suitable organization for the problem-definition phase

Estimate the time required for the study

Identify necessary authority

Identify types of expertise needed in the organization

At the technical policy development function level, the following issues might serve to guide the efforts and activities:

Define waste management

Define waste

Determine the scope of the waste problem to be included in the problem definition phase

Ascertain health and service standards

Define the policies and criteria relative to the acceptability of possible levels of service and location of waste treatment and disposal sites

Determine technical procedures for the study effort

Finally, at the technical function level, the following issues might be pertinent to problem-definition activities:

Performance of technical activities such as sampling, measuring, and estimating

Review pertinent research on the properties of the various waste products

Determine the state of the art on the performance and effectiveness of existing types of waste handling procedures such as collection, transportation, and disposal

Investigate the performance of existing waste disposal, including problems associated with dump and consolidation controls and subsequent seepage and environmental impacts

Identify suitable potential treatment and waste disposal sites

Any one of the typical question issues raised in this or other phases in waste management solution process can be addressed in a manner that requires the development of models, techniques, and procedures based on the concepts and material presented in this book.

The challenge to the engineer is to determine the boundaries within which the functional responsibilities must be performed and to establish the approach to be taken in the exercising of the functional responsibilities.

14.6 SUMMARY

The engineering activities matrix can be developed as an overall summary guide in laying out the activities that must be accomplished. At the same time, the detailed formulation of the matrix becomes an aid to the engineer in defining the role of a particular engineer or teams of engineers. The matrix in itself is capable of depicting an organizational system, and the engineer is frequently required to develop the appropriate representation for a specific situation.

The matrix can serve as a guide to defining a proposed effort. While a formal proposal does require a considerable amount of effort, it does reflect an understanding of a problem and the attack suggested to address the problem. A proposal is an important document, whether it is prepared by an engineer for a professional effort or by a student for a required project. In both cases, it is likely that the proposal will give an indication as to the capabilities of the individual or group to successfully address the defined problem.

In all areas of professional effort, engineers must demonstrate an awareness and ability to deal with the complex problems that are presented by the current needs of society. Throughout the engineering process, systems concepts can be applied to the technical as well as the managerial aspects of professional activities. The successful application of these concepts is contingent upon the abilities of the individual to understand the problem involved and its solution.

PROBLEMS

P14.1. Prepare an engineering activities matrix for the following types of projects.
   (a) Selecting and purchasing a piece of equipment
   (b) Extending a utility service to a small subdivision
   (c) Expanding a warehouse
   (d) Constructing a major regional engineering works, such as an irrigation system, urban freeway, or a power distribution system

P14.2. Visit an engineering office and develop an understanding of their functional role, organizational structure, and location within the problem definition and solution implementation process. Based on this understanding, prepare an engineering activities matrix for the functional responsibilities of that office.

P14.3. Select a traffic- or transportation-oriented problem. Prepare a proposal that defines an effort to investigate alternative solutions for that problem.

P14.4. You are a plant engineer in charge of expanding the production facilities of the plant. How would you approach the expansion problem in order to increase manufactured output by
   (a) 10 percent
   (b) 40 percent
   (c) 100 percent
   How would the project plans, focus, decision variables, and constraints change as the size of the expansion increases?

P14.5. Any work that an engineer performs is not very useful unless it is conveyed to other people. It may be necessary to transmit the information to sell the project or to get the project built after it is designed. The acceptability of the information is often influenced by the manner in which it is presented. The information is most often presented in the form of a report. The report outline might be the following.

   1. Purpose of the report
   2. Need for the proposed alternative
   3. A general statement of the resources required by the alternative
   4. General design concepts
   5. Capital cost breakdown

6. Operational cost breakdown

7. Anticipated occupancy and income profile

8. Work sheets

An architectural-engineering firm is reporting to the client the results of their design study for a high-rise building. Comment on the format of the report. What do you think would be included in each of the sections of the report as outlined above? Where would you obtain the information you include? How might the report change to meet the requirements of decision makers at different levels?

P14.6. Davis (1968) presented an example of how to approach a planning study of a water-quality problem in a river or estuary. Read the preface and Chapter 1 of that reference. On pages viii, ix, and 12, the author indicates the personnel and time involved in the study. Look through the other chapters.

Complete an engineering activities matrix that indicates the study activities that would correspond to each component of the matrix. Indicate who might perform each of these activities. For example, the principal investigator may have decided to use simulation to analyze the dissolved oxygen levels in the river and estuary but probably did not develop the computer program to do the computations. Where would the decision to use simulation be located in the engineering activities matrix? Where would the development of a computer program to do the analysis be located in the matrix, and who would develop the computer program?

P14.7. Large engineering projects in heavily populated and utilized areas inevitably affect many government, private, and local agencies and pose considerable site, access, and delivery problems. In order to cope with these issues in the decision process, extensive organizational structures and planning procedures are necessary.

(a) Discuss the organizational structure and planning procedures that would have to be initiated in a project as large as a large building or bridge.

(b) Identify site, access, and delivery problems for this project.

P14.8. The engineer is usually not told how to use the various approaches and techniques discussed in the text. Because problems are generally complex and the portion of the problem addressed by a given technique is limited, the engineer is faced with deciding where to use the different techniques and how to integrate them into a study of the overall problem. It may be possible, for example, for one technique to be relevant at several levels of the overall problem. The manner in which the engineer visualizes what the problem actually is, defines it, and gives it some structure is the hard part of most practical engineering problems. This is engineering and requires judgment, study, and often several attempts before the real problem is solved.

There are several possible ways of approaching the following problem; however, the questions are arranged so as to stimulate your thinking concerning some of the possible ways in which previously discussed techniques might be used and incorporated into an overall study of the problem.

A large construction project is located in a remote area and requires the development of a self-contained community that will support 5000 residents to ensure availability of labor. One aspect of the development and operation of this community is handling and processing solid wastes (i.e., garbage).

(a) Assuming that you are the engineer responsible for designing and planning this aspect of the community, pose the problem as a systems problem identifying components, identifying the requirements of that component, the basic issues associated with meeting those requirements, and their interaction within the system structure.

(b) Linear graphs can be used in two different ways in the problem-solving process. There are certain problem-solving techniques associated with linear graphs, such as cuts-sets or maximum flow–minimum cut theorem. In using these techniques, it is usually necessary to draw the graph even though it represents only the connectivity of the components. Sometimes this picture of the connectivity is all that can be obtained from the graph, since there is no solution technique associated with it for the particular problem. Nevertheless, the graph serves a very useful purpose in showing interdependence of the components. Identify and discuss where linear graph concepts could be used in this problem.

(c) In order to use mathematical programming, the engineer must recognize problem areas over which there is some control and for which the possibility of obtaining an improved solution exists. The engineer must be able to formulate objectives and identify decision variables and constraints so that a mathematical programming formulation becomes possible. Where do you think mathematical programming concepts could be usefully applied in this problem? What might be the decision variables and constraints involved? How would you go about quantifying the mathematical formulation?

(d) An engineer must often make a decision now from among a set of alternatives that influence future actions and whose future outcomes are uncertain. Furthermore, the engineer may have to document the basis for this decision. Among these decisions are those regarding the type and depth of study required as well as which alternative is best. In the text, decision tree analysis, and simulation were proposed as possible techniques for use in decision making. Discuss how these techniques might be used in this problem and how you would obtain the data necessary to apply these techniques. How would you handle situations in which no data exist and the need for immediate decisions precludes the possibility of gaining further data?

(e) A basic problem in any project is identifying tasks, assigning resources to the tasks, and scheduling the work tasks. Project network models can be developed for areas with clearly defined sequential logic. Assigning resources and scheduling work tasks must often be made subjectively. You can solve the problem if you have certain information about the tasks, the order in which they can be performed, and the times required to do them. Some of this knowledge comes with experience on the job and cannot be taught. Most design and construction offices keep careful records of times required to do each job so that future estimates will be based on sound information. What is the relevance of project network models to this problem?

(f) Indicate how the studies cited above can be integrated into an overall study of the solid waste problem for the proposed community. Include:

1. The type of information obtained from each step or study and how it would be used.
2. A description of how these various phases or steps would interact (i.e., the way in which the results of one phase would affect the study being performed in another phase). Interaction may also result from the fact that another study is being performed even before results are available.
3. The manner in which the validity of the results would be evaluated for both the individual steps as well as for the overall project. This part is designed to indicate what is involved in the complete planning and execution of a project. It is perhaps the most important part of the project because it integrates the isolated components of the approach into the problem.

P14.9. Community life raises many general problem areas that engineers must face; for example housing, food and water distribution, waste disposal, transportation, and others. The acceptance of particular solutions are influenced by the standards, magnitude, and level of social awareness that exists and that contributes to the environment in which the problem arises. The more complex the environment, the more forces, issues, and decision areas that must be considered. This requires a broader based interdisciplinary approach. Contrast your approach to the design and planning of a new integrated waste disposal system for a large existing city with the approach used in P14.8 for a remote area.

Bibliography

THE SYSTEMS APPROACH TO ENGINEERING PROBLEMS

Anonymous, "The Future of the Super Hi-Rise Building," *Modern Steel Construction,* First Quarter, 1972.

Asimow, M., *Introduction to Design.* Englewood Cliffs, N.J.: Prentice-Hall, Inc., 1962. A philosophy of engineering design is presented. This philosophy is comprised of three parts: a set of consistent principles; an operation discipline that leads to action; and a critical feedback apparatus that measures the advantages, detects the shortcomings, and illuminates the directions of improvement.

Blanchard, B. S., and W. J. Fabrycky, *Systems Engineering and Analysis.* Englewood Cliffs, N.J.: Prentice-Hall, Inc., 1981. Contains a general overview of systems and analyses that can be applied in the design of new systems and the analysis of existing systems.

Churchman, C. W., *The Systems Approach.* New York: A Delta Book, Dell Publishing Company, Inc., 1968. Considers an approach that characterizes the nature of a system in such a way that decision making can take place in a logical and coherent manner.

Heathington, K. W., and R. B. Bunton, "Systems Approach and the Civil Engineer," Engineering Issues, *Jour. Professional Activities, ASCE,* **97,** No. PP1, October 1971, 65–82, with discussion by T. C. Kavanagh.

Katz, D. L., R. O. Goetz, E. R. Lady, and D. C. Ray, *Engineering Concepts and Perspectives.* New York: John Wiley & Sons, Inc., 1968. A problem-oriented approach is used to stimulate engineering students to relate theory and formulations to a variety of real-world physical situations: automobile engines, rockets, and electrical and control systems.

KAUFMANN, A., *The Science of Decision Making*. New York: McGraw-Hill Book Company, 1968. Presents procedures by which the factors in a given situation can be analyzed and the result can be used to suggest the best course of action to be taken. The mathematical and graphical aspects of rather complex questions are alluded to, but only elementary problems are treated.

KRICK, E. V., *An Introduction to Engineering and Engineering Design* (2nd ed.). New York: John Wiley & Sons, Inc., 1969. This book presents a description of engineering practice, an introduction to important abilities of engineers, and a motivating description of the fields in which they can profitably apply their talent.

MESAROVIĆ, M. D., D. MACKO, AND Y. TAKAHARA, *Theory of Hierarchical Multilevel Systems*. New York: Academic Press, Inc., 1970. In Part I of this book hierarchical systems are discussed in terms of levels of abstraction, levels of complexity of decision making, and levels of priority in a multiunit decision system. A mathematical theory of coordination is developed in Part II.

SAGE, A. P., *Methodology for Large-Scale Systems*. New York: McGraw-Hill Book Company, 1977. Presents the methodologies and modeling that are applicable to addressing large-scale systems problems.

WILSON, W. E., *Concepts of Engineering Systems Design*. New York: McGraw-Hill Book Company, 1965. The text assumes that the engineering design of a system is of major interest to students, and thereby introduces them to the concepts of the profession and the function inherent in the design.

WOODSON, T. T., *Introduction to Engineering Design*. New York: McGraw-Hill Book Company, 1966. The book is written for engineering students who are beginning their design experience, and it guides them through all stages of an authentic design project.

LINEAR GRAPH MODELING AND ANALYSIS OF SYSTEMS

AU, T., AND T. E. STELSON, *Introduction to Systems Engineering, Deterministic Models*, Chapter 6. Reading, Mass.: Addison-Wesley Publishing Company, Inc., 1969. Presents procedures for determining the maximum flows in networks with capacity constraints.

BUSACKER, R. G., AND T. L. SAATY, *Finite Graphs and Networks: An Introduction with Applications*. New York: McGraw-Hill Book Company, 1965. Graph theory is presented in the setting of a relatively informal discussion of central ideas that are amplified and illustrated by a variety of applications.

FRANK, H., AND L. T. FRISCH, "Network Analysis," *Scientific American*, July 1970. Describes the basic principles of graphical network analysis and their application in designing gas pipelines for offshore drilling and in reliability analysis of electrical networks.

JENSEN, P. A., AND J. W. BARNES, *Network Flow Programming*. New York: John Wiley & Sons, Inc., 1980. Methods and algorithms are presented for the solution of network flow problems.

MARSHALL, C. W., *Applied Graph Theory*. New York: John Wiley & Sons, Inc., 1971. This is an introduction to graph theory and its application. Illustrative applications are given from the social sciences, physics, operations research, and related fields.

ORE, O., *Graphs and Their Uses*. New York: Random House, Inc., 1963. Presents an introduction to the kind of analyses that can be made by means of graphs and some of the problems that can be attacked by such methods. Only some of the simplest problems from graph theory are treated, so little technical knowledge is needed to understand the material.

PHILLIPS, D. T., AND A. GARCIA-DIAZ, *Fundamentals of Network Analysis*. Englewood Cliffs, N.J.: Prentice-Hall, Inc., 1981. This books contains a thorough and comprehensive discussion of network theories and models that relate to network analysis.

MATHEMATICAL MODELING OF ENGINEERING SYSTEMS

AU, T., AND T. E. STELSON, *Introduction to Systems Engineering, Deterministic Models*, Chapters 1 and 2. Reading, Mass.: Addison-Wesley Publishing Company, Inc., 1969. Presents the principles underlying different mathematical methods of analysis, with simplified examples for illustration.

BUSACKER, R. G., AND T. L. SAATY, *Finite Graphs and Networks*, Chapter 7. New York: McGraw-Hill Book Company, 1965. Presents graphical and mathematical methods for analyzing network flows.

FENVES, S. J., AND F. H. BRANIN, "Network-Topological Formulation of Structural Analysis," *Jour. Structural Division, ASCE*, **89**, No. ST4, August 1963, 483–514.

FORD, L. R., JR., AND D. R. FULKERSON, *Flows in Networks*. Princeton, N.J.: Princeton University Press, 1962. Methods are presented for dealing with a variety of problems that have formulations in terms of flows in capacity-constrained networks. The problems discussed range from practical ones to more purely theoretical ones.

HAITH, D. A., *Environmental Systems Optimization*. New York: John Wiley & Sons, Inc., 1982. An introduction to the application of systems analysis, mathematical modeling, and optimization techniques is presented.

JENSEN, P. A., AND J. W. BARNES, *Network Flow Programming*. New York: John Wiley & Sons, Inc., 1980. Methods and algorithms are presented for the solution of network flow problems.

KESAVAN, H. K., AND M. CHANDRASHEKAR, "Graph–Theoretical Models for Pipe Network Analysis," *Jour. Hydraulics Division, ASCE*, **98**, No. HY2, February 1972, 345–364. Models based on concepts from linear graph theory are developed for analyzing nonlinear pipe networks.

MARSHALL, C. W., *Applied Graph Theory*. New York: John Wiley & Sons, Inc., 1971. This book is an introduction to graph theory and its application. Illustrative applications are given from the social sciences, physics, operations research, and related fields.

SESHU, S., AND M. B. REED, *Linear Graphs and Electrical Networks*. Reading, Mass.: Addison-Wesley Publishing Company, Inc., 1961. A rigorous mathematical treatment of linear graph theory and its application to electrical networks. This book is aimed primarily at advanced graduate students who already have some fundamental knowledge of linear network analysis.

WYMORE, A. W., *A Mathematical Theory of Systems Engineering: The Elements* (3rd ed.). New York: John Wiley & Sons, Inc., 1977. Presents the modeling of systems, with an interesting example on the mathematical modeling of an open-pit copper mine.

OPTIMIZATION

ACKOFF, R. L., S. K. GUPTA, AND J. S. MINAS, *Scientific Method: Optimizing Applied Research Decisions*, Chapters 2 and 3. New York: John Wiley & Sons, Inc., 1962. Presents the meaning of an optimal solution and the approach to problem formulation.

BEIGHTLER, C. S., D. T. PHILLIPS, AND D. J. WILDE, *Foundations of Optimization* (2nd ed.). Englewood Cliffs, N.J.: Prentice-Hall, Inc., 1979. This book presents a comprehensive, unified treatment of the field of optimization theory.

BEVERIDGE, G. S. G., AND R. S. SCHECHTER, *Optimization: Theory and Practice*. New York: McGraw-Hill Book Company, 1970. A broad but unified coverage of optimization is presented. The interrelationships between the various facets of optimization are demonstrated.

HALL, W. A., AND J. A. DRACUP, *Water Resources Systems Engineering*, Chapters 3 and 4. New York: McGraw-Hill Book Company, 1970. This book presents fundamentals of the systems approach to complex water resources problems. Coverage includes the nature and objective functions of water resources systems; investment timing; large-scale, complex, multiple-purpose systems; groundwater systems; and water-quality systems. Chapter 3 deals with the principles of systems analysis and Chapter 4 discusses the development of objective functions for water resources development.

HESTEMUS, M. R., *Optimization Theory, The Finite Dimensional Case*. New York: John Wiley & Sons, Inc., 1975. An introduction to variational theory and finite-dimensional optimization theory, including unconstrained and constrained extrema and the use of the Lagrange multiplier.

HITCH, C. J., "On the Choice of Objectives in Systems Studies," in *Systems: Research and Design*, ed. D. P. Eckman. New York: John Wiley & Sons, Inc., 1961.

KAUFMAN, A., *The Science of Decision Making*, Chapters 2 and 3. New York: McGraw-Hill Book Company, 1968. Chapter 2 suggests the use of the mathematical theory of sets as a basis for establishing the order of preference or value function. Chapter 3 discusses the use of linear programming models for optimization and the problems of suboptimization and sensitivity analysis.

WOODSON, T. T., *Introduction to Engineering Design*, Chapters 13 and 15. New York: McGraw-Hill Book Company, 1966. Chapter 13 presents a discussion of criteria functions. Chapter 15 describes the basic principles and methods of optimization with an example on the optimum design of a bucket conveyor for a concrete mixing plant.

MATHEMATICAL PROGRAMMING

COHEN, C., AND J. STEIN, *Multipurpose Optimization System: User's Guide* (*Version 3, Manual 320*). Evanston, Ill.: Vogelback Computing Center, Northwestern University, 1976. Instructions are given for using an integrated system of computer programs to solve optimization problems on the CDC 6000/CYBER computers.

DANTZIG, G. B., *Linear Programming and Extensions.* Princeton, N.J.: Princeton University Press, 1963. A comprehensive treatment of the theoretical, computational, and applied areas of linear programming.

GARFINKEL, R. S., AND G. L. NEMHAUSER, *Integer Programming.* New York: John Wiley & Sons, Inc., 1972. A comprehensive treatment of theory, methodology, and application of integer programming is presented. Many examples and exercises, ranging from numerical calculation to research problems, are provided and techniques are illustrated by example.

GASS, S. I., *Linear Programming* (4th ed.). New York: McGraw-Hill Book Company, 1975. This is a basic presentation of the theoretical, computational, and applied areas of linear programming.

GREENBERG, H., *Integer Programming.* New York: Academic Press Inc., 1971. Presents theory and examples of integer programming and methods of solving practical problems.

HILLIER, F. S., AND G. J. LIEBERMAN, *Introduction to Operations Research* (3rd ed.). San Francisco: Holden-Day, Inc., 1980. Includes an introduction to linear programming, dynamic programming, integer programming, nonlinear programming, and simulation.

KUESTER, J. L., AND J. H. MIZE, *Optimization Techniques with FORTRAN.* New York: McGraw-Hill Book Company, 1973. Twenty-six FORTRAN-coded optimization algorithms are presented with a description and example for each.

LLEWELLYN, R. W., *Linear Programming.* New York: Holt, Rinehart and Winston, 1964. An introduction to the formulation and solution of linear programming problems. The model formulation and solution procedures for networks with multiple sources and sinks are presented clearly at an introductory level.

SALKIN, M., *Integer Programming.* Reading, Mass.: Addison-Wesley Publishing Company, Inc., 1975. Presents integer programming techniques and applications as well as computational experiences on various algorithms and problems.

WAGNER, H. M., *Principles of Operations Research: With Applications to Managerial Decisions* (2nd ed.). Englewood Cliffs, N.J.: Prentice-Hall, Inc., 1975. An extensive coverage of the formulation and solution of linear models at an introductory level. Sensitivity testing and duality concepts are also covered.

ZIONTS, S., *Linear and Integer Programming.* Englewood Cliffs, N.J.: Prentice-Hall, Inc., 1974. Coverage of linear and integer programming, including such topics as sensitivity analysis, network flow methods, and zero–one problems.

ORGANIZATIONAL NETWORKS

ANTILL, J. M., AND R. W. WOODHEAD, *Critical Path Methods in Construction Practice* (3rd ed.). New York: John Wiley & Sons, Inc., 1982. Presents the concepts and procedures of this method of construction planning and project control. Discusses the use of CPM as a practical system for integrating project development and management.

CLOUGH, R. H., *Construction Contracting* (3rd ed.), Chapter 12. New York: Wiley-Interscience, 1975. The chapter centers about a discussion of the critical path method and its associated applications of least-cost expediting and resource leveling.

PEURIFOY, R. L., *Construction Planning, Equipment and Methods* (3rd ed.), Chapter 2. New York: McGraw-Hill Book Company, 1979. Discusses the application of the critical path method in job planning and management. A practical example on a highway project is used to illustrate the use of CPM to determine the duration of the project; to schedule materials, equipment, and labor; and to schedule the amount and duration of the financing required for the project.

SHAFFER, L. R., J. B. RITTER, AND W. L. MEYER, *The Critical Path Method*. New York: McGraw-Hill Book Company, 1965. Explains the purpose and function of the Critical Path Method. Emphasizes the application of the method through the use of practical examples taken from the construction industry.

Symposium, Verrazano-Narrows Bridge, *Civil Engineering*, **34**, No. 12, December 1964, also *Jour. Construction Division, ASCE*, **92**, No. CO2, March 1966.

WIEST, J. D., AND F. K. LEVY, *A Management Guide to PERT/CPM* (2nd ed.). Englewood Cliffs, N.J.: Prentice-Hall, Inc., 1977. Presents the basic ideas of PERT and CPM scheduling techniques and the variety of management problems to which they may be applied.

DECISION ANALYSIS

BARISH, N. N., *Economic Analysis for Engineering and Managerial Decision Making* (2nd ed.), Chapters 20 and 21. New York: McGraw-Hill Book Company, 1978. A discussion of different criteria and methods of analysis for decision making under risk and uncertainty.

BENJAMIN, J. R., AND C. A. CORNELL, *Probability, Statistics, and Decisions for Civil Engineers*. New York: McGraw-Hill Book Company, 1970. A book on applied probability and statistics with the illustrations and problems taken from the civil engineering field.

FISHBURN, P. C., *Decision and Value Theory*, Chapters 2, 3, and 4. New York: John Wiley & Sons, Inc., 1964. These chapters deal with the decision structure and problem formulation, basic decision models, and the measurement of relative values (utility).

FISHBURN, P. C., *Utility Theory for Decision Making* (2nd ed.). New York: John Wiley & Sons, Inc., 1979. Presents a unifying upper-level treatment of preference structures and numerical representations of preference structures.

RAIFFA, H., *Decision Analysis: Introductory Lectures on Choices under Uncertainty*. Reading, Mass.: Addison-Wesley Publishing Company, Inc., 1968. Decision tree is presented as an organizational scheme that can be used to help select the action to be taken when faced with a situation that requires a decision. The consequences of the chosen action may be uncertain. The Bayesian viewpoint, which uses both utilities and subjective probabilities, is utilized.

TRIBUS, M., *Rational Descriptions, Decisions, and Designs*, Chapter 8. Elmsford, N.Y.: Pergamon Press, Inc., 1969. In addition to decision tree and utility value, this chapter also deals with the decision problems relating to the design of experiments and with sequential testing.

SYSTEM SIMULATION

BALINTFY, J. L., D. S. BURDICK, K. CHU, AND T. H. NAYLOR, *Computer Simulation Techniques*. New York: John Wiley & Sons, Inc., 1966. Discusses the rationale for computer simulation, the formulation of simulation models, as well as the design of simulation experiments. Also included are chapters on simulation languages and techniques for generating random numbers.

HUFSCHMIDT, M. M., AND M. B. FIERING, *Simulation Techniques for Design of Water-Resource Systems*. Cambridge, Mass.: Harvard University Press, 1966. Presents the various steps and procedures required to institute a simulation study of a water resource system, including procedures for collecting and organizing hydrologic and economic data, and for developing the necessary logic and detailed computer code using the Lehigh River basin in Pennsylvania as an example.

MAASS, A., M. M. HUFSCHMIDT, R. DORFMAN, H. A. THOMAS, JR., S. A. MARGLIN, AND G. M. FAIR, *Design of Water-Resource Systems*. Cambridge, Mass.: Harvard University Press, 1962. The results of a large-scale research program devoted to the methodology of planning and designing complex, multiunit, multipurpose water resource systems. Discusses techniques of systems analysis appropriate for preliminary screening and the detailed analysis of alternatives.

MIDDLETON, J. T., "Planning against Air Pollution," *American Scientist*, **59**, March–April 1971, 188–194.

RAND CORPORATION, *A Million Random Digits with 100,000 Normal Deviates*. Glencoe, Ill.: The Free Press, 1955.

SHANNON, R. E., *Systems Simulation: The Art and Science*. Englewood Cliffs, N.J.: Prentice-Hall, Inc., 1975. Provides an overview of the development and use of simulation models together with the analysis and application of the results from the models.

STEPHENSON, R. E., *Computer Simulation for Engineers*. New York: Harcourt Brace Jovanovich, 1971.

TAKEDA, T., M. A. SOZEN, AND N. N. NIELSEN, "Reinforced Concrete Response to Simulated Earthquakes," *Jour. Structural Division, ASCE*, **96**, No. ST12, December 1970, 2557–2573.

SYSTEM PLANNING

BARISH, N. N., *Economic Analysis for Engineering and Managerial Decision Making* (2nd ed.). New York: McGraw-Hill Book Company, Inc., 1978. This is a technique-oriented presentation of the basic reasoning and methodology of economic analyses that are important in decision making.

DE NEUFVILLE, R., AND J. H. STAFFORD, *Systems Analysis for Engineers and Managers*. New York: McGraw-Hill Book Company, Inc., 1972. Presents an extensive economic treatment of systems problems.

HILL, M., "A Goals-Achievement Matrix for Evaluating Alternative Plans," *Journal, American Institute of Planners*, **34**, No. 1, January 1968, 19–29. The goals-achievement matrix is postulated as a technique to be used in evaluating alternative plans. It is considered in terms of the requirements of the rational planning process.

JAMES, L. D., AND R. R. LEE, *Economics of Water Resources Planning*, Chapters 1 through 9, and 22. New York: McGraw-Hill Book Company, 1971. Presents the basic concepts of engineering economy, microeconomics, the criterion of economic efficiency, and financial analysis.

MASSE, P., *Optimal Investment Decisions: Rules for Action and Criteria for Choice*. Englewood Cliffs, N.J.: Prentice-Hall, Inc., 1962. A unifying treatment of the theory of investment that includes linear programming, probability theory, and stochastic processes, and applications of these techniques to problems in decision making under both certainty and uncertainty.

THUESEN, H. G., W. J. FABRYCKY, AND G. J. THUESEN, *Engineering Economy* (5th ed.). Englewood Cliffs, N.J.: Prentice-Hall, Inc., 1978. An introduction to the principles and techniques required for evaluating engineering alternatives in terms of worth and cost.

Water Resources Council, "Proposed Principles and Standards for Planning Water and Related Land Resources," *Federal Register*, Part II, **36**, No. 245, Washington, D.C., December 21, 1971, 24144–24194. Proposed objectives for water resources development and a method of analysis of alternative plans (the goals-achievement matrix).

PROJECT MANAGEMENT

BLANCHARD, B. S., *Engineering Organization and Management*. Englewood Clifis, N.J.: Prentice-Hall, Inc., 1976. Focuses on concepts related to the life cycle of systems and the integration of engineering disciplines and administration functions.

CLOUGH, R. H., *Construction Project Management*. New York: John Wiley & Sons, Inc., 1972. Deals with the estimating, planning, scheduling, and management of construction projects.

FARKAS, L. L., *Management of Technical Field Operations*. New York: McGraw-Hill Book Company, 1970. Treats technical field operations as a special area of management with special practices, problems, and solutions.

FORRESTER, J. W., *Industrial Dynamics*. Cambridge, Mass.: The MIT Press, 1961. Presents a computer-based methodology for the modeling and analysis of industrial management systems.

HACKNEY, J. W., *Control and Management of Capital Projects*. New York: John Wiley & Sons, Inc., 1965. The author describes all aspects of controlling cost, time, and value and the interpersonal relationships that form the ever-present background for all capital projects.

HALPIN, D. A., AND R. W. WOODHEAD, *Design of Construction and Process Operations*. New York: John Wiley & Sons, Inc., 1976. Introduces modeling methods for describing process components in terms of a system. Includes numerous examples of the application of such models in the construction area.

Rubey, H., J. A. Logan, and W. W. Milner, *The Engineer and Professional Management* (3rd ed.). Ames, Iowa: The Iowa State University Press, 1970. Considers the fundamentals of management and its role in engineering and engineering services.

Starr, M. K., *Production Management: Systems and Synthesis* (2nd ed.). Englewood Cliffs, N.J.: Prentice-Hall, Inc., 1972. Discusses production management in a systems context. Includes management techniques, mathematical models, and the application of the synthesis concept as a management tool.

STATE CONCEPTS OF SYSTEMS THEORY

Ashby, W. R., *An Introduction to Cybernetics*, Chapter 3. New York: John Wiley & Sons, Inc., 1963. Presents the concept of system states clearly at an introductory level.

Bellman, R., *Dynamic Programming*. Princeton, N.J.: Princeton University Press, 1957. Presents an introduction to the mathematical theory of multistage decision processes (dynamic programming).

Bellman, R. E., and S. E. Dreyfus, *Applied Dynamic Programming*. Princeton, N.J.: Princeton University Press, 1962. Describes how the theory of dynamic programming can be applied to the numerical solution of optimization problems in connection with satellites and space travel, the determination of trajectories, feedback control and servomechanism theory, inventory and scheduling processes, allocation of resources, and the determination of prices.

Heidari, M., V. T. Chow, P. V. Kokotovic, and D. D. Meredith, "Discrete Differential Dynamic Programming Approach to Water Resources Systems Optimization," *Water Resources Research*, 7, No. 2, April 1971, 273–282.

Nemhauser, G. L., *Introduction to Dynamic Programming*. New York: John Wiley & Sons, Inc., 1966. The theory and computational aspects of dynamic programming are presented at an introductory level.

Wagner, H. M., *Principles of Operations Research with Applications to Managerial Decisions*, (2nd ed.). Englewood Cliffs, N.J.: Prentice-Hall, Inc., 1975. An introduction to dynamic optimization model formulation and solution procedures with examples of dynamic programming for bounded and unbounded horizons.

SYSTEMS CONCEPTS IN ENGINEERING

Brandt, C. T., *A Systems Study of Soft Ground Tunneling*. Tulsa, Oklahoma: Fenix and Scisson, Inc., May 1970. National Technical Information Service Accession No. PB 194 769. Describes a study to investigate new ideas and radical concepts for soft-ground tunneling. The project was supported jointly by the Office of High Speed Ground Transportation and the Urban Mass Transportation Administration.

Chen, K. (ed.), *Urban Dynamics—Extensions and Reflections* (University of Pittsburg School of Engineering Publications Series 3). San Francisco: San Francisco Press, 1972. Contains a collection of papers by various authors which pertain to the concepts and ideas put forth by Forrester in the book *Urban Dynamics*.

DAVIS, R. K., *The Range of Choice in Water Management*. Baltimore, Md: The Johns Hopkins Press, 1968. Economic-engineering analysis is used to explore alternatives and to provide the information needed to make comparisons and choices in the interests of serving seciety's preferences for the case of water-quality planning for the Potomac estuary.

DAWES, J. H., "Tools for Water-Resource Study," *Jour. Irrigation and Drainage Division, ASCE*, **96**, No. IR4, December 1970, 403–424. Presents cost versus size relationships suitable for preliminary planning studies for reservoirs, water-transmitting wells, pumps, municipal sewage treatment, and water pumping in Illinois.

DE NEUFVILLE, R., *Airport System Planning*. Cambridge, Mass.: The MIT Press, 1976. Presents a rather comprehensive discussion of airport planning and critically examines the traditional techniques and practices that have been utilized. The first chapter emphasizes the need for the systems approach and the recognition of functional use of a facility.

DE NEUFVILLE, R., J. SCHAEKE, JR., AND J. H. STAFFORD, "Systems Analysis of Water Distribution Networks," *Jour. Sanitary Engineering Division, ASCE*, **97**, No. SA6, December 1971, 825–842. The systems analysis methodology is applied to the design and planning for the $1 billion Third City Tunnel for New York City.

GIBSON, J. E., *Designing the New City: A Systematic Approach*. New York: John Wiley & Sons, Inc., 1977. Applies systems analysis methodology to urban development with an examination of the interaction of the various elements of the infrastructure.

LA PATRA, J. W., *Applying the Systems Approach to Urban Development*. New York: McGraw-Hill Book Company, 1977. Presents a rather extensive discussion of systems theories and analysis techniques in terms of urban problems.

MEREDITH, D. D., AND B. B. EWING, "Systems Approach to the Evaluation of Benefits from Improved Great Lakes Water Quality," *Proceedings*, 12th Conference of Great Lakes Research, International Association of Great Lakes Research, 1969, 843–870. Outlines a systems approach for evaluating benefits that would accrue due to an improvement in the quality of water in the Great Lakes. Presents a mathematical model that can be solved to determine the benefits from a change in water quality.

MESAROVIC, M. D., AND A. REISMAN, *Systems Approach and the City*. Amsterdam: North-Holland Publishing Company, 1972. Discusses a range of urban problems and the application of the systems approach to these problems.

Regional Planning Council, "A Consistent Trade-Off Approach to Rapid Transit System Planning," Baltimore, Maryland, February 1970. National Information Service Accession No. PB 192 692. Describes an application of linear programming to rapid-transit-system planning.

SAGE, A. P., *Methodology for Large-Scale Systems*. New York: McGraw-Hill Book Company, 1977. Presents the methodologies and modeling that are applicable to addressing large-scale systems problems. The book also contains a discussion of the definition of activities associated with the solution of these problems.

SHEMDIN, O. H., "River-Coast Interaction: Laboratory Simulation," *Jour. Waterways, Harbors and Coastal Engineering Division, ASCE*, **96**, No. WW4, November 1970, 755–766.

Random Numbers

APPENDIX A

258164	244733	824904	959712	284925	062825
547250	466759	943814	751744	707634	376550
279794	797398	656465	505360	241001	256756
676883	778968	934335	028735	444391	538814
056700	668517	599657	172246	663342	229231
339846	006566	593875	032328	975552	373848
036783	039384	559225	193777	846672	240567
220480	236066	351556	161368	074279	441791
321406	414815	106967	967134	445197	647755
926274	486088	641104	796227	668169	882135
551342	913235	842276	771953	004479	286810
304312	473198	047928	626475	026876	718933
823825	835986	287273	754598	161107	308715
937351	010233	721707	522461	965570	850209
617730	061361	325338	131225	786849	095472
702187	367781	949838	786484	715749	572211
208356	204205	692568	713559	289632	429389
248744	223866	150708	276511	735843	573432
490798	341698	903251	657207	410058	436704
941463	047882	413364	938779	457579	617269
642372	286994	477391	626291	742379	699424
849870	720032	861112	753498	449229	191795
093443	315302	160820	515872	692334	149489
560052	889689	963853	091735	149304	895946
356517	332082	776563	549817	894838	369583
136699	990251	654104	295173	362940	215001
819290	934772	920183	769050	175190	288566
910170	602271	514838	609073	049977	729456
454833	609543	085541	650304	299551	371782

725920	653122	512693	897409	795288	228180
350587	914302	072686	378353	766325	367552
101159	479593	435653	267561	592743	202833
606294	874310	610972	603571	552441	215643
633650	239915	661686	617332	310901	292418
797598	437881	965626	699801	863313	752542
780166	624326	787185	194055	174009	510141
675692	741722	717763	163035	042897	057390
049565	445296	301705	977129	257123	343977
297081	668767	808201	856124	541013	061544
780488	008061	843715	130923	242413	368876
677624	048345	056556	784673	452850	210769
061141	289772	338980	702709	714037	263205
366460	736682	031592	211482	279375	577461
196284	415086	189369	267476	674370	460850
176396	587709	134951	603059	041642	761984
057200	922951	808817	614263	249601	566725
342842	531427	847407	681412	495933	396506
054743	184960	078683	083843	972241	376355
328117	108529	471591	502517	826831	255591
966490	650465	826352	011698	955365	531832
792363	898373	952494	070140	725692	187387
748793	384127	708486	420392	349224	123071
487669	302166	246104	519512	092989	737614
922713	810962	474975	113551	557332	420673
530000	860263	846638	680563	340216	521195
176412	155727	074070	078756	039000	123638
057296	933327	443946	472031	233763	741017
343418	593614	660673	828993	401014	441071
058191	557661	959553	968321	403375	643445
348781	342185	750789	803337	417523	856303
090331	050801	499633	814559	502313	131999
541400	304492	994413	881820	010477	791123

Discount Factors

Single payment compound amount factor, CAF:

$$(CAF, i, n) = (1 + i)^n$$

Single payment present worth factor, $PWSP$:

$$(PWSP, i, n) = \left(\frac{1}{1 + i}\right)^n$$

Uniform series compound amount factor, $USCA$:

$$(USCA, i, n) = \frac{(1 + i)^n - 1}{i}$$

Uniform series sinking fund factor, SFF:

$$(SFF, i, n) = \frac{i}{(1 + i)^n - 1}$$

Uniform series present worth factor, $PWUS$:

$$(PWUS, i, n) = \frac{(1 + i)^n - 1}{i(1 + i)^n}$$

Uniform series capital recovery factor, CRF:

$$(CRF, i, n) = \frac{i(1 + i)^n}{(1 + i)^n - 1}$$

Discount Factors

	Single Payment		Uniform Series			
N	Compound Amount Factor, CAF Given P, Find F	Present Worth Factor, PWSP Given F, Find P	Compound Amount Factor, USCA Given A, Find F	Sinking Fund Factor, SFF Given F, Find A	Present Worth Factor, PWUS Given A, Find P	Capital Recovery Factor, CRF Given P, Find A
			1% Discount Factors			
1	1.0100	0.99010	1.000	1.00000	0.9901	1.01000
2	1.0201	0.98030	2.010	0.49751	1.9704	0.50751
3	1.0303	0.97059	3.030	0.33002	2.9410	0.34002
4	1.0406	0.96098	4.060	0.24628	3.9020	0.25628
5	1.0510	0.95147	5.101	0.19604	4.8534	0.20604
6	1.0615	0.94205	6.152	0.16255	5.7955	0.17255
7	1.0721	0.93272	7.214	0.13863	6.7282	0.14863
8	1.0829	0.92348	8.286	0.12069	7.6517	0.13069
9	1.0937	0.91434	9.369	0.10674	8.5660	0.11674
10	1.1046	0.90529	10.462	0.09558	9.4713	0.10558
11	1.1157	0.89632	11.567	0.08645	10.3676	0.09645
12	1.1268	0.88745	12.683	0.07885	11.2551	0.08885
13	1.1381	0.87866	13.809	0.07241	12.1337	0.08241
14	1.1495	0.86996	14.947	0.06690	13.0037	0.07690
15	1.1610	0.86135	16.097	0.06212	13.8651	0.07212
16	1.1726	0.85282	17.258	0.05794	14.7179	0.06794
17	1.1843	0.84438	18.430	0.05426	15.5623	0.06426
18	1.1961	0.83602	19.615	0.05098	16.3983	0.06098
19	1.2081	0.82774	20.811	0.04805	17.2260	0.05805
20	1.2202	0.81954	22.019	0.04542	18.0456	0.05542
25	1.2824	0.77977	28.243	0.03541	22.0232	0.04541
50	1.6446	0.60804	64.463	0.01551	39.1961	0.02551
100	2.7048	0.36971	170.481	0.00587	63.0289	0.01587
			1.5% Discount Factors			
1	1.0150	0.98522	1.000	1.00000	0.9852	1.01500
2	1.0302	0.97066	2.015	0.49628	1.9559	0.51128
3	1.0457	0.95632	3.045	0.32838	2.9122	0.34338
4	1.0614	0.94218	4.091	0.24444	3.8544	0.25944
5	1.0773	0.92826	5.152	0.19409	4.7826	0.20909
6	1.0934	0.91454	6.230	0.16053	5.6972	0.17553
7	1.1098	0.90103	7.323	0.13656	6.5982	0.15156
8	1.1265	0.88771	8.433	0.11858	7.4859	0.13358
9	1.1434	0.87459	9.559	0.10461	8.3605	0.11961
10	1.1605	0.86167	10.703	0.09343	9.2222	0.10843
11	1.1779	0.84893	11.863	0.08429	10.0711	0.09929
12	1.1956	0.83639	13.041	0.07668	10.9075	0.09168
13	1.2136	0.82403	14.237	0.07024	11.7315	0.08524
14	1.2318	0.81185	15.450	0.06472	12.5434	0.07972
15	1.2502	0.79985	16.682	0.05994	13.3432	0.07494
16	1.2690	0.78803	17.932	0.05577	14.1313	0.07077

	Single Payment		Uniform Series			
	Compound Amount Factor, *CAF* Given *P*,	Present Worth Factor, *PWSP* Given *F*,	Compound Amount Factor, *USCA* Given *A*,	Sinking Fund Factor, *SFF* Given *F*,	Present Worth Factor, *PWUS* Given *A*,	Capital Recovery Factor, *CRF* Given *P*,
N	Find *F*	Find *P*	Find *F*	Find *A*	Find *P*	Find *A*
17	1.2880	0.77639	19.201	0.05208	14.9076	0.06708
18	1.3073	0.76491	20.489	0.04881	15.6726	0.06381
19	1.3270	0.75361	21.797	0.04588	16.4262	0.06088
20	1.3469	0.74247	23.124	0.04325	17.1686	0.05825
25	1.4509	0.68921	30.063	0.03326	20.7196	0.04826
50	2.1052	0.47500	73.683	0.01357	34.9997	0.02857
100	4.4320	0.22563	228.803	0.00437	51.6247	0.01937

4% Discount Factors

N						
1	1.0400	0.96154	1.000	1.00000	0.9615	1.04000
2	1.0816	0.92456	2.040	0.49020	1.8861	0.53020
3	1.1249	0.88900	3.122	0.32035	2.7751	0.36035
4	1.1699	0.85480	4.246	0.23549	3.6299	0.27549
5	1.2167	0.82193	5.416	0.18463	4.4518	0.22463
6	1.2653	0.79031	6.633	0.15076	5.2421	0.19076
7	1.3159	0.75992	7.898	0.12661	6.0021	0.16661
8	1.3686	0.73069	9.214	0.10853	6.7327	0.14853
9	1.4233	0.70259	10.583	0.09449	7.4353	0.13449
10	1.4802	0.67556	12.006	0.08329	8.1109	0.12329
11	1.5395	0.64958	13.486	0.07415	8.7605	0.11415
12	1.6010	0.62460	15.026	0.06655	9.3851	0.10655
13	1.6651	0.60057	16.627	0.06014	9.9856	0.10014
14	1.7317	0.57748	18.292	0.05467	10.5631	0.09467
15	1.8009	0.55526	20.024	0.04994	11.1184	0.08994
16	1.8730	0.53391	21.825	0.04582	11.6523	0.08582
17	1.9479	0.51337	23.698	0.04220	12.1657	0.08220
18	2.0258	0.49363	25.645	0.03899	12.6593	0.07899
19	2.1068	0.47464	27.671	0.03614	13.1339	0.07614
20	2.1911	0.45639	29.778	0.03358	13.5903	0.07358
25	2.6658	0.37512	41.646	0.02401	15.6221	0.06401
50	7.1067	0.14071	152.667	0.00655	21.4822	0.04655
100	50.5049	0.01980	1237.624	0.00081	24.5050	0.04081

6% Discount Factors

N						
1	1.0600	0.94340	1.000	1.00000	0.9434	1.06000
2	1.1236	0.89000	2.060	0.48544	1.8334	0.54544
3	1.1910	0.83962	3.184	0.31411	2.6730	0.37411
4	1.2625	0.79209	4.375	0.22859	3.4651	0.28859
5	1.3382	0.74726	5.637	0.17740	4.2124	0.23740
6	1.4185	0.70496	6.975	0.14336	4.9173	0.20336
7	1.5036	0.66506	8.394	0.11914	5.5824	0.17914
8	1.5938	0.62741	9.897	0.10104	6.2098	0.16104
9	1.6895	0.59190	11.491	0.08702	6.8017	0.14702
10	1.7908	0.55839	13.181	0.07587	7.3601	0.13587

	Single Payment		Uniform Series			
N	Compound Amount Factor, CAF Given P, Find F	Present Worth Factor, $PWSP$ Given F, Find P	Compound Amount Factor, $USCA$ Given A, Find F	Sinking Fund Factor, SFF Given F, Find A	Present Worth Factor, $PWUS$ Given A, Find P	Capital Recovery Factor, CRF Given P, Find A
11	1.8983	0.52679	14.972	0.06679	7.8869	0.12679
12	2.0122	0.49697	16.870	0.05928	8.3838	0.11928
13	2.1329	0.46884	18.882	0.05296	8.8527	0.11296
14	2.2609	0.44230	21.015	0.04758	9.2950	0.10758
15	2.3966	0.41727	23.276	0.04296	9.7122	0.10296
16	2.5404	0.39365	25.673	0.03895	10.1059	0.09895
17	2.6928	0.37136	28.213	0.03544	10.4773	0.09544
18	2.8543	0.35034	30.906	0.03236	10.8276	0.09236
19	3.0256	0.33051	33.760	0.02962	11.1581	0.08962
20	3.2071	0.31180	36.786	0.02718	11.4699	0.08718
25	4.2919	0.23300	54.865	0.01823	12.7834	0.07823
50	18.4202	0.05429	290.336	0.00344	15.7619	0.06344
100	339.3021	0.00295	5638.368	0.00018	16.6175	0.06018
		8% Discount Factors				
1	1.0800	0.92593	1.000	1.00000	0.9259	1.08000
2	1.1664	0.85734	2.080	0.48077	1.7833	0.56077
3	1.2597	0.79383	3.246	0.30803	2.5771	0.38803
4	1.3605	0.73503	4.506	0.22192	3.3121	0.30192
5	1.4693	0.68058	5.867	0.17046	3.9927	0.25046
6	1.5869	0.63017	7.336	0.13632	4.6229	0.21632
7	1.7138	0.58349	8.923	0.11207	5.2064	0.19207
8	1.8509	0.54027	10.637	0.09401	5.7466	0.17401
9	1.9990	0.50025	12.488	0.08008	6.2469	0.16008
10	2.1589	0.46319	14.487	0.06903	6.7101	0.14903
11	2.3316	0.42888	16.645	0.06008	7.1390	0.14008
12	2.5182	0.39711	18.977	0.05270	7.5361	0.13270
13	2.7196	0.36770	21.495	0.04652	7.9038	0.12652
14	2.9372	0.34046	24.215	0.04130	8.2442	0.12130
15	3.1722	0.31524	27.152	0.03683	8.5595	0.11683
16	3.4259	0.29189	30.324	0.03298	8.8514	0.11298
17	3.7000	0.27027	33.750	0.02963	9.1216	0.10963
18	3.9960	0.25025	37.450	0.02670	9.3719	0.10670
19	4.3157	0.23171	41.446	0.02413	9.6036	0.10413
20	4.6610	0.21455	45.762	0.02185	9.8181	0.10185
25	6.8485	0.14602	73.106	0.01368	10.6748	0.09368
50	46.9016	0.02132	573.770	0.00174	12.2335	0.08174
		10% Discount Factors				
1	1.1000	0.90909	1.000	1.00000	0.9091	1.10000
2	1.2100	0.82645	2.100	0.47619	1.7355	0.57619
3	1.3310	0.75131	3.310	0.30211	2.4869	0.40211
4	1.4641	0.68301	4.641	0.21547	3.1699	0.31547

	Single Payment		Uniform Series			
	Compound Amount Factor, CAF Given P,	Present Worth Factor, PWSP Given F,	Compound Amount Factor, USCA Given A,	Sinking Fund Factor, SFF Given F,	Present Worth Factor, PWUS Given A,	Capital Recovery Factor, CRF Given P,
N	Find F	Find P	Find F	Find A	Find P	Find A
5	1.6105	0.62092	6.105	0.16380	3.7908	0.26380
6	1.7716	0.56447	7.716	0.12961	4.3553	0.22961
7	1.9487	0.51316	9.487	0.10541	4.8684	0.20541
8	2.1436	0.46651	11.436	0.08744	5.3349	0.18744
9	2.3579	0.42410	13.579	0.07364	5.7590	0.17364
10	2.5937	0.38554	15.937	0.06275	6.1446	0.16275
11	2.8531	0.35049	18.531	0.05396	6.4951	0.15396
12	3.1384	0.31863	21.384	0.04676	6.8137	0.14676
13	3.4523	0.28966	24.523	0.04078	7.1034	0.14078
14	3.7975	0.26333	27.975	0.03575	7.3667	0.13575
15	4.1772	0.23939	31.772	0.03147	7.6061	0.13147
16	4.5950	0.21763	35.950	0.02782	7.8237	0.12782
17	5.0545	0.19784	40.545	0.02466	8.0216	0.12466
18	5.5599	0.17986	45.599	0.02193	8.2014	0.12193
19	6.1159	0.16351	51.159	0.01955	8.3649	0.11955
20	6.7275	0.14864	57.275	0.01746	8.5136	0.11746
25	10.8347	0.09230	98.347	0.01017	9.0770	0.11017
50	117.3909	0.00852	1163.909	0.00086	9.9148	0.10086
			12% Discount Factors			
1	1.1200	0.89286	1.000	1.00000	0.8929	1.12000
2	1.2544	0.79719	2.120	0.47170	1.6901	0.59170
3	1.4049	0.71178	3.374	0.29635	2.4018	0.41635
4	1.5735	0.63552	4.779	0.20923	3.0373	0.32923
5	1.7623	0.56743	6.353	0.15741	3.6048	0.27741
6	1.9738	0.50663	8.115	0.12323	4.1114	0.24323
7	2.2107	0.45235	10.089	0.09912	4.5638	0.21912
8	2.4760	0.40388	12.300	0.08130	4.9676	0.20130
9	2.7731	0.36061	14.776	0.06768	5.3282	0.18768
10	3.1058	0.32197	17.549	0.05698	5.6502	0.17698
11	3.4785	0.28748	20.655	0.04842	5.9377	0.16842
12	3.8960	0.25668	24.133	0.04144	6.1944	0.16144
13	4.3635	0.22917	28.029	0.03568	6.4235	0.15568
14	4.8871	0.20462	32.393	0.03087	6.6282	0.15087
15	5.4736	0.18270	37.280	0.02682	6.8109	0.14682
16	6.1304	0.16312	42.753	0.02339	6.9740	0.14339
17	6.8660	0.14564	48.884	0.02046	7.1196	0.14046
18	7.6900	0.13004	55.750	0.01794	7.2497	0.13794
19	8.6128	0.11611	63.440	0.01576	7.3658	0.13576
20	9.6463	0.10367	72.052	0.01388	7.4694	0.13388
25	17.0001	0.05882	133.334	0.00750	7.8431	0.12750
50	289.0022	0.00346	2400.018	0.00042	8.3045	0.12042

	Single Payment		Uniform Series			
	Compound Amount Factor, *CAF* Given *P*,	Present Worth Factor, *PWSP* Given *F*,	Compound Amount Factor, *USCA* Given *A*,	Sinking Fund Factor, *SFF* Given *F*,	Present Worth Factor, *PWUS* Given *A*,	Capital Recovery Factor, *CRF* Given *P*,
N	Find *F*	Find *P*	Find *F*	Find *A*	Find *P*	Find *A*
14% Discount Factors						
1	1.1400	0.87719	1.000	1.00000	0.8772	1.14000
2	1.2996	0.76947	2.140	0.46729	1.6467	0.60729
3	1.4815	0.67497	3.440	0.29073	2.3216	0.43073
4	1.6890	0.59208	4.921	0.20320	2.9137	0.34320
5	1.9254	0.51937	6.610	0.15128	3.4331	0.29128
6	2.1950	0.45559	8.536	0.11716	3.8887	0.25716
7	2.5023	0.39964	10.730	0.09319	4.2883	0.23319
8	2.8526	0.35056	13.233	0.07557	4.6389	0.21557
9	3.2519	0.30751	16.085	0.06217	4.9464	0.20217
10	3.7072	0.26974	19.337	0.05171	5.2161	0.19171
11	4.2262	0.23662	23.045	0.04339	5.4527	0.18339
12	4.8179	0.20756	27.271	0.03667	5.6603	0.17667
13	5.4924	0.18207	32.089	0.03116	5.8424	0.17116
14	6.2613	0.15971	37.581	0.02661	6.0021	0.16661
15	7.1379	0.14010	43.842	0.02281	6.1422	0.16281
16	8.1372	0.12289	50.980	0.01962	6.2651	0.15962
17	9.2765	0.10780	59.118	0.01692	6.3729	0.15692
18	10.5752	0.09456	68.394	0.01462	6.4674	0.15462
19	12.0557	0.08295	78.969	0.01266	6.5504	0.15266
20	13.7435	0.07276	91.025	0.01099	6.6231	0.15099
25	26.4619	0.03779	181.871	0.00550	6.8729	0.14550
50	700.2330	0.00143	4994.521	0.00020	7.1327	0.14020
16% Discount Factors						
1	1.1600	0.86207	1.000	1.00000	0.8621	1.16000
2	1.3456	0.74316	2.160	0.46296	1.6052	0.62296
3	1.5609	0.64066	3.506	0.28526	2.2459	0.44526
4	1.8106	0.55229	5.066	0.19738	2.7982	0.35738
5	2.1003	0.47611	6.877	0.14541	3.2743	0.30541
6	2.4364	0.41044	8.977	0.11139	3.6847	0.27139
7	2.8262	0.35383	11.414	0.08761	4.0386	0.24761
8	3.2784	0.30503	14.240	0.07022	4.3436	0.23022
9	3.8030	0.26295	17.519	0.05708	4.6065	0.21708
10	4.4114	0.22668	21.321	0.04690	4.8332	0.20690
11	5.1173	0.19542	25.733	0.03886	5.0286	0.19886
12	5.9360	0.16846	30.850	0.03241	5.1971	0.19241
13	6.8858	0.14523	36.786	0.02718	5.3423	0.18718
14	7.9875	0.12520	43.672	0.02290	5.4675	0.18290
15	9.2655	0.10793	51.660	0.01936	5.5755	0.17936
16	10.7480	0.09304	60.925	0.01641	5.6685	0.17641
17	12.4677	0.08021	71.673	0.01395	5.7487	0.17395
18	14.4625	0.06914	84.141	0.01188	5.8178	0.17188

	Single Payment		Uniform Series			
	Compound Amount Factor, CAF Given P,	Present Worth Factor, PWSP Given F,	Compound Amount Factor, USCA Given A,	Sinking Fund Factor, SFF Given F,	Present Worth Factor, PWUS Given A,	Capital Recovery Factor, CRF Given P,
N	Find F	Find P	Find F	Find A	Find P	Find A
19	16.7765	0.05961	98.603	0.01014	5.8775	0.17014
20	19.4608	0.05139	115.380	0.00867	5.9288	0.16867
25	40.8742	0.02447	249.214	0.00401	6.0971	0.16401
			18% Discount Factors			
1	1.1800	0.84746	1.000	1.00000	0.8475	1.18000
2	1.3924	0.71818	2.180	0.45872	1.5656	0.63872
3	1.6430	0.60863	3.572	0.27992	2.1743	0.45992
4	1.9388	0.51579	5.215	0.19174	2.6901	0.37174
5	2.2878	0.43711	7.154	0.13978	3.1272	0.31978
6	2.6996	0.37043	9.442	0.10591	3.4976	0.28591
7	3.1855	0.31393	12.142	0.08236	3.8115	0.26236
8	3.7589	0.26604	15.327	0.06524	4.0776	0.24524
9	4.4355	0.22546	19.086	0.05239	4.3030	0.23239
10	5.2338	0.19106	23.521	0.04251	4.4941	0.22251
11	6.1759	0.16192	28.755	0.03478	4.6560	0.21478
12	7.2876	0.13722	34.931	0.02863	4.7932	0.20863
13	8.5994	0.11629	42.219	0.02369	4.9095	0.20369
14	10.1472	0.09855	50.818	0.01968	5.0081	0.19968
15	11.9737	0.08352	60.965	0.01640	5.0916	0.19640
16	14.1290	0.07078	72.939	0.01371	5.1624	0.19371
17	16.6722	0.05998	87.068	0.01149	5.2223	0.19149
18	19.6733	0.05083	103.740	0.00964	5.2732	0.18964
19	23.2144	0.04308	123.414	0.00810	5.3162	0.18810
20	27.3930	0.03651	146.628	0.00682	5.3527	0.18682
25	62.6686	0.01596	342.603	0.00292	5.4669	0.18292
			20% Discount Factors			
1	1.2000	0.83333	1.000	1.00000	0.8333	1.20000
2	1.4400	0.69444	2.200	0.45455	1.5278	0.65455
3	1.7280	0.57870	3.640	0.27473	2.1065	0.47473
4	2.0736	0.48225	5.368	0.18629	2.5887	0.38629
5	2.4883	0.40188	7.442	0.13438	2.9906	0.33438
6	2.9860	0.33490	9.930	0.10071	3.3255	0.30071
7	3.5832	0.27908	12.916	0.07742	3.6046	0.27742
8	4.2998	0.23257	16.499	0.06061	3.8372	0.26061
9	5.1598	0.19381	20.799	0.04808	4.0310	0.24808
10	6.1917	0.16151	25.959	0.03852	4.1925	0.23852
11	7.4301	0.13459	32.150	0.03110	4.3271	0.23110
12	8.9161	0.11216	39.581	0.02526	4.4392	0.22526
13	10.6993	0.09346	48.497	0.02062	4.5327	0.22062
14	12.8392	0.07789	59.196	0.01689	4.6106	0.21689
15	15.4070	0.06491	72.035	0.01388	4.6755	0.21388
16	18.4884	0.05409	87.442	0.01144	4.7296	0.21144

	Single Payment		Uniform Series			
N	Compound Amount Factor, *CAF* Given P, Find F	Present Worth Factor, *PWSP* Given F, Find P	Compound Amount Factor, *USCA* Given A, Find F	Sinking Fund Factor, *SFF* Given F, Find A	Present Worth Factor, *PWUS* Given A, Find P	Capital Recovery Foctor, *CRF* Given P, Find A
17	22.1861	0.04507	105.931	0.00944	4.7746	0.20944
18	26.6233	0.03756	128.117	0.00781	4.8122	0.20781
19	31.9480	0.03130	154.740	0.00646	4.8435	0.20646
20	38.3376	0.02608	186.688	0.00536	4.8696	0.20536
25	95.3962	0.01048	471.981	0.00212	4.9476	0.20212
22% Discount Factors						
1	1.2200	0.81967	1.000	1.00000	0.8197	1.22000
2	1.4884	0.67186	2.220	0.45045	1.4915	0.67045
3	1.8158	0.55071	3.708	0.26966	2.0422	0.48966
4	2.2153	0.45140	5.524	0.18102	2.4936	0.40102
5	2.7027	0.37000	7.740	0.12921	2.8636	0.34921
6	3.2973	0.30328	10.442	0.09576	3.1669	0.31576
7	4.0227	0.24859	13.740	0.07278	3.4155	0.29278
8	4.9077	0.20376	17.762	0.05630	3.6193	0.27630
9	5.9874	0.16702	22.670	0.04411	3.7863	0.26411
10	7.3046	0.13690	28.657	0.03489	3.9232	0.25489
11	8.9117	0.11221	35.962	0.02781	4.0354	0.24781
12	10.8722	0.09198	44.874	0.02228	4.1274	0.24228
13	13.2641	0.07539	55.746	0.01794	4.2028	0.23794
14	16.1822	0.06180	69.010	0.01449	4.2646	0.23449
15	19.7423	0.05065	85.192	0.01174	4.3152	0.23174
16	24.0856	0.04152	104.935	0.00953	4.3567	0.22953
17	29.3844	0.03403	129.020	0.00775	4.3908	0.22775
18	35.8490	0.02789	158.405	0.00631	4.4187	0.22631
19	43.7358	0.02286	194.254	0.00515	4.4415	0.22515
20	53.3476	0.01874	237.989	0.00420	4.4603	0.22420
25	144.2101	0.00693	650.955	0.00154	4.5139	0.22154

Index

H

I

S